Happy
Father's
Day

DADA !

♡ Sarasen

WEYERHAEUSER ENVIRONMENTAL BOOKS

William Cronon, Editor

Weyerhaeuser Environmental Books explore human relationships with natural environments in all their variety and complexity. They seek to cast new light on the ways that natural systems affect human communities, the ways that people affect the environments of which they are a part, and the ways that different cultural conceptions of nature profoundly shape our sense of the world around us.

WEYERHAEUSER ENVIRONMENTAL BOOKS

LANDSCAPES
OF CONFLICT

The Oregon Story, 1940–2000

WILLIAM G. ROBBINS

Foreword by William Cronon

UNIVERSITY OF WASHINGTON PRESS

Seattle and London

Landscapes of Conflict: The Oregon Story, 1940–2000 has been published with the
assistance of a grant from the Weyerhaeuser Environmental Books Endowment,
established by the Weyerhaeuser Company Foundation, members of the Weyerhaeuser family, and Janet and John Creighton.

University of Washington Press, P.O. Box 50096, Seattle, WA 98145
www.washington.edu/uwpress

Library of Congress Cataloging-in-Publication Data
Robbins, William G., 1935–
Landscapes of conflict : the Oregon story, 1940-2000/
William G. Robbins ; foreword by William Cronon.
p. cm.—(Weyerhaeuser enviromental books)
Includes bibliographical references and index.
ISBN 0-295-98442-2 (acid-free paper)
1. Oregon—Enviromental conditions—History.
I. Title.   II. Weyerhaeuser enviromental book.
GE155.07R58   2004   333.7'09795—dc22   2004052613

Cover image: Painting by Andrew Vincent, *The Boegli Ranch (Crooked River at Cove
Palisades)*, c. 1960. Oil on canvas, 25 1/2 x 33 in. Collection of Michael Parsons.

FOR KARLA

Best friend and companion

through three decades of time

CONTENTS

FOREWORD

STILL SEARCHING FOR EDEN

AT THE END OF THE OREGON TRAIL

William Cronon

William Robbins' *Landscapes of Conflict: The Oregon Story, 1941–2000* brings to completion the magisterial two-volume survey of Oregon environmental history that he began in 1997 with the publication of *Landscapes of Promise: The Oregon Story, 1800–1940*. There are very few books like this one. Although we have a number of fine studies that examine different aspects of environmental change and politics at the state level—Arthur McEvoy's pioneering account of legal regulation in the California fisheries, Mark Fiege's richly nuanced interpretation of irrigated agriculture in Idaho, Thomas Huffman's analysis of bipartisan environmental politics in Wisconsin during the 1960s, to name just three distinguished examples—rarely has a senior historian offered a full, broad-gauged synthesis of environmental change and political controversy for an entire state from frontier times to the present.[1] Not only will these two volumes instantly become the standard works on Oregon environmental history for the foreseeable future; they should inspire others to see the advantages of state-level geographical scales for understanding the complicated ways that political economy and social change interact with natural environments to produce the landscapes and communities we inhabit today.

Perhaps because so many of the key laws that now govern environmental politics in the United States were passed by Congress in the 1960s and 1970s, it is easy to imagine that the federal government is where the action is if one wants to understand the history of environmentalism in the second half of the twentieth century. Certainly that is the way the subject is most commonly taught in undergraduate courses. Political historians are often drawn to the national scale because it is so much easier to study one body of federal law than it is to grapple with the vast array of separate statutes produced by the fifty states, to say nothing of all the counties and municipalities that lie still further down the federal hierarchy. And certainly one should never downplay the impact of federal laws like the Clean Air and Water Acts, the Wilderness Act, the Endangered Species Act, or the National Environmental Policy Act, not least for their cascading consequences at the state and local levels.

But in fact, many of the most important environmental changes and controversies take place far below the national scale, and even most federal laws find their ultimate expression in enforcement regimes operating at the local level. To take just one obvious example, the laws and funding mechanisms to promote American hydropower in the first half of the twentieth century may have been debated and passed in Washington, D.C., but the impacts of those laws occurred on rivers like the Tennessee, the Colorado, and the Columbia. Moreover, one cannot even begin to comprehend the complexity of those impacts if one limits oneself to the bird's-eye views from Capitol Hill or the White House. Only on the ground can one see the power lines and irrigation ditches and monocultural hybrid crops and pesticides and invasive species and farms and clearcuts and highways and factories and cities and landfills and sewage systems—to say nothing of simple human homes—that together make up the landscapes and political economies and cultures we actually inhabit. As Robbins masterfully demonstrates in this book, the great beauty of a state-based environmental history is that all these phenomena and many more can fit within the single narrative and interpretive frame of a state called Oregon. Although it is true that one cannot understand local history without placing it in its national and global context, it is no less true that one cannot understand national or global history without grounding them in local places where the seemingly abstract effects of much larger systems and processes become real. These systemic environmental interconnections have been William Robbins's scholarly quarry for his entire career, and the care and nuance with which he pursues them in these two volumes make *Landscapes of Promise and Conflict* his masterpiece.

In *Landscapes of Promise*, Robbins carried his story down to the eve of the Second World War. His narrative essentially traced a transformation from a colonial settlement frontier to the maturing agro-industrial economy of a settled region. As I wrote in my foreword to that volume, Oregon became in the early nineteenth century an icon for the land of journey's ending, the destination at the far end of an overland trail which despite all its hardships might nonetheless deliver migrating families to a place of milk and honey where it seemed that the mythic promises of Eden might finally be fulfilled. Robbins's own love for the land and people of his adopted state—he first moved to Oregon as a graduate student, and has been there ever since— mean that he fully appreciates the stunning beauty and rich cultural complexity that can so easily make Oregon appear to be a promised land. But perhaps for this very reason, Robbins is equally conscious of the ways in which Eden has also been eroded and betrayed over the long sweep of the state's history. By the end of his first volume, he had demonstrated how the hoped-for landscape of Jeffersonian yeoman farmers that had drawn pioneers to the Oregon Trail in the 1840s had by the early twentieth century become increasingly controlled by large corporations and bureaucratic institutions that threatened environment and community alike.

Now, in *Landscapes of Conflict*, Robbins shows that nothing turned out quite as it seemed in the days when dreams of progress drove pioneers and capitalists alike to harvest the bounty of the Oregon environment for their own profit. The law of unintended consequences meant that the very technologies which made the harvest possible also began to put it at risk. Hydroelectric dams might fuel urban growth and regional prosperity, but they also blocked the spawning runs on which the state's fisheries depended. Pesticides might seem like miracle solutions to the invasive pests fostered by vast acreages of monocultural crops, but their hidden effects on wildlife and human health would become one of the most notorious themes of the nascent environmental movement of the 1960s. Logging might be the lifeblood of dozens of small rural communities, but the changing economics of the lumber industry gradually rendered those communities unsustainable at the same time that environmental regulations and spotted owls could serve as convenient scapegoats for those deeper structural causes. And so on and on. Precisely because Robbins is able in this book to juxtapose so many different aspects of the Oregon environment that underwent transformation during the post-World War II era, he reveals their interconnections in ways that make this book invaluable even for readers with no special interest in Oregon.

Among the most important and revealing of these interconnections are those between economic and technological change on the one hand, and political institutions and controversies on the other. The period covered by this volume of course witnessed the emergence of modern environmentalism, which has been a significant player in the conflicts that give this book its title. Whereas the conservation movement had focused to a considerable degree on the wise use of natural resources through the elimination of waste via the application of sound scientific knowledge by an enlightened managerial state (a theme that has long been associated in the field of environmental history with the classic work of Samuel P. Hays[2]), environmental politics after the Second World War became much more complicated as new concerns about pollution, toxicity, biological diversity, recreation, and the health of natural ecosystems began to call into question the primary commodity production that had fueled so much of Oregon's prewar economy.

At the same time, the growth of metropolitan and suburban communities—especially in the vicinity of Portland, which during this period emerged as one of the preeminent "green cities" of the United States—helped generate an electorate that became ever more sharply divided between rural and small-town working class communities on the one hand, and metropolitan, suburban, and exurban professional communities on the other. The rise of the service sector along with the growing importance of tourism and recreation meant that growing portions of the state's economy and the political concerns of its voters no longer looked to ranching, mining, farming, and fishing as the main sources of jobs and economic prosperity. Quite the contrary. Especially when coupled with technological innovations that eliminated jobs and increased environmental impacts, industries focusing on primary commodity production became ever more controversial with the passage of time.

Federal laws and agencies were certainly crucial to this story. One cannot understand the environmental history of twentieth-century Oregon without coming to terms with the Bonneville Power Administration, the U.S. Forest Service, the U.S. Fish and Wildlife Service, the Bureau of Land Management, and other such bureaucracies. Robbins gives all of them the attention they deserve. But he situates them always in relationship to state and local politics, so that one can suddenly see western public lands in their surrounding context. Controversies over municipal sewage systems created by sprawling suburban developments in the Willamette watershed, for instance, become as important to the Oregon story as dams on the Columbia or federal fisheries policies on the open seas. Here, one finally sees them in relationship to each other.

Moreover, Robbins is able to show how a politician like Richard Neuberger applied the lessons of his Oregon experience not just to defending the landscapes of the state he loved, but to the broader application of those lessons at the national level during his career in the United States Senate, though Robbins makes clear the less attractive sides of Neuberger's personality and politics as well. Likewise, Oregon Governor Tom McCall would emerge during the late 1960s and early 1970s as a visionary figure who promoted aggressive environmental and land-use regulations at the state level and so inspired similar actions elsewhere in the nation. By this reading, the history of environmental politics in Oregon is not just about how one state *responded* to an emerging national movement, but how the nation was in part shaped and led by the state. American federalism is a two-way street.

William Robbins would be the last person to hold up Oregon as a state that solved the environmental challenges of the twentieth century, or that somehow escaped the environmental crises and controversies that have characterized the modern era. He is as critical of his home state as he is of the larger culture and political economy of which it is a part. The central arc of his narrative, after all, repeatedly shows promises giving way to conflicts. But one nonetheless comes away from this book impressed by the strong sense of place that seems so often to have motivated Oregonians (including Robbins himself) to struggle with and against one another for the future of the lands and homes they loved.

Robbins certainly doesn't say that this shared love of the state's land-scape will be enough to save it in the long run, and he clearly fears that the greatest threat of modernity is the rootlessness that robs us of the moral and political commitment to struggle for justice and to take collective responsibility for the natural and human communities in which we dwell. Despite these dark misgivings, though, Robbins clearly believes—and I agree with him—that it is difficult to imagine any future solutions to our present environmental dilemmas that do not somehow invoke this special sense of place, this special human commitment to caring for the lands and crea-tures and fellow human beings with which we share our homes. Such emo-tions are hardly unique to Oregon, of course, and are as much under siege there as anywhere else. But that makes this history of one special place all the more suggestive and poignant for its relevance to all those other spe-cial places, all those other Edens at the ends of all those other trails, in which each of us tries to make a home.

NOTES

1. Arthur F. McEvoy, *The Fisherman's Problem: Ecology and Law in the California Fisheries, 1850–1980* (New York: Cambridge University Press, 1986); Mark Fiege, *Irri-gated Eden: The Making of an Agricultural Landscape in the American West* (Seattle: University of Washington Press, 2000); and Thomas R. Huffman, *Protectors of the Land and Water: Environmentalism in Wisconsin, 1961–1968* (Chapel Hill: Univer-sity of North Carolina Press, 1994). For examples of other state-level environmental histories comparable to Robbins in scale if not so much in approach, see Mark Derr, *Some Kind of Paradise: A Chronicle of Man and the Land in Florida* (New York: William Morrow, 1989); and Jan Albers, *Hands on the Land: A History of the Vermont Landscape* (Cambridge: MIT Press, 2000).

2. Samuel P. Hays, *Conservation and the Gospel of Efficiency: The Progressive Con-servation Movement, 1890–1920* (Cambridge: Harvard University Press, 1959).

PREFACE

The Oregon that we know today had its beginnings in mythic nineteenth-century stories of a restless immigrant population on the move to a proverbial green land of promise, a sweep of country that required only the dedication and hard work of a committed people to build a truly progressive society. Steeped in romantic stories about the region's potential, the newcomers who came in increasing numbers in the 1830s and 1840s took advantage of nature's wealth to establish an economy based largely on natural-resource extraction. As they ruthlessly pushed the native inhabitants aside, the growing number of immigrants and their successor generations set about the arduous task of establishing for themselves a celebrated place in Oregon history. The striking historical features underscored in my 1997 book, *Landscapes of Promise: The Oregon Story, 1800–1940*, addressed the dramatic transformation of Oregon from the Indian period to the onset of the Second World War. The deliberate and accidental introduction of exotic flora and fauna, new economic reckonings, the commodification of natural phenomena, the damming and diversion of waterways, and ever more intensive agricultural practices refashioned Native American ecosystems into settings better suited to commercial enterprise.

Those early comers and their successor generations plowed the soil, hewed the timber, and fished the region's streams with neither caution, introspection, nor reflection. For several decades Oregon's natural bounty yielded a modest living for many and significant riches for a few. Although glutted markets and depressed prices periodically slowed the state's agricultural and lumber sectors—and especially so during the great crash and economic desolation of the 1930s—true believers always held to the faith that Oregon remained a land of possibilities and opportunity. This book argues that the mythical promise of the Oregon Trail lived on following the Second World War. A broadly shared consensus existed among newspaper editorialists, planners, and magazine writers who argued that still abundant timber supplies, a thriving agricultural sector, and a booming California market promised a bright future. If the war's beginning signaled an abrupt shift from peace to wartime conditions, the end of the conflict meant a transition to a very different world, one that Americans hoped would bring prosperity and improved living standards. The victories in Europe and Asia, new developments in science and engineering, and Oregon's wealth of natural resources suggested a future free from want and deprivation.

The combination of full employment, deferred purchasing, and savings accumulated during the war unleashed unprecedented spending and extraordinary economic growth after 1945. For the resource-dependent Northwest, a booming economy meant accelerated timber harvests, expanding water-development projects, and unparalleled growth in agriculture and other extractive activities. Economic forces were primary during the postwar decades. The movement to the suburbs, America's love affair with the automobile, and a growing affluence and interest in outdoor recreation profoundly reshaped Oregon's landscape after 1945. With the exception of environmental restraints—most of them coming after 1970—market-driven forces remained the critical determinant in the conduct of human activity. Until the imposition of state land-use restrictions, federal environmental regulations, Endangered Species Act requirements, and the regulation of field-burning and chemical use, alterations to natural systems were treated as externalities rather than part of the cost of industrial activity.

When I arrived in Oregon in the early 1960s, loggers cut trees to water's edge, "cat skinners" drove bulldozers through streambeds, and some of the largest timberland owners were indifferent to reforesting cutover land. Willamette Valley farmers plowed from fence row to riverbank, removed hedgerows, and drained sloughs to create ever larger fields, all in the interest of economies of scale. Monocropping and petrochemical treatments

fueled the productivity of the big fields. In the Deschutes, Klamath, Umatilla, and Malheur-Owyhee basins, the Bureau of Reclamation expanded
its system of canals and laterals to bring more irrigated land into production. The notion that enough water should remain in streams to support
healthy and sustainable fish populations did not fit into the agency's calculations. While the human imprint on Oregon's landscape was already
considerable at the end of the Second World War, the next half century
would witness a vast acceleration of those forces.

Although Oregon is a politically bounded place, it has always enjoyed
economic, demographic, and ecological exchanges and associations with
more distant environments. As a subordinate component of a global
superpower in the postwar era, the state's economy and politics have been
partners to geopolitical strategies worked out in Washington, D.C. National
and international events and circumstances have influenced dam-building
projects, timber harvests, and agricultural production. Columbia River
dams, proponents argued, would control flooding, enhance inland transportation, provide irrigation water for the arid Columbia Plain, and—most
important—produce abundant hydropower to drive the region's aerospace
industry. If there was a sacred canon in the postwar era, it was vested in
erecting an infrastructure to power economic development. A growing population in the Pacific Coast states meant expanding markets for forest and
agricultural products and, increasingly, for surplus Columbia River
hydropower.

The sections of this book address the interrelated stories of postwar optimism, natural-resource extraction, extensive water-development projects,
and the growing numbers and affluence of Oregon's population. The great
spurt of dam building, ushered in with great enthusiasm immediately after
the war, eventually generated opposition from sports and commercial
fishers, Indian tribes, and citizens who wanted to protect anadromous fish
runs and to preserve scenic streams such as the McKenzie and Deschutes
Rivers. In the production-driven worlds of forestry and agriculture, policymakers and industry leaders promoted efficiency, economies of scale, and
the use of an ever increasing array of new chemicals. By the late 1960s, however, a growing environmental community challenged those industrial inno-

vations through legislative restraints and regulations, court challenges, and appeals to health, quality-of-life issues, and aesthetic considerations.

Two well-known journalists-turned-politicians, Richard Neuberger and Tom McCall, provide contrasting styles and ideas about conservation and environmentalism. Each was sometimes contradictory and ambivalent; both held a deep affection for Oregon and the Northwest. Through his prolific writings, the gifted Neuberger was more responsible than anyone for giving spectacular Hells Canyon its name. As a U. S. senator, Neuberger also supported the construction of a huge federal dam that would have flooded much of the canyon. While Tom McCall took courageous action against polluters to clean up the Willamette River and to enact Oregon's pioneering land-use legislation, the governor and his equally progressive political friend Robert Straub never questioned nuclear-power development or heavy timber harvests, issues that subsequently came to bedevil Oregon's politics and its environment.

If this book begins at a brighter, more optimistic moment in Oregon's history than the previous volume, it concludes on a more ambiguous note. Rivers that seemingly had been restored and cleaned up have once again been sullied and spoiled through human activity. Trace elements of DDT and other toxic chemicals in river-bottom sediments remind us of an earlier day when the use of such agents was considered a sign of progress and forward thinking. Water-development projects once deemed symbols of progress for improving navigation and producing hydropower are now dubbed salmon killers, destroyers of a regional icon. Efforts to rectify these and other environmental stresses have divided rural and urban Oregon, citizen from citizen, and worsened class schisms.

I agree with the writer Evan Eisenberg, who urges us to pay closer attention and to manage better "our cities, our factories and our farms."[1] Exercising greater concern for our own behavior—or at least giving it as much attention as we do wilderness—would help us build and maintain healthy human and natural systems. It is my hope that the larger narrative thrust of *Landscapes of Conflict* will point to those individuals and groups who valued civic commitment and stewardship over raw profiteering, who sought to build a society that promised something more than the reduction of all human behavior to market transactions. As the prospect of global warming looms before us, it is obvious that the heavy hand of human agency is literally everywhere. We ignore such indicators at our peril.

ACKNOWLEDGMENTS

The credits for the making of this book include many of the same people who provided suggestions, encouragement, and assistance in the research and writing of its companion volume, *Landscapes of Promise: The Oregon Story, 1800–1940* (1997). David Brooks, a social scientist with the Pacific Northwest Research Station of the United States Forest Service, first invited me to consider writing a regionally based environmental history in 1992. Don Wolf and Amy Seivers Fackler—two Oregon State University graduate students at that time— collected valuable documents for the first volume. Many of the research files they gathered—federal and state reports, census data, newspaper and magazine searches—provided important grounding for *Landscapes of Conflict*. My daughter, Aubrey Robbins, an undergraduate student at Oregon State University from 1996 to 1999, proved cheerfully expert at reading microfilm and ferreting out important articles from the *New York Times,* the Portland *Oregonian,* and other state and regional newspapers.

Oregon State University's always courteous Valley Library staff guided me to sometimes difficult-to-find sources, especially obscure state and regional scientific reports. Good friend and University Librarian Karyle Butcher kindly listened to my grumblings about occasionally malfunctioning microfilm machines. University Archivist Larry Landis and Senior Staff Archivist Elizabeth Nielsen offered critical direction and guided me through the intricacies of the Oregon State University Archives. Staff Archivist Karl R. McCreary provided valuable assistance with the unit's large photo collections. Former graduate student Linda Hahn early on pointed me to valuable additional materials housed in the University Archives. The computer-generated maps of Oregon and the Klamath Basin are the good work of OSU geographer Jon Kimerling. Todd Welch, an archivist with the Oregon Historical Society, offered helpful assistance and unfailingly had manuscripts ready for perusal during my frequent treks to Portland. The Northwest Coalition for Alternatives to Pesticides (NCAP) in Eugene made available important scientific reports and congressional hearings on pesticide issues. The Oregon Council for the Humanities and its executive director, Christopher Zinn, provided financial support that enabled me to travel to archival repositories in Bend, Eugene, Portland, and southern Oregon.

Perhaps my greatest debt is to the many professional associates and friends whose books and articles, interpretive ideas, insights, and suggestions helped shape this study. I always value Bill Cronon's patience and exceptional work in guiding this series and his constructive advice and wise counsel in how to properly frame a book. I also treasure the ongoing discussions of things historical and contemporary with Steve Haycox, a long time friend who occasionally ventures "outside" (from Anchorage) to attend conferences and symposia. This book has benefited from historian Katrine Barber's insights on the cultural context of the flooding of Celilo Falls. Thanks, as well, to colleague Ron Doel for guiding me to important secondary sources on postwar science.

My special gratitude to Marianne Keddington-Lang, the talented editor of the *Oregon Historical Quarterly,* for her continued encouragement and support. Bill Lang deserves foremost thanks for his insights, perceptive questions, and critical reading of the manuscript. Bill's impressive and wide-ranging reading interests, his abiding intellectual curiosity, and his own research on Columbia Basin water-development are reflected in some of the narrative argument in this book. Although we frequently disagree about issues such as Oregon exceptionalism, I have always delighted in the ongoing exchanges with Bill.

Urban historian Carl Abbott's writings on a variety of Oregon-related subjects prompted me to rethink approaches to the state's political history. Carl's work on the state's pioneering land-use system also suggested an important theme for the book. I have always valued occasional conversations with historians Eckard Toy and Mark Spence. Mark and I shared several "railroad seminars" during regular train trips to Portland to work at the Oregon Historical Society in the fall of 1999. Chris Mundt, a plant pathologist at the Oregon State University, directed me to appropriate specialists on a variety of scientific issues. Bonnie Hill, a good friend for nearly forty years, provided a critical reading of chapter Nine and steered me to several scientific reports on the issue of herbicide spraying in forest environments. My thanks also to running friends Irv Horowitz, Steve Payne, and Jim Ridlington for patiently listening to my stories about the making of this book. As always, Kay Schaffer and Paul Farber, administrators *extraordinaire*, have offered their encouragement and support for this project. I am especially indebted to Paul for making office space available after my retirement in June 2002. The always supportive and helpful History Office staff, Ginny Domka and Marilyn Bethman, graciously provided infrastructure support (computer needs, office supplies, and mailings) for the duration of this venture.

Judy Austin's copy editing talents helped to improve the clarity of the narrative and saved me from egregious stylistic lapses. Julidta Tarver, my long time editorial connection with the University of Washington Press, is patient beyond words, suffering through my occasional frustrations over relatively minor issues. Lita's quiet encouragement and persistence also kept me in motion to finish the volume. Lita, I greatly value our friendship. Finally, my wife Karla's professional work with the Benton County Juvenile Department serves as a constant reminder that historians should never take themselves too seriously.

Relief map of Oregon. Courtesy Jon Kimerling.

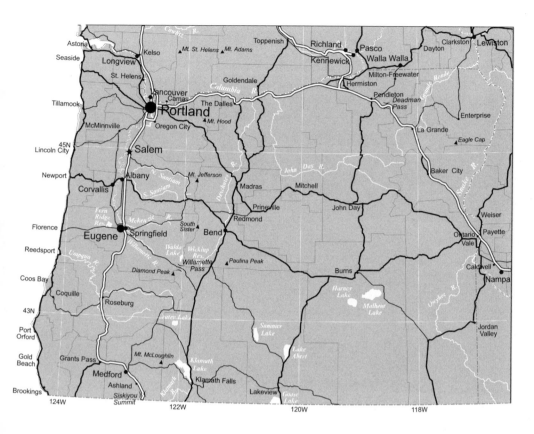

Selected cities, towns, and highways of Oregon. Courtesy Jon Kimerling.

LANDSCAPES OF CONFLICT

The Oregon Story, 1940–2000

PROLOGUE

A TIME TO REMEMBER

The 20th century is the American Century—HENRY R. LUCE[1]

O n Wednesday, August 15, 1945, the *New York Times* declared the end of the war in the Pacific with eight-column headlines across the front page:

JAPAN SURRENDERS, END OF WAR!

EMPEROR ACCEPTS ALLIED RULE;

M'ARTHUR SUPREME ALLIED COMMANDER;

OUR MANPOWER CURBS VOIDED

Arthur Krock, a *Times* columnist, wrote the same day: "The bloody dream of the Japanese military caste vanished in the text of a note to the Four Powers accepting the terms of the Potsdam Declaration of July 26, 1945." The previous evening, after Japan's surrender announcement flashed on the electric sign at the Times Building at 7:03 P.M., 2 million people gathered in New York City's Times Square. In a front-page story the following day, the *Times* reported that the victory roar greeting the end of the war "beat upon the eardrums until it numbed the senses. For twenty minutes wave after wave of that joyous roar surged forth."[2] In his best-selling book *The Crucial Decade*, historian Eric Goldman recounted the city's festivities, the waves of cheering that stretched from the East River to the Hudson: "The hoopla swirled on into the dawn, died down, broke out again the next afternoon, finally subsided only with another midnight."[3] The long,

3

wearying, and sometimes uncertain war years had finally come to an end with the explosion of two atomic bombs over Hiroshima and Nagasaki. In New York City and elsewhere across the country, the American people collectively celebrated this unique national moment.

Three thousand miles away at the western edge of the continent, where the war's end held even greater significance because of the region's proximity to the Pacific theater of operations, residents of the small coastal logging town of Coos Bay, Oregon, hastily organized a V-J Day parade to celebrate the end of "their" war. The next day the city was calm, according to the local newspaper, "like a ship at sea after a storm." The Coos Bay *Times* described a night of "riotous celebrations," with blowing whistles and horns and torn pieces of paper strewn everywhere. From a police standpoint, however, the celebrating brought little trouble: only two people were arrested on charges of drunkenness.[4] Further north, in the lower Columbia River port of Astoria, "business and industry all but stood still" as citizens recovered from the previous night's celebration. The *Astorian Evening Budget*'s Harold Haynes celebrated the end of hostilities in his daily op-ed piece: "The tumult and the shouting is still ringing in our ears," he wrote, but at this moment "we are bringing you greetings in a new world of peace and comparative quiet" with both the Germans and the Japanese "down for the count in a world that looks good again." Although Oregon's governor, Earl Snell, had proclaimed August 15 and 16 state holidays, the *Budget* noted that local canneries were limiting time off to one day because of the perishable nature of fish.[5]

Although the celebrations in those coastal settlements were part of a larger national euphoria, they were also vivid reminders that the Pacific Northwest was physically closer to the Pacific war zone than any other part of the forty-eight states. The Japanese bombing of Pearl Harbor brought instant fear and rumor to West Coast communities. Civilian volunteers and military personnel patrolled strategic communications and transportation sites to guard against the threat of a Japanese attack. The United States Navy placed submarine nets in Puget Sound, mined the mouth of the Columbia River, and the Boeing Company stretched a huge camouflaged net over its sprawling building complex to resemble a residential neighborhood from the air. On Oregon's Coos Bay, as early as Sunday evening on December

7, 1941, a small self-appointed defense force was stopping and questioning travelers crossing the new McCullough Bridge over the bay. Those anxieties persisted through the war years, occasionally heightened when a small Japanese submarine shelled Fort Stevens at the mouth of the Columbia River on June 12, 1942, and again later that summer when the same boat launched a small plane that dropped incendiary bombs on forest land in Curry County.[6] The Japanese surrender in Tokyo Harbor ended those fears.

Elsewhere in the state the Pendleton *East Oregonian* exclaimed, "The war is over and a new era is at hand." The newspaper's editor attributed American success in the war to the nation's values of "freedom, democracy, and opportunity." It was American policy "not to enslave the human mind." The great successes at the nearby Hanford nuclear reservation could not be attributed solely to brilliant scientists and the DuPonts: "Back of them were the American people with a strength that is great because our policies promote advancement, not stagnation." A few days after the battleship *Missouri* floated into Tokyo Bay to conclude the formal conditions of surrender with Japan, the Labor Day edition of the *Medford Mail-Tribune* praised the productivity of the American free-enterprise system. With labor-saving devices and mechanical energy enhancing human muscle power, the nation had proved it could produce "in conflict as well as in tranquility." Citizens could point with pride to a capitalist economy that encouraged individual enterprise and provided "rightly earned rewards and incentives." Victory in Europe and in the Pacific, the *Mail-Tribune* continued, would strengthen the belief that thrift, research, and high productivity were the bedrock of material progress. "Free laborers and unshackled captains of industry," together the centerpiece of the free-enterprise system, made the nation omnipotent and prosperous and had contributed to winning "the greatest war in history." Although Americans might not be a chosen people, there was no secret about the nation's economic and political success.[7]

"Oregon Folk Moving to Great Outdoors in First Heavy Holiday Rush in Years," the *Oregonian* exclaimed on the eve of the first peacetime Labor Day weekend in four years. People left for coastal beaches or mountain retreats to celebrate the return to normal. Although newspapers heralded the long weekend event as an opportunity to picnic and enjoy the great outdoors, the end-of-summer holiday was a rather tame affair by prewar standards. The Associated Press reported a relatively peaceful weekend nationwide, with only 187 fatalities from automobile accidents, drowning, and other causes. That figure compared favorably with the 626 violent deaths recorded in 1941 on the last prewar Labor Day weekend. According to the *Oregonian,* Portland's Union Station was so busy that clerks had no time to answer telephones; dispatchers at the local bus terminal also reported heavy crowds of travelers on Saturday and Sunday, and servicemen "doing the city on three-day passes" filled Portland hotels to capacity.[8]

One week removed from the big Labor Day weekend, the city of Pendleton observed its thirty-fourth annual cowboy event, dubbing this one the "Victory Round-Up." The list of participants in the various roping and riding activities bristled with the big names of former champions, and the first capacity opening night attendance in many years—3,700 people—turned out for the inaugural events. An even larger crowd, between 4,000 and 5,000, participated in the big Friday parade through Pendleton's streets, and "magnificently costumed" Indians brought "a brilliant climax to that mighty cavalcade." While it boasted about the Round-Up's record attendance, the local newspaper also pointed with pride to Pendleton's contribution to the war effort: more than 10 percent of its young men and women served in the armed forces; nearby Pendleton Field sent forth bomber crews to batter enemy forces; and the new community at Ordnance provided one of the three greatest storage areas in the nation for bombs, mines, ammunition, "and other articles of war."[9] Those articles, which included poison gases, would eventually become a frightening and toxic nightmare for nearby residents.

The reflections of Oregon newspapers differed little from the sense of buoyant optimism that existed in virtually every corner of the United States as the Pacific war reached its climactic end. In those critical autumn months of 1945, the nation stood powerful and confident. In a brief period of time it had built up an enormously productive industrial capacity, and as historian Howard Zinn observed, the United States entered the war declaring the right of all peoples to be free of external control, and in the end it had defeated both Germany and Japan. For most Americans, August 15, 1945, marked the conclusion of the most just of wars—"the good war,"

Studs Terkel would later call it. At no time since have the American people been as united and convinced that they were fighting a "good war" to defeat unmitigated evil.[10] Eric Goldman captured the spirit of the return to peace: "The America of V-J was prosperous, more prosperous than the country had been in all its three centuries of zest for good living. The boom rolled out in great fat waves, into every corner of the nation and up and down the social ladder."[11]

The Second World War was a profoundly dramatic and transforming event for the American people. For one thing, the onset of war in Europe and Asia brought an abrupt end to years of unemployment and economic stagnation. People who had been out of work for long stretches during the 1930s suddenly found themselves in a sellers' market. Despite the long workdays (during peak production, my father worked a twelve-hour shift seven days a week for the American Brass Company in Torrington, Connecticut), despite rationing, food shortages, and military conscription, the wartime period meant dramatically improved economic conditions for most, a heady experience for families who had suffered through a decade or more without steady work. Even before Pearl Harbor, the construction of defense plants in California helped to revitalize western Oregon's moribund lumber industry. Once the war was underway, the region's leading newspapers printed extra pages of help-wanted ads—a striking departure from the Depression experience.[12]

The war years revolutionized the American workforce in several other ways. Thousands of women took wage-earning jobs for the first time. Between 1941 and 1945, the number of women taking wage-earning jobs increased by 57 percent, and the percentage of the entire workforce made up by women increased from 25 percent to 36 percent at the close of hostilities. At the peak of Boeing's wartime production effort, women made up 46 percent of its labor force. Women with no work experience outside the home could earn $200 to $250 a month at Boeing or in the shipyards. To accommodate families with both parents at work or in the armed forces, day-care centers became important features of urban life for the first time. Such services made it possible for women to take wage-labor jobs traditionally the province of males. At Evans Products on Coos Bay, females numbered 800 of the 1,200 employees, and the local union forced the company to pay the same wage for people doing the same job, irrespective of gender.[13]

The war scrambled communities and reshaped virtually every region of the country; it uprooted people from home and sent them to work in distant defense industries or into the armed forces. In many instances, before young men and women were shipped overseas, they were sent to training camps far from home. One such person was young, red-headed James Brewer from Winfield, Louisiana, who enlisted in the Army in 1943 and was ordered to report to Oregon's Camp Adair, a mid-Willamette Valley Army training base. There he met Patricia Skaling, an Oregon College of Education student in nearby Monmouth. Before Jim took ship in the summer of 1944 for the European theater of operations, the two were married in a quiet ceremony in Salem. When Jim mustered out of the Army in 1946, they lived for a time in his native Louisiana; but on Christmas Day 1947 the two returned to Salem, where they decided to make their home and raise a family. One of their seven children, red-haired Karla, subsequently became my wife (after I, too, discovered Oregon via military service in the 1950s).

Such personal stories can be repeated thousands of times across the United States, with the most prominent movement of people being from east to west. "It was as if someone had tilted the country," historian Richard White writes, and "people, money, and soldiers all spilled west."[14] Although many war workers came from outside the Northwest, residents within the region made up the largest percentage of the wartime labor force. An exodus of people from the countryside to booming urban centers contributed to a dramatic reshuffling of the region's population, and many of the more rural districts actually declined in population during the war. The dynamics of Idaho's demographic profile provide a striking example: the state's resource-dependent population decreased by 15,000 between 1940 and 1945.[15]

Individual stories of movement and relocation also resonate with testimony to Oregon's promise and abundance, the popular belief that the postwar years would usher in the good life for those willing to work hard and take advantage of new opportunities made available through cheap hydropower, the opening up of newly reclaimed agricultural lands, and jobs created through the region's expanding industrial base. The watchword for the new world emerging in Oregon and the American West centered on the desirability of uninhibited economic development, the widespread belief that the West stood at the threshold of an explosive growth in material wealth. An atmosphere of expectancy and faith in the future, that tomorrow would witness the realization of those dreams, pervaded discourse in virtually every western state as the war drew to a close.[16]

War-related labor shortages in the Pacific Northwest's natural resource industries accelerated the move to adopt mechanized production processes. Gasoline- and diesel-powered tractors replaced horses, mules, and oxen on farms and ranches across the region. In a fit of urban arrogance, the *Oregonian* stumbled badly in late August 1945 when it ridiculed a U.S. Senate committee's fear that rural workers would refuse to return to farm life when the shipyards and other war industries closed. Although farm laborers might want "to play around a while" before they returned to "the pig and the plow," the newspaper did not share the committee's apprehension that unemployed workers would flood the nation's cities. Transplanted farmers "know, as well as we do, that the farm is the best place of all."[17]

While the Portland newspaper may have over-idealized agrarian life, it did not anticipate the role that technology would play in displacing workers in extractive industries. An even greater dearth of workers in woods operations had a similar effect in accelerating the mechanization of timber harvesting. Logging bosses placed a premium on labor-saving devices and other innovative techniques to boost production. On southern Oregon's Coos River, the Irwin-Lyons Company turned the waterway into an industrial arteriole, building first one and then a second hydraulically operated splash dam to move huge timbers downstream to tidewater. Dow Beckham, one of the key men responsible for "splashing" logs on the river, helped the firm build a second splash dam in 1944 so that its "very small crew" could better avoid the constant need to fight logjams.[18] Although the dams provided an efficient means to move the logs to tidewater, daily flooding scoured the riverbed and further degraded the waterway's ecology. Beckham recognized later in life that the effects of the constant flushing caused damage to downstream plant and aquatic life.[19]

The most visible signs of change in the Pacific Northwest involved the two principal centers of war manufacturing, the Portland-Vancouver area and

Seattle-Tacoma vicinity. The Boeing Company employed 50,000 people during 1944, its peak production year. The story of the company's enormous
assembly-line production records, the contributions of the region's alu-
minum-reduction plants, and the cheap hydropower provided by Bonneville and Grand Coulee Dams have become part of the region's celebrated
folklore of winning the war. The manufacture of Boeing's B-17, the famous
Flying Fortress, reached an astounding sixteen planes every twenty-four
hours in 1944. The company was also capable of producing a B-29 in five
days at each of its nearby Renton assembly lines. Working on cost-plus contracts, Boeing put up immense buildings, larger than anyone had ever imagined, to house the long rows of planes under construction. By the close of
hostilities, Seattle ranked among the top three cities in the United States
in per-capita war orders.[20] The region's huge defense plants were directly
related to the great Columbia River hydropower projects and seemed to
promise the long-standing objective of diversifying the Northwest economy. Although those industrial structures reflected the special needs of a
wartime economy, they also harbored the roots of future social and environmental debates.

Equally renowned and perhaps even more impressive in reshaping
the regional landscape were the three big Kaiser shipyards in the Portland-
Vancouver area: the Oregon Shipbuilding Corporation, on the Willamette
River near Portland's St. John's Bridge; Swan Island, farther upriver; and
a third construction facility in Vancouver, Washington, near the Columbia River Bridge. The three federally subsidized yards eventually employed
as many as 120,000 workers, with an additional 40,000 people engaged in
related jobs. The metropolitan area's population grew by nearly a third at
the peak of war-related employment. This sudden influx of huge numbers
of people created an immediate housing crisis. Existing homes were converted into apartments; trailer camps sprouted everywhere; owners reno-
vated and rented dilapidated homes; sheds and outbuildings were patched
up as temporary dwellings; and some workers lived in tents or in their automobiles. To alleviate the housing emergency, industrialist and shipbuilder
Henry J. Kaiser used Federal Public Housing Administration loans to purchase 650 acres of flood plain, slough, and pastureland on the Oregon side
of the Columbia River. The low-lying area had been subjected to repeated

flooding until the Army Corps of Engineers built a series of high dikes to keep the Columbia's spring rise at bay.[21]

Kaiser's Oregon Shipbuilding Company eventually constructed a sufficient number of units to house 42,000 residents, making "Vanport City" the second largest community in Oregon at its peak in 1944. Supported by wooden blocks, with fiberboard walls, the instant residential city was the nation's largest single wartime housing project—but one that would prove vulnerable to the forces of nature. The Vanport setting became home to a mix of people, including a sizable percentage of African Americans who made the move west to work in the shipyards. Vanport also underscores another reality linking race and class to place of residence. By war's end, African Americans made up 35 percent of Vanport's population, a much larger percentage than anywhere else in the state of Oregon. As long as real-estate covenants prohibited them from living in other sections of Portland, they would remain—along with their white working-class neighbors—in the temporary community adjacent to the volatile waters of the Columbia.[22]

During the war years, Portland and its suburbs (including Vancouver) gained more than 250,000 residents, and the state's growth rate for the entire decade was nearly 40 percent (or 432,000 people). At the peak of wartime production in early 1944, approximately 140,000 defense workers lived in the Portland metropolitan area. The big Columbia River shipyards accelerated the urban concentration of Oregon's population, especially in the Willamette Valley. Whereas every county in western Oregon sustained population increases during the 1940s, four counties east of the Cascade Mountains (Baker, Gilliam, Sherman, and Wallowa) experienced net losses. Moreover, in western Oregon, the three Portland-area counties (Clackamas, Multnomah, and Washington) grew by 37 percent, with the greatest increases taking place in the two non-urban counties (Clackamas, 52 percent, and Washington, 56 percent). Nearly half the state's population lived in the greater Portland metropolitan area by 1950,[23] a trend that would continue during the second half of the century.[24]

Seemingly remote and otherwise isolated places in the Northwest also underwent spectacular change during the war years. The most notable by far—and ultimately the one with the most lasting environmental influence—was the quiet central Washington farming village of Hanford, where the federal government erected a top-secret, multimillion-dollar facility adjacent to the Columbia River to produce plutonium for the first atomic bombs. The United States Army removed 1,500 residents from the 670–square-mile reservation in 1943 and then repopulated the area with a force of more than 45,000 laborers, turning Hanford into one of Washington's largest settlements virtually overnight. The assembled workers set about building several gigantic nuclear reactors, in the process creating an "atomic space" half the size of Rhode Island along the shores of the Columbia. The Hanford site was selected for its special qualities: isolation and a dependable volume of fresh water to cool the reactors and the availability of cheap hydropower produced upriver at Grand Coulee Dam. The mid-Columbia setting also became home to an irony of sorts: the three reactors built during the war produced a form of energy that had no social utility other than mass destruction.[25] Although the Hanford complex is beyond Oregon's borders, the developments that took place there would have long-range implications for both the Columbia River and all living things downwind and downriver from the reservation.

Decisions made at the federal level during the Second World War reshaped the Pacific Northwest landscape in still other ways. The wars in Europe and Asia had far-reaching effects on the region's forest-products trade and its dependent communities. After decades of seasonal and market-induced unemployment, timber districts in the Northwest—especially those with huge stands of old-growth forests—entered a period of sustained expansion that brought full employment and regular paydays for workers. The rekindling of patriotism and self-sacrifice brought an end to federal threats to regulate harvest practices as industry leaders responded to wartime production objectives. With self-sacrifice as their rallying cry, lumber-trade officials informed the Forest Service that trees were less important than human lives and that the United States should forego "future needs for immediate demands." Appeals to sustainable-forestry practices and community welfare, they argued, had to give way to increased production. Throughout Oregon's forested districts, the heavy demand for lumber—new barracks, temporary housing for war workers, government construction projects, and wood materials for Pacific islands such as Hawaii and Midway—placed great strains on the sharply diminished workforce.[26]

Labor was not the only item in short supply in the lumber industry. In its annual report for 1940, the U.S. Forest Service emphasized the critical role of wood fiber in wartime. The armed forces would need forest products for "pontoon bridges and railroad ties, gunstocks, ships and docks, barracks and cantonments, crates, mess halls, hospitals, and post exchanges." Under the leadership of acting chief Earl Clapp, the agency cautioned that war production would place strains on states such as Washington and Oregon, "the Nation's wood basket." Two years later, Clapp declared that pressures to supply war-related demands had contributed to "a new wave of destructive cutting," especially in the old-growth stands of the western states. He pointed out that national-forest harvests had increased sevenfold over the 1939 cut, with most of the sales in the Pacific coastal states. The acting chief worried that "excessively concentrated and unnecessarily destructive practices" in the Pacific Northwest were "jeopardizing opportunities for sustained-yield operations." In the not too distant future, he feared, many communities would be suffering. But Clapp also saw an upside to the rapid liquidation of old-growth timber, an argument that would become an article of faith to most foresters: the huge volume being sent to sawmills had the advantage of "transferring static virgin areas to vigorous second growth."[27] That line of reasoning would become the great rallying cry to progressive foresters in the postwar years, driving the scientific-ideological justification for converting "decadent" and dying old forests into vital, fast-growing second-growth timber.

At the outset of hostilities, the leading lumber trade organization in the country, the National Lumber Manufacturers Association (NLMA), used the war crisis to urge the federal government to avoid regulatory "bottlenecks" and to defer "controversial and explosive issues" until the war was over.[28] For Oregon and Washington, the leading lumber-producing states in the country, federal regulation of private timberlands had special significance. The NLMA and its regional affiliate, the powerful West Coast Lumbermen's Association, fought all federal attempts to regulate harvesting practices on private timberlands. Finally bending to the influential lumber-trade organizations, President Franklin Roosevelt decreed that until the war was over, "there should be no Federal legislation providing for Federal regulation of forestry practices on private lands."[29]

But the regulation issue would not go away, and most of the discussion took place in the Pacific coastal region. Midway through the war, Lyle Watts—Clapp's replacement and former head of Forest Service Region Six (the Pacific Northwest)— cited again the "urgent need" for "public regulation to stop destructive cutting" and to accelerate the program of acquiring private timberlands. His travels about the Pacific Northwest convinced him that "aggressive action rather than complacency" was called for to assure continued forest production. "The public interest in private forest lands," he observed, "is not open to question." Because of shortages in privately held timber, the national forests were playing an increasingly critical role in meeting wartime lumber needs. In his annual report for 1944, Watts announced that national forests now supplied 10 percent of all timber harvested in the United States. If that trend continued in the heavily timbered Pacific Northwest, he feared, local foresters might "exceed sustained-yield cutting budgets." He warned: "There must be no yielding to such pressures."[30]

Changes introduced during the war years would have lasting influences on the landscape and ecology of Northwest forest lands. Gasoline-driven chain saws and the increasing use of diesel-powered yarding machines vastly accelerated both the pace and the scale of disturbance in the forest environment. The special nature of the new machines gave priority to clearcutting practices, harvesting methods that became standard everywhere in Northwest timber districts following the war. Those new and more powerful technologies, and the increasing miles of logging roads that penetrated ever deeper into the forested hills of western Oregon, also compounded erosion and landslide problems and accelerated the spread of a root fungus in the Port Orford (white) cedar district in southwestern Oregon. When the domestic savings accumulated between 1941 and 1945 precipitated a prolonged postwar construction boom, the technological advances enabled the industry to meet the increased demand for lumber. Few in that time and place questioned the essential rightness of the massive clear-cutting, road building, and stream channeling that helped sustain the boom.

Shortly before the United States became actively involved in the wars in Europe and Asia, writers were offering up familiar images about Oregon's quality of life, indulging in what the British writer Denis Brogan referred to as painting word pictures of "America the golden," the deeply ingrained habits of the booster. One of the more widely read and glowing descriptions appeared in the state's contribution to the American Guide Series, published in 1940. "It is easy to write about Oregon," the authors observed. With its "lovely dappled" up-and-down slopes and its "ever-green beauty as seductive as the lotus of ancient myth," the state possessed qualities that inspired great affection. Still unspoiled and uncluttered, Oregon was "a green land of forests and grassy wilderness teeming with wild life," a place with deep-soil valleys giving birth to golden harvests of grain, a land rich in minerals and fast-flowing streams with vast hydropower potential. The Oregon guide referred to the state—with its farmlands subject to "scientific scrutiny," its federal rangelands under grazing regulation, and its national forests serving as models of selective cutting—as a leader in conservation. In a foreword to the volume, Governor Charles Sprague applauded the book for portraying the state to outsiders and presenting "Oregon to Oregonians."[31] Hyperbole aside, the guide was perpetuating Eden-like and mythical images of promise deeply rooted in the region's historical literature.

In a January 1940 editorial celebrating western American attractions, the Southern Pacific Railroad's venerable *Sunset* magazine was lyrical in its praise of Oregon's outback, places beyond pavement's end where trails follow "the quiet peace of high mountains; along fish-filled lakes and rivers; through the green jungles of salal and vine maple." *Sunset* applauded highways that stretched out to reach fruit orchards and fields of grain, that allowed citizens to explore mountains, to travel alongside rivers, to witness the ocean from the advantage of a spectacular overlook. "Our West is many things" large and small, the editorial continued, "yet, one thing is before us always—the fertile earth." The rhythms of the West were of the earth and

in the end offered up a way of living that was different. "This is our West."[32]
The magazine's reference to the outdoor splendors of the West anticipated
the recreational boom that would take place following the war when new
automobiles, highways, and expanded incomes accelerated the American
love affair with the outdoors. *Sunset's* rhetoric would appear to confirm
Brogan's observation that Americans had always been given to overstate-
ment, exaggeration—practicing the habits of boosterism.[33]

The Spectator, a Portland monthly that would not survive the opening
months of the war, presented an equally romantic if less flowery vision of
abundance and prosperity for the region, much of its praise centered on
the Bonneville project and the promise of cheap hydropower. Oregon stood
at the "economic crossroads," writer Rex Eastman reported in January 1940.
Up to that moment, the state's economy had been "frontier or colonial,"
with agriculture and lumbering leading the way. As a consequence of
dependency, Oregon's people "sell cheap and buy high." But the prospects
for relief, especially new industries for the city of Portland, appeared promising because of Bonneville's splendid system of water transportation and
power development. Inexpensive electricity, newly discovered chemicals
and minerals, and the new "miracle metal," aluminum, would provide the
region with the opportunity to escape its colonial shackles. Because of its
proximity to excellent water transportation and cheap electricity, Eastman
predicted, Portland could become "the center of an enormous aluminum
industry."[34] The promise of the city and the larger region was, to this way
of thinking, vested in new forms of energy, new industries, an increased
ability to manipulate the landscape, and the introduction of increasingly
powerful chemical compounds and synthetics in the environment. With
Depression-era suffering ever present in their minds, Northwesterners were
eager to pursue experiments in social and environmental engineering.

Historian James Patterson has called the Second World War the "central
convulsion" of the twentieth century.[35] It was a time of endings and beginnings; it signaled an end to the Great Depression; and for the thousands of
people who moved from farm and ranch to urban settings, it brought to a
close the steady pace of life and the intimate associations with the seasonal
rhythms of summer and winter. For Oregonians, the war meant a sharp

increase in demand for the state's natural resources, market conditions that accelerated when peacetime returned. A people and an economy that had entered the war years under animal and human power exited the conflict with gasoline-powered engines (especially in agriculture) and new visions for a world free from want and deprivation. War-related industries brought new ideas and new and more ethnically diverse people to Oregon and the Northwest, the seeds of a population influx that in time would change the region's demographic mix. And then, with the surrender of Germany and Japan and with sizable savings accumulated during the war, citizens looked forward to material rewards for their sacrifices.

To tell the story of the modern age, Eric Hobsbawm urges historians "to concentrate on the dynamics of global transformation," especially for the years following the Second World War when a world capitalist economy centered on the United States. The boom that took place was unprecedented, unlike anything the world had experienced, with vast flows of internal migration from country to city, from agriculture to seemingly more promising economic opportunities.[36] That human floodtide leaving rural settings for urban living represented the most striking demographic movement in modern American history. That same story spilled out all across the American West, but especially in California, the spatial playing field for the aerospace industry and its defense-sector affiliates. Those booming markets provided resource-dependent states such as Oregon with an expanding outlet for wood products and agricultural goods. Global events associated with the Cold War, when the United States faced down the Soviet Union, China, and their respective client states in a series of international crises, fueled those expansive and turbulent conditions.

The United States was a nation in perpetual motion in the years following the Second World War. The explosion in construction activity, the rapidly expanding highway system, and the omnipresent need for material goods to fuel the expansion kept the old primary-products sector of the western American economy operating at a frenzied pitch for more than three decades.[37] The boom broadened in the late 1940s, accelerated in the 1950s, and then achieved unimaginable rates of growth during the 1960s, an unprecedented pace of economic expansion that astonished all observers. With 7 percent of the world's population and 42 percent of its income, the

United States generated about half the total manufacturing output and produced 80 percent of the world's autos by the late 1940s. Those were years,
James Patterson argues, of even "grander expectations about the blessings
of science, technology, and expertise in general." The field of scientific
inquiry—in physics, chemistry, biology, and medicine—seemed nearly
omnipotent. The development of penicillin and streptomycin in the 1940s
was followed by antihistamines, cortisone, and new and different antibiotics in the years to follow.[38] It was truly a time to remember.

I
POSTWAR
PROMISE

1

THE GREAT HOPE

FOR THE NEW ORDER

In the hurried, troubled, heartsick days of war, all of us looked back
upon peace and yearned for it. But we can't have it back exactly as it
was in 1940. The stage setting has changed in dozens of little, unsettling
ways, new players are on the scene, and we find that the script has been
altered.—E. R. JACKMAN[1]

It may be said with some truth that the post-industrial age began on
August 6, 1945, when three American B-29s took off from Tinian Island
in the Marianas and directed their course 1,500 miles north to Japan. One
of the best-known planes ever to fly, the *Enola Gay,* carried in its bomb bay
the five-ton uranium bomb known as Little Boy. Shortly before 9:00 A.M.
the Japanese port city of Hiroshima on southwestern Honshu Island was
incinerated in a great spiraling cloud of fire, smoke, and dust. Later that
same day, the United States government released the news that an American airplane had dropped an atomic bomb on Hiroshima with the power
of more than 20,000 tons of TNT. The White House announcement concluded with the terse statement: "It is a harnessing of the basic power of
the universe."[2] There is little question that the awful destruction unleashed
first on Hiroshima and then Nagasaki signaled that the world had crossed
a threshold of sorts, one in which human ingenuity had appropriated an
awesome power—indeed, the seeming capacity to end human life itself. In
Oregon the Medford *Mail-Tribune* thought the date of the bomb's delivery would be etched in human history "along with 1215, 1492, 1776—a date

for school boys to remember, as marking the end of one great epoch and the start of another."[3]

While the bomb suggested an obvious path to doomsday, many journalists and popular writers praised the wondrous potential that atomic energy and other scientific advances would bring to humankind. The benefits of science in the postwar era promised to bring ancient millennial dreams to reality, to bring forth an Eden on earth, to finally make humans ascendant in the physical world. The war itself served as a great divide of sorts, an historical transition separating the evolution of basic science that preceded the 1940s with the technological explosion that followed. Prewar scientific advances, especially in physics and chemistry, created the conditions for the massive reordering and manipulation of nature. And then, under the pressure of urgent military requirements and with huge infusions of federal monies, those revolutionary scientific advances were put to the task of developing new technologies and more productive enterprises. One of the most expensive (and expansive) of those undertakings was the Hanford nuclear complex that took shape on the Columbia River just beyond Oregon's northern border. The great outburst of technological innovation that followed the war included commercial sales of a vast array of synthetics such as DDT, new products that would eventually pose a serious threat to conditions of life in both urban and rural environments.[4]

The new world of miracle chemicals, however, was only one component in a wealth of scientific advances that occurred in the years after 1945. As the wars in Europe and Asia were winding down, President Franklin D. Roosevelt encouraged Vannevar Bush, head of the Office of Scientific Research and Development, to use that agency's expertise in the "war of science against disease." In his now famous report *Science: The Endless Frontier* (1945), Bush argued that progress against disease, the development of new products, new industries, and more jobs required "continuous additions to knowledge of the laws of nature" and application of those findings to practical purposes. Full employment, new jobs, and inexpensive consumer goods must be "founded on new principles and new conceptions which in turn result from basic scientific research." The vast advances in scientific research during the war, Bush predicted, marked only the beginning of even more spectacular developments "if we make use of our scientific resources." Although earlier frontiers had disappeared, the scientific one remained. "Without scientific progress," the nation's health would deteriorate, the standard of living would stagnate, and the jobless rate would increase.[5]

All signs indicated that the federal government would be a full participant in shaping social, economic, and scientific policy following the war. In a 16,000–word text delivered to Congress in September 1945, President Harry Truman offered a twenty-one-point program designed to smooth the way toward "high prosperity" and "a better life here at home and a better world for future generations to come." Among the president's recommendations were several New Deal-type programs: increases in unemployment compensation, fair labor standards, full-employment legislation, farm-price subsidies, federal housing support, and funding for scientific research, public-power development, and expanded highway construction. But the focal point of the president's recommendation centered on a national commitment to build a better future through a vast public-works program and through promoting research in the basic sciences, medicine, and public health.[6]

As the war was drawing to a close, the American media joined the chorus of voices offering up visions of a future free from want and worry. A General Motors advertisement in *Life* magazine proclaimed the United States ready for a "Journey into Tomorrow," a world in which scientists were "moving forward with new methods and improved products" that would bring a wealth of new blessings to humankind. This most American of corporations was, in fact, promoting the fantasy that future happiness could be purchased in the marketplace. But if those songs of optimism were exaggerated and overblown, there was a substance of truth to such commentaries. Discovery and invention were the order of the day—the development of the electron microscope, fiberglass, wonder fibers and plastics, light and handy Styrofoam containers, nylon apparel, cheap vinyl floor coverings, inexpensive plastic food wraps, and latex foam used in mattresses.[7]

The new homes that went up after 1945 differed from their prewar counterparts in several ways. They were often constructed from prefabricated materials and had central heating units, with oil, natural gas, or electricity replacing coal and wood. Those dwellings were also filled with a wide array of modern appliances and gadgets, from cooking stoves to refrigerators, vacuum cleaners, and washing machines. Rapidly expanding suburbs meant outdoor barbecues and instant markets for a wide range of goods— including hotdogs and other foods—and an ever increasing flow of throw-away materials.[8] That brave new world of plastics and synthetic materials spilled a river of consumer goods everywhere across the United States, from the graying factory towns of the eastern seaboard to the expansive aero-space communities of the Far West, and from Rocky Mountain mining centers to the cities and small towns of the Pacific Northwest.

From the vantage of time, however, it is clear that the key innovations of the immediate postwar years—those with particularly long-range consequences—were in the chemical and pharmaceutical fields. "Wonder drugs" such as penicillin and streptomycin, developed during the war, became readily available to the public by the late 1940s, and other new anti-biotics entered the market during the next few years. Congress increased appropriations for the chronically underfunded National Institute of Health, and in 1948 it renamed the agency's disease-specific units the National Institutes of Health. The two most spectacular medical break-throughs in the postwar years included the identification of DNA (deoxyri-bonucleic acid) in 1953, an event that vastly accelerated research in genetics and molecular biology, and the development of the Salk vaccine in the early 1950s, the miracle drug that provided immunity from the dreaded crippling disease poliomyelitis.[9]

The most infamous miracle chemical compound to emerge from the war was the chlorinated hydrocarbon DDT, "the atomic bomb of insecti-cides, the killer of killers, the harbinger of a new age in insect control." First used during the war to combat insects carrying infectious diseases and then made available for civilian use on August 1, 1945, DDT was wildly popular, and few questioned its use until the late 1950s. Because it was highly toxic to insects, inexpensive, and easily adapted to aerial spraying, one economic entomologist referred to DDT as "a miraculous insecticide" that promised to "banish all insect-borne diseases from earth." The only time most Americans thought about DDT, according to historian Thomas Dunlap, was when the spray truck made its annual rounds during the summer months. American farmers became heavily dependent upon such persistent chemicals in the years following the war, and DDT was by far the most popular of them all, enjoying wide public approval until the publication of ecologist Rachel Carson's *Silent Spring* in 1962.[10]

The tremendous productivity of the American economy contributed to one of the central pillars of postwar optimism: an unwavering belief in continued economic growth and unbounded prosperity. It was widely assumed that the country could engineer into existence ever increasing wealth, security, and financial well-being, that the American people could expect a future in which all things were possible. Economist Robert Samuelson underscored the grand postwar vision captured in what he called "the American Dream": "We didn't merely expect things to get better. We expected all social problems to be solved. We expected business cycles, economic insecurity, poverty, and racism to end." The American Dream, he concluded, became "the American Fantasy." Truman Moore, who grew up in Myrtle Beach, South Carolina, during the 1940s, believed that new inventions and unending discoveries would "bring perfection to humankind," that the further into the future one could conjecture, "the better things would be." Synthetics would replace exhausted natural resources, thereby providing "an artificial cornucopia to pour forth abundant substitutes for any shortages." Moore understood the postwar promise to mean a world free from flies and mosquitoes, because DDT was available for civilian use. "I liked the smell of it," he remarked.[11]

At the half-century mark *Life* assured its readers that the United States was leading the world into the scientific age. "These have indeed been decades of dazzling achievement," the magazine declared. A triumphant science had now given humans "the temerity to challenge even the speed of the earth's turning." Moreover, a life expectancy once beyond the reach of human control had been dramatically advanced by the development of cures for certain diseases.[12] Some of that new world of technological wonders seeped into my own life during my growing-up years in rural Connecticut. I vividly remember, as a young high-school student in the early 1950s, the family physician prescribing penicillin for my seasonal bouts with earaches. And as a summer wageworker on my grandfather's dairy farm, one of my responsibilities was to spray the cattle with a DDT insecticide bomb before turning them out to pasture following the evening milking.

The preeminence of science in the immediate postwar years fit well with the nation's overweening confidence as it stood astride a world recently devastated by war. Citizens of the United States were living amidst one of the great myths to come out of the Second World War—Henry Luce's notion that the twentieth century would be "the American Century." Luce asserted—first in a *Life* editorial in 1940—that if the twentieth-century world was to achieve its full vigor and health, it "must be to a significant degree an American Century." Luce based his thesis on the assumption that the United States would prevail because the vibrant strength of American business and the free-enterprise system would combine to make the nation a leader in the world economy. In the coming decades, congressmen and even presidents played upon the American Century idea. In his book *The American Century*, Donald White points out that following the war American ships and aircraft carried United States commerce to distant cities, and everywhere the dollar served as the medium for international transactions. American bank credit reached to the far corners of the globe: "Food, clothing, movies, machines, and science made the American name known throughout the world."[13]

A good case can be made that Oregon and the Pacific Northwest represented a microcosm of the general optimism that prevailed around the country. Writing for the *New York Times Magazine* in late 1945, the Northwest's best-known journalist, Oregon's Richard Neuberger, referred to the region as "The Land of New Horizons." Northwesterners were "buoyant and cheerful" and looked forward to living up to Thomas Jefferson's prediction of "a great, free and independent empire on the Columbia River." Having experienced its greatest industrial development during the war years, the region now confronted a frontier of modern technology that brightened the "prospects for fulfilling Jefferson's vision." Neuberger praised the natural attributes of the Columbia River country: its rich "treasure-trove of hydroelectricity"; its expanding acreages of irrigated land; its future in forest products; and its proximity to Alaska and the prospects of an expanding trade with that distant territory. "But the grandest boast of the Northwest," he wrote, "is that it is 'God's Country.'" In the midst of the region's scenic majesty, the average citizen "enjoys privileges reserved in other places for only a few." Although the Northwest confronted problems with the rapid liquidation of standing timber, Neuberger was confident that sustained-yield forestry and good stewardship practices assured a promising future.[14]

The joyous celebrations that took place in Oregon during that first peacetime holiday season appeared to reflect Neuberger's sense of optimism. Portland reported record sales in consumer goods as Oregonians crowded through stores and shops before Christmas. Because of a transportation bottleneck, more than 10,000 soldiers and sailors added to the crush of people walking downtown streets on the day before Christmas; long queues at state liquor stores promised a "spirited eve," according to the *Oregonian.* Elsewhere in the state, citizens looked forward to an encouraging future; the Roseburg *News-Review* promised that 1946 would be a "banner year, both for the erection of homes and business properties." The newspaper expected "new industries and a score of new business firms" to open early in the year. The Pendleton *East Oregonian* looked forward to "a brand, spanking new year full of rich achievements." While the area faced significant local problems, Pendleton and Umatilla County could look forward to good times "if we want to make it so." With Bonneville and Grand Coulee dams already in operation and McNary Dam under construction, the area's "power problem is virtually solved." In southern Oregon the Medford *Mail-Tribune* praised the bright prospects for agriculture and business, especially activities related to lumber, construction, and real estate.[15]

Beneath the surface of those postwar celebrations and merriment, however, the nation confronted serious problems, especially those associated with the transition from war to peacetime production. With the social and economic chaos of the Great Depression as recent memories, "reconversion" became the popular buzzword for federal and state officials, agency heads, newspaper journalists, and developers who stood to benefit from the shift to a peacetime economy. In an *Oregonian* column published a few days after the Japanese surrender, the normally exuberant Neuberger wrote that he thought the Pacific Northwest would face a supreme test in the postwar period. The civilian population had increased 17 percent since the defense emergency—most of the newcomers working in shipyards and aircraft plants—so reconversion would have "quite a different meaning on the Pacific Coast." Before the war, there were few durable-goods industries in the Far West. Toward what end, Neuberger asked, "will the ship-yards be converted?" And there were still other problems: the Northwest's forests, which supplied half the nation's timber, were overcut; conservation practices had been abandoned; and "the policy of sustained yield—making the annual cut equal new growth—has been forgotten." The usually upbeat Neuberger predicted that Washington and Oregon sawmills might have to reduce rather than expand their postwar workforces.[16]

While Richard Neuberger fretted over the future of the region's economy, Oregon's Postwar Readjustment and Development Commission (PRDC), first appointed in 1944, was already attempting to craft a blueprint for the future. Although its monthly reports repeatedly emphasized the virtues of private enterprise, the Oregon commission (and similar advisory bodies in other western states) echoed some of the old New Deal passion for planning. According to historian Gerald Nash, the effort to develop postwar plans reflected the western states' confidence that the region was "on the threshold of a new age." California governor Earl Warren told a group of executives in 1944 that business and industrial leaders sensed an "atmosphere of expectancy, faith and determination" in the future. When Warren appointed the California Reconstruction and Reemployment Commission, he expected the body to adopt a pragmatic strategy in which the state's government would cooperate with private enterprise to plan for the postwar era. The Oregon commission's first report to Governor Earl Snell placed a similar emphasis on "public and private projects" that would be ready to "go" when hostilities ended.[17]

During its five-year existence, the Oregon commission produced monthly reports that reflected its hopes and aspirations for the future, a body of writing that was decidedly more conventional than radical. Appointees to the commission included utility executives, public officials, state and federal agency representatives, and other prominent individuals from the business community. As the collapse of Germany appeared imminent, the commission reported that the Oregon business community was "becoming restless and anxious to put its postwar plans into effect." It was important that free enterprise not lose time "in preparing for peacetime business." The commission saw problems ahead in pursuing postwar projects: although architects and engineers had already prepared blueprints and designs and some firms had engaged contractors, the War Production Board was refusing to free essential war materials. "At this point free enterprise is stymied," the commission declared in April 1945.[18]

From the vantage point of the twenty-first century, the PRDC's monthly statements read like a prescription for the massive reordering of nature. Although it championed free enterprise and emphasized that public works were "only a stop-gap," its reports highlighted the federal government's role in building the huge dams planned for the Columbia and Snake rivers and the smaller but still significant multipurpose dams associated with the Willamette Valley Project. Its monthly releases praised the reclamation plans for the middle Deschutes system, the Klamath Basin, and the Ontario and Umatilla districts in far eastern Oregon. The reports also called for federal funding to support housing loans; to expand the state's highway system; to "straighten" rivers, lengthen jetties, and deepen harbors; and to subsidize forestry and agricultural research.[19] In all of those varied projects, federal monies would fund the infrastructure for construction activities that would serve as the foundation for all subsequent development, enterprises that ultimately would result in systemic alterations to the region's landscape.

To be fair, not all of the commission's reports focused on uninhibited natural-resource development. Its monthly communiqués occasionally mentioned deforested lands, water pollution, and declining salmon runs. Because the government had declared timber an essential war material, the commission noted on one occasion that War Department pressures had forced loggers to put aside sustained-yield practices "in the frantic endeavor to produce more and more lumber." As a consequence, Oregon's harvest volume was greater than that of any other state and contributed to practices that exacted a heavy price from its chief natural resource. To ensure that there would be forests for future generations, the commission recommended that both federal and state governments undertake extended planting programs. Once industries began "tooling up for peacetime production," federal and state funds should be made available to put "unskilled laborers" to work on reforestation projects, a policy that would provide jobs for those who might otherwise be unemployed.[20]

The commission's reports also made frequent references to stream pollution, especially in the densely populated Willamette Valley. Within a month of Germany's surrender, the Oregon State Sanitary Authority financed a series of surveys to determine pollution levels in the state's waterways. The agency was primarily interested in streams flowing through urban settings to determine proper designs for sewage treatment works. In a scathing section of its February 1946 report, the commission called pollution in the Willamette River system "a State shame." Sewage and industrial wastes, which did not affect "trash fish" such as carp and bullhead, were threatening to exterminate salmon and trout. Citing an Oregon State College Engineering Experiment Station bulletin, the commission declared the Willamette to be "an open sewer," with an especially lethal area between Newburg and the Columbia River polluted with "sulfite liquors, cannery refuse, creamery wastes, sawdust, bark, flax retting waters and alcohol plant discharges." The report cited other problems: unscreened and improperly screened diversion ditches; overflow pools that trapped small fish; and malfunctioning fishways. And, seemingly far ahead of its time, the commission criticized the practice of planting hatchery-reared fish from the Willamette system in other waterways.[21] Polluted waterways and related fishery issues would continue to attract regional newspaper headlines into the next millennium.

Perhaps the most impressive postwar phenomenon was the booming home-construction industry, a widely acknowledged consequence of wartime savings and deferred housing purchases. Burgeoning automobile sales aided and supported the expansion of suburban development, especially among the rapidly growing far western states where many ex-servicemen and defense workers had decided to stay on and enjoy the blessings of a moderate climate. In the first ten years following the war approximately 15 million housing units, most of them single-family homes, went up around the country. There were 114,000 single-family units started in 1944, an additional 937,000 in 1946, and close to 1.7 million in 1950. The ready availability of GI loans, personal savings, and a dramatic increase in the number of marriages forced the market demand for new homes to an all-time high in the years just after the war. Long Island's famous Levittown, a mass-production, assembly-line, instant community of 17,000 homes and 80,000 people, set the standard for the quick construction of cheap, affordable housing. Although the Levitt brothers, Alfred and William, attracted considerable public attention, they were only the best known of what was unquestionably a developer's dream in the postwar years.[22]

When Bill McKenna left Coos Bay to join the Army in 1940, unemployment lay heavily on the area and cheap, vacant houses were abundant. After his discharge, he spent a year in Portland and then returned to the Bay communities in 1946 to find dramatically different conditions. "Times were pretty good, and everybody was working," he remembered. But what most impressed him was the housing market, "because all of a sudden, here's this sudden shortage."[23] McKenna's immediate postwar experiences were repeated everywhere across Oregon. Indeed, when President Harry Truman sent his twenty-one point program to Congress in September 1945, he urged lawmakers to provide funds for public housing, recommending that "over the next ten years, there should be built an average of from a million to a million and a half homes a year."

Truman's message resonated well with Oregon newspaper editorial writers. Some, such as Sheldon Sackett of the Coos Bay *Times,* had anticipated the expanded trade opportunities that would follow the war. Because of their proximity to Pacific waters, Sackett believed that Oregon's mills were well positioned to take advantage of marketing opportunities in Asia. And then, in phrasemaking reminiscent of an earlier day, the garrulous publisher and editor called for open season on the region's natural resources: "If we're going to have full employment and full production . . . after the war, we're not going to make 'reserves' out of great resources of raw materials and foods, we're going to look for even more."[24]

Sackett was not alone in anticipating the lively lumber market that would follow the war. The Postwar Readjustment and Development Commission predicted in the autumn of 1944 that logging and lumbering would be "the outstanding exception to a requirement for reconversion." With the passing of "cut and run" practices and the implementation of sustained-yield policies, the commission expected stabilized employment, permanence for the industry, and an end to ghost towns. With the ending of hostilities, it predicted, the lumber business would operate at full capacity to supply both domestic and foreign markets. In the month of Germany's surrender, the planning agency reported that the state's housing shortage had reached "the

acute stage," much of it attributable to the absence of construction "for
ten years prior to the war."[25]

Oregon newspapers confirmed the critical housing shortage when war
veterans began returning home in droves and placing heavy strains on available living spaces. In the central Oregon town of Bend, the mayor and city
commissioners held September hearings on a postwar construction program, with a special eye toward the housing crisis. Mayor A. T. Niebergall
told the gathering: "No matter how small the start, let's get going." The
Bend *Bulletin* reported that the community was "suffering from one of the
worst housing shortages in its history, . . . a demand that [was] out of all
proportion to supply." For their part, the community's two large pine mills,
Brooks-Scanlon and Shevlin-Hixon, announced that "quality Ponderosa
Pine will again be available for home and other construction purposes."[26]
What the joint advertisement failed to reckon with were the diminishing
stands of privately held timber in proximity to the mills, a matter that would
lead to the sale of the Shevlin-Hixon properties to Brooks-Scanlon within
the decade.

The central Oregon community's immediate problems were similar those
of cities and towns elsewhere. On the southern Oregon coast, the *Port
Umpqua Courier* reported that the national housing crisis was evident in
the mill town of Reedsport—"an influx of population with nowhere to place
them." In Medford, Jackson County's housing agency announced that the
National Housing Authority had authorized the construction of fifty family dwelling units, and the authority's regional representative reported that
federal funds would be made available for fifty units in the Salem area and
twenty-five in the Coos Bay district. Pendleton's *East Oregonian* cited housing as the area's most "troublesome problem." The prospect that construction would soon resume cheered the newspaper's editorial writers;
materials were now becoming available and the time had ended "when those
who wished to buy something were regarded as supplicants." Although
prices might be higher, Veterans' Administration and Federal Housing
Administration (FHA) loans made it possible for returning servicemen and
women to make down payments.[27]

If, as an Associated Press story claimed, the "acute housing shortage"
was the state's most critical problem, the emergency also held the promise

of putting people to work. The need for new homes meant heavy construction activity, jobs for loggers and mill workers, and an expanded service sector. In the monthly reports of the PRDC, the influx of people meant development and a growing consumer market: "It means more housing, more furniture, more household and kitchen equipment, more automobiles, more clothing, more food." An increase in population also required additional services and professional opportunities, including hospitals and schools. Although municipalities would be hard pressed to provide expanded water and sewage systems, electrical and telephone facilities, and police and fire departments, creating that infrastructure would boost economic development.[28]

Nowhere did prospects appear brighter than in Salem, the state's capital city. Citing building permits and other data, the local chamber of commerce predicted that employment in the greater urban area might double by the end of 1946. The *Statesman* estimated that the big Willamette Valley Project and the construction of nearby Detroit Dam would employ as many as 12,500 workers and bring an "influx of people and money." Building the dam and related highway relocation work would mean "three times the amount of living quarters and three times the increase in the present service force." In addition to work on the project, a new battery-manufacturing plant in West Salem and an American Can Company facility on the east side of the Willamette River would require 500 workers during construction. Two automobile dealerships, including Douglas McKay Chevrolet, were putting up new buildings, anticipating expanded auto sales. Several downtown businesses—furniture stores, hotels, the riverside pulp and paper mill, the two local hospitals, and the Greyhound bus station—were already in various stages of construction. State government officials augmented the construction frenzy in the capital city, authorizing $4,400,000 to complete the planning and construction of new state buildings. Those combined projects, the *Statesman* observed, called for immediately building 1,000 new homes in Salem and the surrounding area.[29]

The frenetic construction work that took place in Oregon following the Second World War was only a small-scale version of California's burgeoning home-building market but the bread and butter for Oregon's still heavily dependent forest-products economy. While states west of the Mississippi increased from 32 million to 45 million people between 1945 and 1960, California's growth was the most spectacular by far, mushrooming from 9 million to 19 million in the same period. With federally guaranteed FHA and veterans' loans available, housing and industrial construction, the production of consumer goods, and the booming defense industry were the driving force behind California's remarkable economic growth. For nearly two decades after 1945, lumber orders from the greater San Francisco Bay area and southern California propelled north Pacific slope timber districts through periods of high employment, the easy availability of good-paying jobs, and a steady influx of people from all parts of the country.[30]

Federal forestry officials welcomed the bright promise of the postwar era. Although new industries had temporarily surpassed forest products during the war, Robert Cowlin, an economist with the Pacific Northwest Forest and Range Experiment Station, predicted that with the end of hostilities logging and lumbering would once again "resume top rank by a wide margin." Writing for the *Oregon Business Review,* Cowlin pointed out that Oregon had emerged as the leading lumber-producing state in the nation in 1938 and was quickly distancing itself from Washington—where standing timber had been severely depleted. The industry's major trade organization, the West Coast Lumbermen's Association (WCLA), also followed the new center of production south of the Columbia River, moving its headquarters from Seattle to Portland.[31]

But it was the future that most concerned Cowlin. Because Oregon held the largest volume of standing timber in the United States, the state could expect to supply the greatest bulk of marketable timber with the return to peace. The Forest Service economist worried over forested areas within Oregon that were already cut over, especially the once heavily timbered districts on the lower Columbia River and around the Klamath Basin. The state's national forest lands, southwestern Oregon, and the revested Oregon and California Railroad (O & C) land grant held the greatest volume of old-growth timber. The O & C grant, approximately 2,600,000 acres in western Oregon, was revested to the federal government in 1916 when the U.S. Supreme Court determined that the Southern Pacific Railroad, successor to the O & C, had violated terms of the original land grant. Except for the Coos Bay area, Cowlin reported that there had been relatively little industrial activity in southwestern Oregon until the Second World War, and even then, "a comparatively small number of good-sized companies owned most of the timber." The industry's performance in western Washington, especially around Puget Sound and the Grays Harbor district, also suggested that overexpansion could present dangers for Oregon as well.[32]

For the near future, however, Cowlin predicted that forest products would "provide the major source of employment and of industrial output." Although the postwar era presented both "great possibilities and . . . serious problems" for the state, those difficulties could be avoided by implementing scientific management on forest lands, making full use of wood materials, and adding value through "greater refinement of manufacture." Cowlin's suggestion that scientific management was the answer to cutover districts and social devastation would become the industry's mantra in the decades that followed. In his study of Forest Service policy following the Second World War, historian Paul Hirt argues that the "seeds of modern industrial forestry and the management controversy of the present day germinated in the early post-World War Two era."[33]

Scientific forestry, treating timber as a crop, and intensive forest-management practices became the watchwords for public and industrial representatives alike. H. V. Simpson, the secretary-manager of the WCLA, argued that the lumber industry was in transition from harvesting old-growth timber "to the orderly cropping of forest products" that would be managed to ensure maximum and continued production. Oregon's problem, according to Simpson, was to make certain that its forest lands, "once relieved of stagnant old-growth timber," would immediately begin to grow new trees. Sound forest-management practices in the Douglas-fir region meant promptly replanting "as soon as an acre is relieved of its nonproducing old growth." Simpson's references to "decadent" and "nonproducing" timber stands, which appeared in the *Oregon Business Review* in late 1945, would become the centerpiece of modern scientific forestry and intensive management in the postwar years. Referred to as industrial forestry, the new practices emphasized research and orchestrating the management and growth of timberlands with maximum production as the sole objective. The beauty of scientific forestry, its proponents declared, was its promise that "poorly stocked" lands could be made more productive, and that the Pacific Northwest, "the most favored timber-growing area in the world," was capable of meeting any future market demand.[34]

Technological optimism—the widespread belief that there were scientific and technical solutions to questions about timber production—characterized the emerging world of intensive forest management. And if the promise of forestry had an ideological center, it was vested in sustained-yield management—an always fuzzy concept subject to endless interpretation. The term was included in several pieces of federal legislation, the region's newspapers regularly published articles praising the virtues of sustained-yield management, and the General Land Office (after 1946, the Bureau of Land Management) administered the O & C lands under Public Law 405, the Oregon and California Sustained-Yield Management Act of 1937. Walter Horning, chief forester for the checkerboard 2.5 million acres, viewed the postwar management of the O & C lands as a "grand-scale testing ground" for implementing sustained-yield forestry. Horning argued in a 1946 article that "government forests" provided a good model for forestry practices on private lands. In the case of the O & C forests, they were a "self-sustaining enterprise" providing revenue for county governments. Even more important, Horning pointed out, the O & C legislation promised to protect local communities from timber exhaustion and to invite private timber holders to join with the government in combined sustained-yield cooperative agreements. Because they included both public and private lands, sustained-yield agreements would bring stable manufacturing centers, "settled workers, permanent homes, savings accounts, and all the institutions of modern civilization."[35]

Newspapers in the heavily forested sections of the state also looked to a future where implementing sustained-yield management practices would assure a perpetual supply of timber. Within days of the Japanese surrender the Roseburg *News-Review*'s Charles Stanton predicted that the upper Umpqua Valley would become the center of lumber production on the Pacific Coast and should prepare itself for "increased population, new businesses and new activities." Douglas County would have no conversion problem, because government orders and materials needed for big construction projects at home and overseas made the area's economic outlook "exceptionally favorable." But Stanton was no unabashed worshiper of unlimited harvesting; he urged "full compliance with forest conservation practices," and to avoid the blight and devastation of overcutting, he called for "a realistic view of the forest management problem." Douglas County's timber wealth placed it in an advantageous situation: with a larger volume of standing timber and a greater capacity for annual growth than most other areas, the Umpqua region stood at the threshold of a production boom. "On a sustained yield basis," Stanton proclaimed, "Douglas County had a most outstanding advantage."[36]

Beyond Oregon's forest bounty, another culturally scripted landscape was taking shape in the postwar decades. The most ambitious plans for reordering the Northwest were vested in developing the region's waterways, especially the main stem of the Columbia River and major tributaries such as the Snake and the Willamette. Shortly after the German surrender, the *Oregonian* praised the versatility of Oregon's climate and soil conditions and the diversity of its agricultural crops, virtues that were certain to attract servicemen and women and war workers from other states. The Portland newspaper welcomed all farmers who had "the means and ability to establish themselves in agriculture." In midsummer 1945 the Postwar Readjustment and Development Commission reported a "back-to-the-land" movement in western Oregon, where small farms capable of growing specialty crops were attracting new owners.[37] While the incipient Willamette Valley Project unquestionably attracted potential farmers, promoters and developers expressed even greater enthusiasm for the expansion of big irrigation projects in the Klamath Basin, the middle Deschutes, the Owyhee-Malheur district at the eastern edge of the state, and the lower Umatilla drainage.

Development groups along the state's river corridors aggressively lobbied state legislators and congressional delegations to promote their pet projects. The most imaginative and colossal of those proposals were the giant U.S. Army Corps of Engineers and Bureau of Reclamation projects for the Columbia River. Among the several influential groups promoting dams on the Columbia system was the Pacific Northwest Development

Association, a lobbying organization formed in 1945 when its founders realized "the need for singleness of purpose in the development of the resources of the Columbia River Basin." The association announced on the masthead of its first publication that it supported resource conservation and development "through the combined efforts of private enterprise, group initiative, and established government agencies." The association's leadership was also obsessed with federal proposals to establish a Columbia Valley Authority (CVA), similar to its Tennessee counterpart. Although it denounced the CVA proposal as socialistic, "a left-wing scheme," and collectivist in nature, it praised federal agencies such as the Bureau of Reclamation and the Corps of Engineers for their constructive work.[38]

Although it was less extensive than the state's other reclamation undertakings, central Oregon's federally funded "North Unit" project between Bend and Madras, approved by the Bureau of Reclamation in late 1937, attracted great enthusiasm during the closing months of the war. Civilian Conservation Corps camps cleared the site for Wickiup Dam upstream from Bend and carried out much of the ditching and canal construction before the shortage of labor and materials put the project on hold during the war. With the end of the conflict in Asia, the Bureau of Reclamation resumed work on the enterprise and completed the forty-four-mile canal from Bend north to the project area in the spring of 1946. When water gurgled through one of the laterals near the small town of Culver on the afternoon of May 18, Governor Earl Snell and other dignitaries were on hand to celebrate what the *Bend Bulletin* called "the start of a new epoch in Central Oregon irrigation." Snell told the crowd that the water flowing through the canal represented "years of struggle and effort" and marked the beginning of the transformation of arid central Oregon "into a garden spot of rich agricultural productivity."[39]

The *Bulletin* was effusive in its praise for the vision, patience, and hard work that contributed to the success of the undertaking. When the North Unit project was completed, the water flowing through the system would irrigate 50,000 acres:

Deschutes water, carried over lava lands and across the face of volcanic cliffs in a miniature man-made river, pounded canal headgates overlooking the broad Culver basin at noon today, impatient for release over lands which for unknown thousands of years received their moisture from stingy clouds.[40]

Robert W. Sawyer, editor of the *Bulletin* and a nationally recognized reclamation proponent, called the enterprise "a dream of many years . . . in progress of becoming reality." In an area with rich soils but where "the man on the land was engaged in a bitter struggle for survival," the waters moving through the canals and laterals promised to "reward richly the labor of the farmer." Sawyer and the *Bulletin* envisioned "a new agricultural community in the making."[41]

Journalist and future governor Tom McCall, who spent many of his growing-up years along the Crooked River, wrote that the North Unit project would prove to be a great divide in the history of central Oregon. Referring to the spectacular snowcapped mountains to the west, McCall thought there were intangible features to the project beyond the practical factors that would attract prospective farmers: "the beauty of nature's furnishings is an influence not to be underestimated." The North Unit had the virtue of making valuable that which was valueless, of turning the arid plain's failed farming enterprises into success stories. This was a challenge, the loquacious young journalist wrote, "flung by the souls of eyeless, sagging houses and the bright presence of a poppy or a sweet pea among the tumbleweeds and sage brush engulfing a once lovingly-kept yard." But reclaiming the arid sage brush plains explained only part of the North Unit project's potential, according to McCall, because opening the new highway from Madras northwesterly along the southern slopes of Mount Hood would shorten the distance to Portland by thirty-five miles, bringing farmers within easy reach of the state's largest urban market.[42]

Jarold Ramsey, who grew up on his family's dry-farming operation on Agency Plains outside Madras, remembered that his rural school's enrollment swelled as "whole neighborhoods from western Idaho moved to the Plains in advance of the North Unit Irrigation project." Farming also shifted from the more relaxed seasonal rhythms associated with dry farming, Ramsey recalled, to "the physically and economically frantic enterprise it became when irrigation arrived in the late 1940s."[43] The North Unit project brought other changes to central Oregon, especially to the ecology of the

Deschutes River. Two earlier privately financed storage reservoirs had already altered the river: Crescent Lake, capacity 86,050 acre-feet, constructed in 1922–23; and Crane Prairie, capacity 55,210 acre-feet, completed in 1922. Wickiup Reservoir had a storage capacity of 180,000 acre-feet of water. The coordinated release of those waters into the Deschutes River during the summer months was then diverted into three canals, including the main outtake below Bend, thereby reducing the Deschutes' flow to a mere trickle for several miles.[44] As the newest of the postwar reclamation projects in Oregon, the North Unit transformed the arid plateau country of the middle Deschutes from sagebrush plains to verdant fields of alfalfa, corn, potatoes, and peas. Its extensive system of canals also removed a huge volume of water from the river to irrigate crops between Redmond and Madras, thereby altering water and plant life in the central Deschutes region.

If Richard Neuberger set the journalistic standard for public optimism in the immediate postwar years, he enjoyed lots of company. Oregon's business community, often expressing itself through a variety of state-agency bulletins, aggressively promoted the development of waterways; clearing, diking, and draining wetlands; converting "wasted" resources to productive public use; and providing expanded recreational space for tourism. In a ringing endorsement for what Governor Tom McCall (1967–1975) would later call "coastal condomania,"[45] the Readjustment and Development Commission referred to a mad stampede "to buy, build and develop the Oregon coast" to turn it into "a playground, a recreational corridor" from the Columbia River to the California border. What was essentially a real-estate boom, the agency noted in February 1946, rolled on like "the thunder of the surf" as loggers cleared hillsides and bulldozers leveled sand dunes and filled depressions. Taking pride in the powerful and expensive machines, the commission praised their work in opening new beaches and gouging streets through stands of timber. "The trail of the bulldozer is seen almost everywhere," the report continued, especially in areas where syndicates had purchased large acreages along the coast. In a prelude to Mark Hatfield's reference to sprawling Lincoln City's "twenty miserable miles," the commission observed that small communities such as Manzanita, Rockaway, Ocean Lake, Taft, and Depoe Bay were expanding north and south along Highway 101, "some almost merging and forming one long street."[46]

It was obvious to many people in the development community that Oregon's embryonic tourism industry held great promise in the months following the war. Governor Earl Snell told an audience of business groups in August 1945 that "the tourist trade can be made Oregon's top industry" with sufficient planning and preparation. The governor then appointed a state tourist promotion committee to draft plans to advertise Oregon's scenic attractions through a variety of promotional brochures and displays in other states. The tourism committee also cited the need to refurbish and build new hotels and auto courts. The Pendleton *East Oregonian* urged the local business community to take advantage of the city's location as the "natural traffic point" on the great transcontinental highway that Oregonians had been traveling since time immemorial. Although Pendleton was in an ideal situation to profit from motorcar travel, the area badly needed improved facilities, new hotels, and better tourist camps. To make the state's advertising campaign effective, the eastern Oregon area had to provide modern accommodations that would keep tourists from rushing through the region to other places.[47]

The most aggressive of Oregon's promoters and developers were Portland business leaders, the movers and shakers at the heart of the state's economic and political life. Following decades-old practices, the city's moneyed interests supported most large development enterprises: water projects throughout the Columbia Basin, tourism along the coast and through the Columbia Gorge, deepening the Columbia and Willamette river channels, and several big highway jobs. According to J. R. Woodruff, traffic director for the Port of Portland, the city's waterfront provided a "tidewater gateway to the producing areas of Oregon, eastern Washington, Idaho, and adjacent territories." Portland's twenty-five miles of harborside at the heart of its industrial area provided transfer facilities for agricultural produce, finished lumber, and wheat and minerals from the interior. Woodruff paid tribute to the federal government for its dredging, dike construction, and revetment works, efforts that aided "the growth and development of Portland as a world port." Although the government had developed and maintained a channel 500 feet wide and 35 feet deep from the Columbia River bar to the Port of Portland; the Corps of Engineers' recent work in dredging and building jetties at the mouth "resulted in a splendid entrance channel with a depth of 40 feet at low-water level."[48] The Port's efforts to further deepen the Columbia-Willamette shipping channel to Portland would become even more determined (and controversial) in the ensuing decades.

What Portland-area shippers most desired, however, were additional upriver dams, including low-head dams on the lower Snake River to extend inland water navigation to Lewiston, Idaho. H. T. Shaver, a Columbia River barge operator, contended that dams would enable river carriers to "tap the heavy tonnages originating in central and eastern Washington and western Idaho." Once in place, the Columbia and Snake dams would "practically canalize the Columbia River" and give Portland "a decided rate advantage over competitive Pacific Northwest ports." Once the authorized dams were completed, Shaver saw no special difficulties limiting the expansion of transportation. The problems at Celilo Falls and other points on the upper river would be eliminated once the federal government completed its "program for the improvement of the Columbia River."[49] There was nothing new about those postwar ambitions to improve transportation on the Columbia system; Portland business groups had been pursuing such projects since the late nineteenth century. In an earlier time, the city's support had been critical to completing the Cascade Locks and Canal in 1896 and the more ambitious The Dalles-Celilo Canal in 1915.[50] Neither project produced the economic benefits its promoters promised.

The strategic presence of the Bonneville Power Administration (BPA) headquarters in the city also benefitted Portland's business community. Ever since its provisional appointment in 1937, the BPA and its administrators had been central to the city's politics and economic ambitions. Ivan Block, one of the agency's more visionary administrators and an outspoken proponent of industrial development, set the postwar mood for regional development in a wide-ranging Phi Beta Kappa address at the University of Oregon on May 17, 1946. The Pacific Coast, Bloch told his Eugene audience, was one of the few areas in the world "actively looking ahead to greater development of its resources." Although the West had "always been the expanding horizon for our nation," the postwar years offered unparalleled opportunities to continue "the greatest population movement in our national history." Studies of discharged military personnel who had spent time on the West Coast indicated a strong preference to return to the region. "Public polls throughout the country," Bloch remarked, "show that its eyes are 'West.'" The expected population influx presented a "magnificent challenge" to the region to hew closely to "the proper and wise use of the riches that nature has given to this West Coast."[51]

In rhetoric that may have alarmed some of his contemporaries, the BPA official outlined the ground rules for future development: "Our streams must continue to flow pure and undefiled; our forests must become an everlasting crop; our new industries must develop harmoniously with the countryside and the people who work in them." But Bloch's larger ambition was sufficient opportunity for capital and labor to "maintain this better life we now enjoy." Although no "magical hocus pocus" could provide the answer, he thought the big federal projects on the Columbia River would provide new jobs, and the availability of low-cost electricity would continue to attract new industries to the Northwest. To accommodate the growing numbers of newcomers, Bloch suggested three priorities for the region. First, the Columbia country needed to expand and diversify its manufacturing, especially in the chemical and metallurgical industries. The huge supplies of low-cost hydroelectricity in the Northwest provided a giant step in the direction of regional independence, particularly in the production of light metals such as aluminum. The Northwest also needed to increase and broaden its extractive industries, with a special emphasis on extending its irrigated lands to provide another 3.5 million acres for new farms. The availability of low-cost power and aluminum sprinkler systems would decrease the size of farms, diversify crop production, and open new possibilities for the resident population. Finally, because the Pacific Northwest had not yet fully recognized its tremendous scenic and recreational assets, there was great opportunity to develop the service sector.[52]

Bloch closed his talk by calling for more assertive entrepreneurial effort and suggested that aggressive efforts would set off a chain reaction that would reverberate across the Northwest: "We have a region with assets unique in the history of the nation. . . . Are we going to sit idly by? I think not." Others, including Edward Sammons, president of Portland's United States National Bank, shared Ivan Bloch's call for expanded and more ambitious development programs. As executive head of one of the region's most influential banking institutions, Sammons urged the Pacific Northwest to take advantage of the abundance of hydroelectricity at its doorstep. The Columbia River country, he predicted, would soon be a thriving center for chemical-reduction works and plants for fabricating aluminum into kitchen utensils, automobiles, and sleeping cars. Because of its strategic location, Portland would "tap the richest agricultural hinterland on the continent," and its waterpower would turn the city into a processing center for quick-freezing foods.[53]

The Readjustment and Development Commission predicted in January 1948 that the state's development would continue in high gear. The westward population sweep promised to assure Oregon's rank as the "second state in the Union for percentage increase," the same position it held in 1947. In addition to a long list of plant construction and renovation projects, including a Weyerhaeuser Company sulphate plant at Springfield and an Aluminum Company of America facility on the Columbia River near Portland, the commission pointed to the influx of people moving to Oregon to work on government projects, "the dam-building super work, soon to be underway." The most important of those were the Willamette Valley Project's several dams and McNary Dam at Umatilla Rapids on the Columbia River. Collectively, those undertakings would provide employment for thousands of workers and take more than five years to complete. "Not one but several dams will be under way simultaneously," the commission boasted.[54]

In truth, 1948 was an auspicious year in the postwar history of the Pacific Northwest. The state of Oregon and the Columbia River country were poised at the edge of momentous change, developments that would dramatically rearrange the regional landscape, reconfigure waterways, and vastly accelerate cultural disturbances. The region's population in 1947, estimated at 4.4 million, had increased 25 percent since 1940. The PRDC noted the attractions contributing to that amazing movement: a favorable climate and business opportunities, "no hurricanes, tornadoes," nor intense winters or "blistering summers." To effectively meet the challenge to provide jobs for the mass immigration to the state, Oregon's agency heads and its

TABLE 1.1

DATA FOR OREGON FARMS IN 1945

Farms	63,125
Acreage	19,754,257
Automobiles	59,362
Radios	56,254
Electricity	48,390
Running Water	45,998
Telephones	25,643
Average size (acres)	312.9

business leadership should press ahead with the further development of lumber, agriculture, electrical power, chemicals, and mineral manufactures. "As resources are translated into industries," the commission reported, "they in turn create more housing, jobs, and payrolls. It follows the pattern of a chain reaction." At the marketing end of the spectrum, the commission pointed to the expanding population of the Pacific coastal region, outlets that would "continue to grow indefinitely." Pacific Basin countries such as Japan, China, the Philippines, New Zealand, and Australia also had the potential to fire "the imagination of a number of industries."[55]

Despite the influence of the Second World War in boosting industrial production, especially in the aluminum and petrochemical manufactures, prewar economic activity remained dominant in Oregon and Washington. Agriculture remained in first place and forestry-related work second. More important, of the region's "factory workers," forest industries employed more than 60 percent. Although Oregon's farm population declined from 258,751 to 221,399 during the war, great strides in mechanization had taken place on the existing farmsteads. By the fall of 1945, more than 75 percent of farms had electricity, 40 percent enjoyed telephone service, 90 percent had at least one automobile, and 89 percent owned radios. Table 1.1 provides the raw numbers for these and other categories.[56]

Despite the giant hydro facilities at Bonneville and Grand Coulee, Oregon officials cited a dire need for more electrical power. As rural Oregon was in the process of being electrified, the farm population was purchasing electrical pumps, cream separators, milk coolers, and lights for homes and barns. Moreover, urban and farm dwellers alike were purchasing refrigerators, vacuum cleaners, toasters, washing machines, irons, radios, and other domestic electrical appliances coming on the market—all increasingly popular in a region with cheap power rates.[57]

Both domestic and commercial energy consumption would soar in the next two decades as the Bonneville Power Administration and public and private utility firms promoted the use of electricity for virtually every purpose imaginable. Those were also the years in which "the age of dams reached its apogee," according to writer Marc Reisner, "when hundreds upon hundreds of them were thrown up, forever altering the face of the continent." It was also a time when the two principal dam-building agencies, the Army Corps of Engineers and the Bureau of Reclamation, perfected the art of cost-benefit analyses to justify construction projects. Electrical power production, improving navigation, storage reservoirs for irrigation, flood control, recreational values, and pollution abatement were factors written into most of the Northwest proposals submitted to Congress.[58] Despite protestations and assurances to the contrary, protecting the region's once prodigious runs of anadromous fish was not an important agenda item to water-development proponents.

2

INTO THE BRAVE
NEW WORLD

With Bonneville already seeking outlets for power in the postwar period, it is very doubtful whether further hydroelectric development is essential in the immediate future. Flood control is a vital factor in some areas. But in every case where water is to be diverted for some utilitarian purpose, full consideration should be given to the possibility of damaging recreational resources.—Charles Stanton, *Roseburg News-Review*, August 24, 1945

F ew symbols more powerfully evoke the essence and meaning of the Pacific Northwest than salmon. Fisheries biologist Jim Lichatowich argues that salmon are survivors, "living through volcanic eruptions, ice ages, mountain building, fires, floods, and droughts."[1] The great salmon runs provided spiritual reassurance and sustenance for Native people, and for most of the 150 years since Euro-American settlement, salmon have served as physical manifestations of nature's bounty. Shortly before his discharge from the United States Army, Richard Neuberger worried that the once abundant Columbia River chinook were in serious decline. The explanation for their decreasing numbers, he reasoned, were multiple: the extraordinary commercial catch in the lower river; pollution and industrial wastes that had choked salmon habitat; logged-off hillsides and overgrazed grasslands; and dams and irrigation works that blocked access to traditional spawning grounds. Neuberger also noted the fluctuating and generally downward trend of the fish count passing through Bonneville Dam between 1938 and 1944. But it was not too late to save the chinook, Neuberger concluded, because Bonneville's fish ladders worked and Northwest sportsmen were promoting programs to clean up urban pollution. The young journalist pointed out that the new dams proposed for the Columbia would generate "power at the cheapest rates on earth." It was possible, he believed, for the water projects and salmon to coexist.[2]

Despite his occasional bouts of ambivalence, few Northwest writers surpassed Neuberger in praising the region's bright prospects for the postwar era. First in order were the federal government's Columbia River energy projects, especially the hydroelectric power generated at Bonneville and Grand Coulee dams, "the cheapest power on earth." Low-cost electricity offered promising opportunities for aluminum-fabrication factories, industrial employment that would turn metal into pots, pans, and appliances. The waters impounded behind Grand Coulee Dam and pumped into Banks Lake promised to irrigate vast acreages in the Columbia Basin where new farms would directly support an estimated 85,000 people. Neuberger predicted that it would require another 170,000 workers in nearby communities to process foodstuffs and provide essential services. And he would include Grand Coulee's already significant contributions to the Hanford facilities on the Columbia's sagebrush uplands.[3]

John Gunther's best-selling 1946 book, *Inside U.S.A.*, followed Neuberger's lead in celebrating the abundant and cheap power of the Columbia River. Bonneville and Grand Coulee helped create the region's aluminum industry and played an important role in the development of the atomic bomb. Although the bomb was "a kind of apocalyptic, demonic child of the Columbia," Gunther noted that the prospects for the new peacetime waterway represented the industrial and social future of the nation.[4] Gunther's observations mirrored regional boosters and developers who allied themselves with two powerful federal agencies, the Army Corps of Engineers and (to a lesser degree) the Bureau of Reclamation, to promote the huge engineering works on the Columbia River system. The end result of their efforts—after more than four decades of sometimes frenzied construction activity—was a thoroughly rationalized and regulated waterway.[5] The engineered river provided a water highway for barge traffic more than 400 miles to Lewiston, Idaho, powered turbines that produced electricity delivered as far as southern California, furnished irrigation water for thousands of acres, and cooled Hanford's nuclear reactors. Its series of dams evened the annual flow of the main stem to the point that it no longer posed a threat of annual flooding.

The Army Corps of Engineers, the government agency charged with river-basin surveys and drafting blueprints for the big engineering projects, was the principal federal bureau driving Columbia River development. Ralph Tudor, head of the Corps' Portland District, told members of the Portland Rotary in August 1943 that the Columbia River was "the greatest national asset of this or any other nation." He urged Congress to adopt a comprehensive engineering plan for the entire Columbia Basin and cautioned about "the fallacy of piecemeal development." Tudor had every reason to believe that, "if congress so wills it," the Umatilla Rapids Dam and the Willamette Valley flood-control project would begin as soon as the war ended. Like his successors, Tudor worked closely with business and commercial interests, congressmen, and other powerful Northwest public figures to generate support for river development.[6]

The project that captured the most attention at the end of the war was the Corps of Engineers' recommendation to build a giant dam just upriver from The Dalles. The huge structure would flood the ancient Indian fishery at Celilo Falls and a half-dozen rapids further upstream. Corps official Theron Weaver argued that flooding the falls and rapids would bring great economic benefits to the interior Columbia Basin. The Dalles pool, the second in a series of four proposed slack-water lakes, would aid commercial navigation to Lewiston, produce significant amounts of hydropower, provide water for irrigators, and enhance recreation. As for the Indians living at Celilo, Weaver recommended alternative fishing facilities for those who wanted to continue fishing in their traditional ways.[7] In the end, The Dalles Dam would be the most contentious and controversial of all the Columbia River development projects. And the tribes, promised "in-lieu" fishing sites in exchange for building the dams, were left with marginal and limited access to the river. In *Empty Nets*, Roberta Ulrich argues that "the delay in replacing the sites carries all the elements of cultural arrogance, hostility and, at bottom, indifference that has marked so much of white Americans' dealings with those who were here first."[8]

To coordinate development in the Columbia Basin, the Federal Inter-

Agency River Basin Commission established a regional subsidiary, the Columbia Basin Inter-Agency Committee, in 1946. The committee brought together state and federal officials, but it represented federal agendas; state officials had no voting authority at committee meetings. Speaking for the committee, the Corps' W. O. Silverthorn argued that the construction of Bonneville and Grand Coulee Dams indicated the need for rationalized, coordinated planning and cooperation. The Bonneville Power Administration controlled the distribution of power through its transmission lines; but until the appointment of the committee, no one body provided overall administrative oversight of planning and development.[9]

The Columbia River Development Association, local chambers of commerce, the Army Corps of Engineers, and the *Oregonian* presented a consensus in the immediate postwar years, describing the great Columbia River as a shining symbol of hope. In lofty and exaggerated prose, *Oregonian* staffer Paul Hauser heralded "the vast promise of the Columbia" and the move "toward full realization" of the river's potential. In addition to work already under way at McNary, he reported in 1946 that the Corps was planning a dam at The Dalles to "wipe out the dangerous rapids and swift water." In a rhetorical flourish similar in style to Neuberger's, Hauser praised low-cost hydropower:

> [It is] the golden asset of the Pacific northwest. It is its hope and promise for the future. . . . It will be energy that is always there—small snow or big snow, little flood or big—for the storage features of the dams on the main river and on the tributaries will enable the engineers to keep the river at a fairly constant flow. There will be little water flowing to waste over the spillways of Bonneville, the last chance to convert the force of its falling to the sea into power.

The big Columbia projects, Hauser argued, would make it possible for the Pacific Northwest to break free "from old strangle holds" that had kept it dependent on the industrial east.[10]

To satisfy congressional and local critics, the Army Corps of Engineers officials used cost-benefit arguments to justify its river-development proposals. Speaking to the American Society of Civil Engineers in Spokane,

Portland District's Robert Higdon contended that turning the Columbia into a navigable waterway would offer public benefits worth four times the money spent on the projects. Extending the jetties at the river's mouth, dredging the channel, and building dikes along navigable sections had already vastly increased tonnage passing upriver from Vancouver. "The Columbia river system," Higdon told the engineers, "constitutes one of the greatest assets of the Pacific Northwest." Realizing its full potential would "result in a great industrial development and increased movements of commerce." O. E. Walsh, the Corps district engineer, informed the same audience that the Columbia Basin was one component of national flood-control policy that provided the multiple benefits of power development, flood control, irrigation, and navigation.[11]

While Pacific Northwest promoters and developers fought tenaciously to achieve their objectives, sports and commercial fishers, Indian tribes, conservation groups, and others who wanted to protect the scenic and natural qualities of rivers were not without influence. Although boosters vigorously pushed congressional funding for the full network of regional projects, opponents fought several of the multipurpose dam proposals. Corps of Engineers and Bureau of Reclamation officials were optimistic that fish ladders and hatcheries would effectively overcome the problems posed by dams; scientists, fishery interests, and the tribes were more skeptical. Paul Needham, a young aquatic biologist with the U.S. Fish and Wildlife Service, referred to dams as "large-scale experiments" and charged that engineers had mostly neglected fish in their planning. Referring specifically to the Willamette Valley Project in an agency memo, Needham expressed doubt that high dams and healthy salmon populations were compatible. "Maybe our salmon will follow the dodo and the passenger pigeon into the limbo of forgotten and neglected assets." If salmon runs were to be preserved, Needham contended, the fish must be afforded "greater, instead of less protection."[12]

Although time, circumstance, and nature conspired to accelerate the dam-building juggernaut, it was not apparent until sometime in 1948 that the developers would prevail. There were other options, other alternatives, and conflicts over McNary, The Dalles, and dams on the lower Snake River

and in the Willamette Valley continued. Opponents provided stiff opposition to river-development proposals through the media and at congressional and agency hearings.[13] The debate was not new; such discussions had
been in the public domain since the 1938 revision of the Army Corps of
Engineers' 308 Report, the first time the agency joined the lower Columbia and Snake Rivers in a control plan.

Among the notable public figures questioning the big engineering
schemes was the well-known naturalist William Finley, a longtime critic
of water projects that he deemed detrimental to fish and waterfowl. If the
present development of the Columbia River continued, Finley feared,
salmon would be "exterminated." Other groups, increasingly worried that
dams would harm salmon and steelhead runs, joined Finley in opposing
specific projects. Sports and commercial fishery interests formed the
Columbia River Fisheries Development Association, an umbrella organization for sixteen affiliated agencies and private groups from all three northwestern states. In addition to representatives from the Columbia River
Fishermen's Protective Union, the Columbia River Salmon and Tuna
Packers Association, the International Fishermen and Allied Workers, and
the Mid-Columbia Fisheries Association, the association's membership also
counted fishery biologists, Columbia River treaty tribes, and fish and game
officials from the affiliated states. Because the Corps of Engineers, the Bureau
of Reclamation, and business groups were moving aggressively to promote
river-development projects, the Fisheries Development Association passed
a resolution in May 1945 calling upon Congress to appoint a committee to
investigate Columbia River fisheries and the damages the resource would
suffer "from the building of obstructing dams." Speaking for the association, Milo Moore, Washington's director of fisheries, charged that existing dams had already caused great harm. Merle Chessman, a state senator
from Astoria and active in Columbia River issues, asked the federal government to compensate Oregon for the damages to its fishery. A Washington delegate suggested the addition of a second power unit at Grand
Coulee before building any further power dams on the Columbia.[14]

Although the Fisheries Development Association lost the larger struggle to prohibit the construction of additional dams on the main stem of
the Columbia and the lower Snake, the association and allied groups were

successful in forcing a series of public hearings that placed the future of many of the projects in doubt. The formal and informal hearings lasted through 1946 and 1947. Association representatives first took part in a 1945 hearing at The Dalles, criticizing the proposed dam that would inundate Celilo Falls and destroy the most productive Indian fishery on the river. It was more successful in bringing about hearings in 1946 that directly challenged the Corps of Engineers' and the Bureau of Reclamation's grandiose plans for river development.[15] With the benefit of hindsight, it is clear that the years from 1946 to 1950 marked a critical period for Columbia Basin development projects.

Sports-fishing interests also voiced their opposition to river-development projects on the Rogue and Umpqua Rivers, southwestern Oregon's two major waterways. The Bureau of Reclamation and the Corps of Engineers had completed preliminary surveys in both basins. The majestic Rogue had long been touted as a "recreational stream"; the Izaak Walton League warned that storage dams "would destroy a tremendous area of the finest migratory fish spawning grounds in the entire Rogue River"—and severely damage a major industry. The equally famous Umpqua River enjoyed even greater natural advantages from a recreational viewpoint, according to the *Roseburg News-Review*'s Charles Stanton. "Properly developed, advertised and managed," the Umpqua "could produce as much income annually from local tourist and recreational expenditures" as lumber or agriculture. Stanton recognized the Umpqua as a cold-water stream with unlimited possibilities, and the Roseburg editor argued for the river's proper care, "coupled with high-type, adequate accommodations, sufficiently advertised to draw the desired type of patronage." He urged that the Umpqua be retained in its "original recreational capacity through control," and he opposed high dams in the Columbia and Willamette basins "that would destroy recreational values."[16] A conservative anti-New Dealer opposed to "makework" projects, the curmudgeonly Stanton never joined the postwar water-development bandwagon.

The big project already under way in the Willamette Valley attracted some of the earliest and most strident opposition to Corps of Engineers' plans. When the Portland District's Colonel George Zimmerman called a meeting in Eugene in January 1946 to explain the Willamette Valley flood-control program, he was greeted by a storm of protest. Zimmerman opened the lengthy evening meeting by telling the audience, "As I see it, gentlemen, you have a problem, and we have a solution." The flood-control program had already been studied and approved. Although plans were still "in the blueprint stage," the Corps was ready to take action as soon as funds were available. After outlining progress on dams already under construction, Zimmerman cited the additional benefits the projects would bring to the valley: power generation, irrigation, navigation, and pollution abatement. Douglas McKay, a Salem legislator, future governor, and interior secretary, urged Zimmerman to notify congressional representatives to speed appropriations for the dams. Opponents in the audience, especially members of the McKenzie River Protective and Development Association, objected to the project in its present form. William Puustinen, a farmer on the lower McKenzie and a chief spokesman for the group, suggested that the Corps widen and deepen the present river channels to speed the flow of water. Clarence Belknap, speaking for residents of the upper McKenzie, wanted to preserve the river's recreational attractions.[17]

Conservation organizations, sports and commercial fishers, and Columbia River treaty tribes uttered their loudest protests against the multipurpose-dam proposals at a series of hearings in Portland in 1946 (and later in Walla Walla, Washington, the following year). Fishery interests and their allies offered much of the testimony and were sharply critical of the headlong rush to build dams. The dissenters objected to seemingly incessant propaganda coming from the Corps and the Bureau of Reclamation and the agencies' unwillingness to consider alternatives to main-stem dams. They represented a well-organized and skillfully orchestrated attempt to convince the Army Engineers to modify its original proposals. Although the effort failed to slow the frenzy to build dams, it succeeded in bringing some modification in design, especially for the much celebrated Willamette Valley Project.[18]

When the Columbia Basin Fisheries Development Association ran a series of newspaper advertisements in the spring of 1946 strongly objecting to water projects targeted for the Columbia system (including the seven-dam flood control scheme for the Willamette Valley), the Corps of Engineers invited the fisheries groups to an informal hearing in Portland. As one Corps official put it, the meeting was called "in hopes of learning something . . . of your troubles." Those attending included Milo Bell, the hydraulic engineer who worked on Bonneville Dam; Arnie Suomela, master fish warden with the Oregon Fish Commission; Frank Wire, director of the Oregon Game Commission; Paul Needham, now with the Oregon Fish Commission; Henry Niemela and William Puustinen of the Columbia River Fisherman's Protective Union; and Mert Folts of the Oregon Wildlife Federation and the Oregon division of the Izaak Walton League. Folts, who orchestrated the presentations, first addressed the losses in land and taxes that Lane County would incur if the Corps were to dam the main stem of the McKenzie River near the small town of Vida. He pointed out that the Corps' design would destroy a world-famous trout fishery and spectacular "white water boating" and would severely damage the livelihoods of several craftsmen who made their living building the famous McKenzie River boats. He reminded Corps officials that famous people from around the world had fished the McKenzie, including former president Herbert Hoover and current Supreme Court Justice William O. Douglas.[19] Movie actor Clark Gable and writer of Western fiction Zane Grey also fished the river.

Folts then called on Paul Needham, who told the gathering that the proposed reservoirs would be largely "intolerable to fish life" because the water level would fluctuate so dramatically. "It is a shame," the biologist argued, "that any dam of such nature should be used in Oregon streams and particularly the McKenzie River." Because aquatic plants and insect life increase in relation to sunlight, shallows around the shores of lakes normally produced most of the food. Needham feared that the reservoir's seasonal fluctuations would be inimical to plant and insect life. Others in the audience pointed out that the reservoir's shores would never be desirable sites for summer homes because of fluctuating water levels. Hydrologist Milo Bell pointed out that water drawdowns for irrigation would begin in June, making the lake less attractive to vacationists. For the remainder of the six-hour meeting, Folts guided discussions from the "headwaters to the wide ocean" to indicate how fisheries would suffer with the construction of "high dams" on scenic rivers such as the McKenzie. At the close of the hearing, the dissenters requested modifications to the Willamette Valley

Project's original seven-dam design, substituting dams on tributary streams in lieu of projects planned for main-stem waterways. No person was more important to that testimony than William Puustinen, fisherman, Springfield-area farmer, and policy consultant to the Columbia Basin Fisheries Maintenance Program.[20]

Citing his credentials as "a farmer on the McKenzie" and "a commercial fisherman on the Columbia," Puustinen told the Corps that, like his McKenzie River neighbors, he wanted flood control but not at the expense of destroying salmon runs and recreational values. There were alternatives to managing the flood-control problem without destroying vast resources: "The planning is wrong," Puustinen told the gathering. As spokesperson for the group, Folts then submitted a petition for a modified plan for the McKenzie River that included tributary flood-control dams on Blue River and Cougar Creek. Those alternatives, he contended, would avoid costly highway relocation, save valuable tillable land that would otherwise be inundated, avoid dispossessing people who lived along the McKenzie, and reduce the loss of spawning grounds for salmon and trout. The revised plan would also provide flood control for large areas.[21] The remarkable thing about this inconspicuous informal hearing was the Corps' willingness to modify its plans and build dams on tributary streams in lieu of the main stem of the McKenzie. While the Corps of Engineers and its allies were successful in gaining authorization for most river development projects, the modifications to McKenzie's design were unique. Mert Folts later recalled that he became an opponent to building high dams on the Willamette system "at the occasion of the big furor over the rape of the McKenzie River." Based on the strength of the dissenters' testimony, he noted, the Corps "substituted large tributary plans in place of main river works."[22] That the dissenters included a former president of the United States, a sitting U.S. Supreme Court justice, a popular movie actor, and influential McKenzie River landowners strengthened their argument.

William Puustinen was also a principal witness when the House Subcommittee on Merchant Marine and Fisheries held two hearings in Portland. At the initial hearing conducted by California congressman Richard J. Welch, Puustinen accused the Corps of Engineers and the Bureau of Reclamation of carrying out an "insidious public relations" campaign and preventing fisheries people from gaining a fair audience with federal officials. The two agencies, he charged, issued a continuous flow of publicity that grossly exaggerated the benefits to be gained through flood control, improvements in navigation, water for irrigation, and earnings from hydropower production. Nowhere, Puustinen told Congressman Welch, were fisheries mentioned: "The volume of printed, radio, and oral publicity going out on the other features of any dam project . . . soon develops a proportional consciousness of the project in the public mind which grows with the growth of the project until this consciousness finally completely overshadows the subdued fishery features." Although the agencies should not "be charged with viciously neglecting the fisheries," time, funding, and personnel limited their work. The two agencies simply lacked "experience for making an intelligent fisheries study such as lifetime scientists can make." How absurd it was, Puustinen concluded, that one should expect the engineers under such conditions to "develop the same authority for fisheries information and the same wisdom" as such scientists.[23]

Three congressmen—including Washington's rising young politician Henry M. "Scoop" Jackson— conducted the more significant of the House Subcommittee hearings in Portland on August 14, 1946. Astoria's state senator, Merle Chessman, representing commercial fishers, told the panel that his group would not oppose river projects essential to the region's industrial development. He drew the line, however, when it came to public works, "the benefits from which cannot justify the heavy public expenditures." Fishery interests did not oppose the construction of Bonneville Dam, "even though they knew it would be detrimental to fish." But the federal agencies' constant din for additional projects "without any regard to their effect on fish life" bordered on "criminal negligence." If additional dams were built, Chessman insisted, there must be a demonstrated need for more electricity, and the projects should "give maximum protection for the fisheries." He reserved his major criticism for McNary and the lower Snake River dams, facilities that would serve only the interests of barge transportation and inland navigation. Before he left the stand, Chessman entered into the record J. T. Barnaby's report to the director of the U.S. Fish and Wildlife Service, which pointed out that McNary and the four Snake River dams presented "problems which cannot be looked upon optimistically."[24]

In his important testimony before the panel, Izaak Walton League of America director Kenneth Reid observed that the earlier hearing was a revelation to the Engineers. "Our attempt at the time was to have some say in the locating of dams on the McKenzie River," he told the visiting congressmen. The outcome of those deliberations "was to achieve a modification of the plans, . . . the abandonment of a main stream dam." Reid's larger purpose, however, was to use the McKenzie plan as an example of the agency's faulty cost accounting. He chided Corps officials for assuming a scientific approach when in fact they were manipulating words and figures to achieve "a sales piece which by its technicality and voluminous content assures itself a sanctuary from the prying eyes of the public-minded Congressman with an eye to the public pocketbook." Because of the slanted publicity of federal-bureau public-relations departments, Reid said, Congress was being "led by the nose" and "hypnotized by its own Frankenstein creations."[25]

Turning to the larger issue of river development, Reid asked, "why the great haste to dam or divert every river in the country?" Could it be "that we have too many engineers on the Federal payroll?" And then in words dripping with scorn, he criticized the reckless speed to authorize and fund water-development projects:[26]

> Like a horde of locusts, the engineers of the great dam building agencies have descended upon our rivers, mapping every possible dam site with bland unconcern for national need, sound economics, or effect on public aquatic resources. And, when these job hungry engineers connive with selfish local promoters and vote seeking politicians, all afflicted with the dangerous water-borne diseases of "hydro-mania," "reclamania" and "navigamania," which blind them to all values in water except kilowatt hours, acre-feet for irrigation, or avenues for boats, that triumvirate is a tough combination to beat.

In summary remarks to the committee, the League director referred to McNary Dam as a "sorry mistake," because it would destroy a large fall chinook spawning ground and harm steelhead runs in northeastern Oregon. If Oregon senator Charles McNary were alive, Reid speculated, he might object to having his name "attached to a project . . . where the benefits are for the few at the expense of the many." Reid urged Congress to be aware that the Pacific Northwest fishery was "at doom's door" and that lawmakers should "thoroughly probe for the truth."[27]

Ed Averill, representing a Portland sportsmen's club, told the subcommittee that conservationists were not attacking "dams merely because they are dams." Rather, they were asking that fisheries and wildlife resources be given "early and equal consideration in planning stages, with power, irrigation, and flood control." Because energy consumption had declined since its wartime peak, fully developing Rock Island and Grand Coulee dams would make available more than twice the hydropower produced during the war. Averill thought it absurd that dams would, as the Corps put it, provide "steps up and down the river for the traffic of boats and barges . . . as far inland as Lewiston, Idaho." River transport represented backward thinking, "even further behind the times than the proverbial horse and buggy." It was equally ridiculous, Averill argued, to believe that main-stem dams on the Columbia would control floods. To achieve effective flood-control, dams should be numerous but should be placed on headwater streams where their effect on fish life would be limited. Damming the Columbia, he charged, would alter the river's biology and encourage the proliferation of predatory fish that would prey upon salmon fingerlings. Columbia River salmon "should not be sacrificed to provide employment for an army of engineers or a handful of river steamboat men who are still living in the dead and almost forgotten past."[28]

Touting once again his experiences as a farmer, fisherman, and federal appointee to the Columbia Basin Fisheries Committee, William Puustinen drew distinctions among the federal agencies charged with managing the Columbia River. He applauded the U.S. Fish and Wildlife Service for moving "actively and very promisingly" to restore salmon runs on the lower Columbia and Willamette systems. There was a problem, however, because the Army Engineers and Bureau of Reclamation "almost completely ignored" the Fish and Wildlife Service in their development plans for the Columbia River. In the past, Puustinen noted, the Bureau of Reclamation had constructed scores of dams "without a single cent of consideration given to the maintenance of the runs of salmon and steelheads." Dams and other human structures already blocked 75 percent of the Columbia Basin's original spawning grounds, and only recently had the Bureau made an effort to build simple devices to allow salmon to pass through through its dams en route to the remaining spawning grounds. Puustinen then turned his attention to the Army Corps of Engineers, faulting the agency's design for the Willamette system as giving at best "naive and unsatisfactory" attention to fisheries. While he applauded the revisions in the McKenzie River design, he called the changes the "very late result" of the fishers' informal conference with the Corps.[29]

River-basin developers and federal-agency personnel did not let the testimony of fisheries proponents go unchallenged. Writing on behalf of the Willamette River Basin Commission, Ivan Oakes submitted a brief to the subcommittee applauding Corps of Engineers accomplishments in the Willamette Valley. Oakes filed technical data about soil losses from flooding, indicating that bank cutting had already removed thousands of acres of fertile land from cultivation. Those losses meant reductions in snap-bean production, a fact that dam opponents completely ignored. In order to keep a proper balance, Oakes wrote, "it is necessary to construct multipurpose dams." Because the Willamette Valley was "the garden spot of the United States," every means should be taken to preserve and create its recreational potential. He expressed full confidence that "the proper construction of hatcheries . . . and recreational facilities at reservoir sites" would provide ample leisure activity. Oakes cited as proof the visitor counts at two completed projects, Fern Ridge and Cottage Grove.[30]

When the Willamette River Basin Commission issued its first biennial report in 1946, it boasted that in three years of operation Fern Ridge and Cottage Grove Dams had already saved in potential flood damages "more than two-thirds their original cost." In one sense, the biennial report reads like a blueprint to the promised land. When the Willamette system was completed, it would eliminate most flood damage; provide storage for irrigation, domestic use, and power production; enhance navigation; furnish water for pollution abatement during low-flow periods; and expand recreational opportunities. The 1946 report was even more ambitious: the

Willamette Valley Project would "strengthen the economy of the state" and require "further studies of projects that are not now feasible."[31]

Future development in the Columbia Basin appeared to reach a critical juncture in the fall of 1946 when the Fish and Wildlife Service recommended a ten-year moratorium against further dam construction on the Columbia and lower Snake Rivers. The Department of the Interior, which adopted the subordinate agency's proposal as its official position, focused on two critical issues: how to provide safe upstream passage for migrating fish, and how to move downstream migrating smolts safely through the dams on their way to the ocean. Although there was little public information that the dams posed problems for smolts, the Bureau of Reclamation recognized as early as 1934 the difficulties of getting the small fish through dams. In 1947 biologist Harlan Holmes began gathering data on their mortality as they passed through dams, observing that some turbines were "literal sausage grinders." In his study of the four lower Snake River dams, *River of Life, Channel of Death,* historian Keith Petersen argues that "it is more a deception of recent political convenience than a statement of facts known at the time to say that the Corps was unaware of the difficulties its river work caused smolts."[32]

No single set of testimonies is more critical to understanding the conflicting political forces than the Columbia Basin Inter-Agency Committee hearings in Walla Walla, Washington, in June 1947. The hearings were forced when U.S. Fish and Wildlife Service officials complained about the Corps of Engineers' plans for the Basin. Conflicting statements delivered at the meeting provide a window to disputes taking place over water-development projects everywhere. David Charlton, who operated a chemical-materials and ecology testing business in Portland, believed a moratorium against further dam building would allow more time to study the effects of dams on anadromous fish. Charlton wrote Kenneth Reid that every possible effort should be made to get the moratorium idea accepted, although he believed it would "be difficult to sell the other agencies" on the plan.[33]

As might be expected, hydropower, irrigation, and inland-navigation interests presented stiff opposition to the Interior Department's proposal. Interior's plan called for a ten-year delay in dam construction to allow the

Fish and Wildlife Service to carry out further biological studies on the viability of fishways, establishing fish refuges, transplanting migratory fish stocks, and the feasibility of artificial propagation. The Interior's proposal would stop all new dam construction on the Columbia below the Okanogan River and on the Snake below the Salmon River.[34] Three months before the Walla Walla hearings, the Corps' district engineer disputed the notion that additional dams would harm fish. Because the fish facilities at Bonneville Dam were functioning well, Colonel O. E. Walsh wrote, "this office still is of the opinion that the Snake River dams will not eliminate the runs of migratory fish on that stream."[35]

Supporters of the moratorium presented their testimony on Wednesday, June 25, the opening day of the Walla Walla hearings. "Because one power dam is a good thing is no sound argument that six . . . are better," T. F. Sandoz of the Columbia River Packers told the committee. Addressing the economic value of salmon, he noted that "the wealth of the sea" was being sacrificed to augment the region's need for electricity. Oregon Fish Commission master fish warden Arnie Suomela pointed to Fish and Wildlife Service findings that cited the high mortality rates among juvenile salmon passing through Bonneville Dam. And the Fish Commission's Paul Needham criticized the notion that upriver salmon runs could be transplanted into lower Columbia River tributary streams; there was simply no scientific evidence indicating that this could be done successfully. He warned the committee, "You can't have high dams and salmon too." Chief Tommy Thompson of the Wyam band at Celilo closed the day's testimony with an appeal to leave the falls in its natural state: "I don't know how I would live if you put up a dam which will flood my fishing places. . . . It is the only food I am dependent on for my livelihood, and I am here to protect that."[36]

On the second day of the hearings, moratorium opponents stepped forward with their public-relations apparatus operating at full throttle, blasting fishing interests for falsely charging that dams were responsible for declining salmon runs. A representative of the Inland Empire Waterways Association told the Interagency panel that the river needed a ten-year moratorium on fishing "to give the salmon a chance to populate themselves." The owner of Tidewater Barge Lines attributed the salmon's decline to overfishing, greedy fishing interests that had destroyed the historic runs with their big seines and traps. Overfishing and the absence of conservation measures were the real reasons for declining catches, according to Inland Navigation Company representative Gus Carlson. Herbert West, mayor of Walla Walla and an executive with the Inland Empire Waterways Association, charged that a ten-year moratorium would actually mean a twenty-year delay because of the slow planning and authorization process: "This is a well-devised scheme to kill Columbia and Snake river reclamation, power, navigation, irrigation, and flood control development for our time."[37]

Urban representatives also opposed the moratorium. Hillman Luedemann, speaking for Portland's mayor and the city's port-development committee, emphasized the vital need for electrical power. A spokesperson for the Seattle Chamber of Commerce supported full development of the Columbia's power and navigation potential with one caveat: the city might protest freight-rate advantages that Columbia River communities would gain once the series of locks was in place. W. S. Nelson, representing The Dalles Chamber of Commerce, supported the dam program, insisting that construction begin immediately on the project just upriver from his city to remove the "Celilo canal bottleneck" and the problems it posed to river navigation. If river transportation were ever to become "fully economic," it would be necessary to flood the Celilo area. Before the hearing adjourned, a host of farmers, grange members, and port and chamber of commerce people added their collective voices to protest the proposed moratorium. Finally, Colonel Theron Weaver, who chaired the Inter-Agency Committee, told those attending that no immediate decision would be made, because the volume of briefs filed at the hearing would require considerable study.[38]

Oregon newspapers were quick to pass editorial judgment on the Walla Walla hearings. The *Oregonian* accused sports and commercial fishers and Indian tribes of being "allied in an all-out, uncompromising drive" to stop work on McNary and prevent future construction on the river. The newspaper accused commercial fishermen and their Astoria business-community supporters of ignoring "the fact that the Columbia river salmon runs were on a steady decline for years before Bonneville dam was built." Fishers, it charged, lacked interest in building up the salmon runs below McNary dam, "the only practical way to avoid a complete destruction of the fish by the inevitable construction of multipurpose dams." The editorial put an interesting spin on the story when it observed that the Fish and Wildlife Service call for a moratorium "caught both the army engineers and the Bonneville power administration flat-footed." If McNary was stopped, it warned, upriver dam projects would never catch up with the demand for power. McNary was more important than mere power production; it was also a critical element contributing to the navigability of the Columbia and Snake Rivers. But the inland waterway would not be feasible until "The Dalles dam drowns out the Indian treaty fishing grounds at Celilo" and until the construction of the four lower dams on the Snake River. Salmon runs on the lower Columbia could be saved, the *Oregonian* argued, through hatchery production, the removal of stream obstructions on tributaries, control of industrial and municipal pollution, screening of diversion ditches, and establishment of fish refuges.[39] The newspaper hinted that fishing interests should invest their energies toward productive ends rather than opposing dam projects.

The most prescient and fascinating critique of Columbia River development originated in Paul Needham's address to the annual meeting of the Izaak Walton League of America in March 1947. Needham then published his talk, "Dams Threaten West Coast Fisheries Industry," in the June issue of *Oregon Business Review*. The essay opened with the warning that if the Columbia River dam builders had their way, they "will completely ruin for all time some of the richest fishery resources of this nation." Bonneville and Grand Coulee, the latter with its mammoth hatchery on the Wenatchee River, introduced "a new era of potential ruin" to the fisheries of the Great River of the West. Needham judged the hatchery experiment only partly successful, with expenses "out of all proportion to the success." Hatcheries were not the simple and easy solution to saving the runs. "Unfortunately, the teachings of early-day fish culturists still form the basis of public thought on this problem. With salmon and steelhead, it is sadly in error." In seventy years of artificial propagation, Needham declared, hatcheries had not maintained a single major run of salmon or steelhead; and that "gets us back to what nature can and will do for us if given half a chance."[40]

In light of the contentious political debate in our own time over the fate of the lower Snake River dams, Needham's most perceptive assessment addressed water-development projects on the drawing boards of the Corps of Engineers and Bureau of Reclamation. If the present plans were carried through, he predicted, "the rich anadromous fishery resources of the Columbia Basin are doomed." McNary Dam, already authorized and under construction, would be "the beginning of the end" of steelhead and salmon runs on the upper river:

> But the finishing touch will be the four-dam plan now being recommended by the Army Engineer Corps for construction on the Snake River to provide slack-water navigation to Lewiston, Idaho. *All western fishery biologists with whom I have talked agree that this plan, if followed, will spell the doom of salmon and steelhead migrations up the Snake River* as well as up its best tributary, the Salmon River in Idaho [emphasis mine].

Taxpayers wanted "expert, unbiased analyses" of the huge multipurpose dam proposals, Needham insisted. Accepting without question the Corps' recommendations was akin to taking the advice of the first salesman when purchasing a new car. "Maybe we ought to shop around a little," he concluded.[41]

In the aftermath of the Walla Walla hearings, matters moved quickly. Inter-Agency Committee staff sifted through thousands of pages of testimony, briefs, letters, and one motion picture before making a recommendation on the Interior Department's request for a dam-building moratorium. When the committee met in executive session in Portland a month after the hearings, it decided against a moratorium and passed that recommendation to the Federal Inter-Agency River Basins Committee. The parent agency also opposed the moratorium, declaring bluntly: "Facts and evidence presently available do not substantiate the fear that additional dams on the main stem of the Columbia and Snake Rivers will result in major loss or extinction of fish life on these streams."[42]

In declaring against the moratorium, the federal panel ignored the Fish and Wildlife Service declaration that McNary Dam alone threatened all upstream salmon runs with extinction. Building all of the proposed dams, the Service contended, "would literally destroy the valuable Columbia River salmon fishery."[43] But there is still more to this story. One of the key arguments against the moratorium—and part of an agreement the Inter-Agency Committee made in 1947—was the Lower Columbia River Fisheries Plan, in which the federal government promised to protect all fishery interests downstream from The Dalles. To further enforce the lower-river plan, the Washington State Legislature passed a statute in 1949 designating the Cowlitz River a "fish sanctuary" to protect it from future water projects.[44]

While federal agencies and project supporters in the Northwest were sharpening their attacks against commercial fishers and tribal groups, nature was conspiring to create conditions that would place powerful propaganda tools in the hands of Columbia Basin developers. During the winter season of 1947–48 unusually heavy snows blanketed the upland regions of the interior Northwest and the northern Rocky Mountains. Cool early-spring weather, above-normal temperatures in late April and May, and then heavy rainfall between mid-May and mid-June sent a huge volume of water rushing down streams and tributaries to the Columbia. The Kootenai, Clark Fork, Clearwater, and Snake Rivers contributed the greatest quantity of water pouring into the Great River. When the peak of the flood reached the gauging station at The Dalles on the last day of May, more than one million cubic feet per second (cfs) roared through the Gorge into the lower Columbia. "For nearly three weeks," historian William Willingham wrote, "the water flowed not less than 900,000 cfs in comparison to the annual peak discharge on the Columbia of about 583,000 cfs." The deluge created flood conditions everywhere on the lower river and inflicted damages that placed the Northwest in the national headlines.

"Bustling until late yesterday afternoon with a post-war population of 18,500," the *New York Times* reported on May 31, 1948, the "dead city" of Vanport "lay buried today under the yellow, muddy waters of the Columbia River." Under the front-page headline, "LAKE THAT ONCE WAS A CITY COMBED VAINLY FOR ANY DEAD," the newspaper declared that smashed and abandoned buildings were all that was left of the wartime housing project located on a flood plain north of Portland. Fed by melting snow and heavy rains in headwater streams, the Columbia's roiling waters swept brown and fast below Bonneville, inundating low-lying areas, destroying dikes and revetments, and ultimately threatening all flood-plain structures, including Henry Kaiser's wartime settlement at Vanport City.[45]

The *Oregonian's* report differed from that of the *Times* only in providing greater detail and more eyewitness reports. The wartime community was squared on four sides by dikes, a highway, and railroad embankments. When the railroad bed gave way at 4:30 on Sunday afternoon, the breach quickly widened to 600 feet and the water came "in waves like breakers at the seashore." Sloughs between the break and the housing center absorbed some of the inrushing torrent, thereby providing valuable extra minutes for people to flee. Because of the threatening waters of the Columbia, many residents had already left; still others were out of town enjoying the beautiful Memorial Day weather. In the end only fifteen people lost their lives, a remarkably low number considering that, as the *Oregonian* put it, the "flood descended like a tidal wave with virtually no warning." Elsewhere on the lower Columbia, the newspaper reported the failure of a dike below Rainier that flooded farms and dairies on the Oregon side of the river. Three small boys drowned near Portland's Swan Island when their raft disintegrated, and two men drowned on the Washington side while fighting the flood. Portland's small but growing airport was "closed down tight," and staff moved expensive technical equipment out of harm's way.[46]

When the flooding Columbia's waters breached the railroad embankment and destroyed the community, Vanport became an instant symbol of the need for flood control. Indeed, if there was a defining moment in the immediate postwar history of Oregon and the Northwest, it should be attributed to the finicky forces of nature, the great Columbia River flood of 1948. The damage inflicted by the rampaging waters attracted impassioned editorials calling for flood-control projects; it brought the president of the United States to tour the lower Columbia country; it loosened federal purse strings to complete McNary and Hungry Horse Dams; and it provided a big boost for authorizing additional dams. There had been no similar event since the great flood of 1894, when the region's population was much smaller. Although the river crested well below the 1894 level, damage was much greater, causing more than $100 million in property losses.[47]

While the Vanport disaster was complete, it represented only part of the damage inflicted in 1948. For two weeks following the destruction of Vanport, flood conditions on the lower Columbia worsened as the continued pressure on dikes and levees threatened to inundate lowland areas. The Corps of Engineers brought in hundreds of troops for flood-control work and urged all people living behind dikes to evacuate their homes. At the peak of the effort, 2,000 military personnel and 5,000 civilians were involved in the flood-fighting effort. Levees, roadways, and dikes failed, bridges were washed away, and then on June 12 heavy rains temporarily worsened the situation. But the most critical period had passed, and conditions improved to the point that the Corps began to demobilize by the middle of the month. William Willingham's history of the Portland District of the Army Corps of Engineers calls the 1948 flood "the greatest single disaster in the history of the Columbia River Basin." In addition to Vanport's destruction, the uncontrolled waters took nearly forty lives, destroyed 5,000 buildings, and put 50,000 people out of their homes. Add to this the thousands of acres of eroded and submerged land and an additional 15,000 acres of inundated commercial and urban space, including more than 600 Portland city blocks.[48]

Within three days of the great flood, the *Oregonian* asked the federal government to accelerate appropriations for river-development projects. The multimillion-dollar cost of the damages on the Columbia and its tributaries, the newspaper observed in a lengthy editorial, brought "national as well as regional attention to methods of control." If an additional four multiple-purpose dams had been in operation, Army engineers reported, "the present crest would have been lowered at least three feet." Those projects included flood-control dams on the north fork of the Flathead River, the Clark Fork of the Columbia, the Pend Oreille, and in Hells Canyon on the Snake. The last structure alone "probably would have reduced the flood level one foot." The newspaper pointed out that neither Bonneville nor Grand Coulee was intended to control floods: "The truth is that congress did not think much about flood control when Bonneville and Grand Coulee were approved." For effective flood prevention, headwater streams had to be controlled. The dams targeted for the lower Snake and main-stem Columbia (McNary and the proposed John Day and The Dalles projects), the *Oregonian* cautioned, "may be written off as far as flood control is concerned." Although some of the upstream engineering work would require agreements with Canada, Columbia Basin citizens would be fortunate in being able "to combine flood control and power facilities on upstream dams which are least damaging to fish migrations."[49]

As the river was receding, President Harry Truman made an unscheduled visit to the soggy ruins at Vanport and remarked that the scene was "just as bad as I thought it was." In earlier comments in Portland, the president told the audience that he would urge Congress to fund the Columbia River development projects so that "these disastrous floods will never happen again." The president also repeated what was becoming the conventional rationale for Bonneville and Grand Coulee dams: "Had it not been for the immense power dams on the river, it would have been much more difficult for us to win the last war. If they had not been built over the protests of the opposition, World War II would have cost us many more lives." As for the rest of the federal plan for the Columbia system, the blunt-speaking president declared, "I hope to see it outlined and completed." *Oregonian* staff writer Paul Hauser thought the president had shown the way for western Democrats who were expected to "ride the flood for all it's worth." The destruction from the 1948 flood pointed to the "perfectly obvious conclusion that control and development of the river is the region's No. 1 problem and interest."[50] It was left to others, who used more fanciful language and exaggerated rhetoric, to further promote river development and push for additional dams.

Federal officials were already taking aggressive measures to advance Columbia Basin development proposals. The Army Corps of Engineers gave its official sanction to a flood-control and hydropower dam in the Snake River's Hells Canyon. The Corps reported that the dam would reduce flooding on the Columbia and generate one million kilowatts of electrical power. The big project also had the approval of Interior secretary Julius

Krug, who forwarded the Bureau of Reclamation report on the proposed dam to governors and state and federal agencies in the Pacific Northwest. Reclamation commissioner Michael Straus thought the enterprise would "make a substantial contribution to the upstream storage so urgently needed to control the devastating Columbia river floods." Although the Bureau would be charged with actual construction, the Corps supported the dam because of its flood-control potential. The Corps' North Pacific Division head, Theron Weaver, cited engineering studies suggesting that the Hells Canyon dam would reduce peak flooding in the Portland-Vancouver vicinity by 1.2 feet.[51]

Richard Neuberger joined the chorus of voices supporting additional multipurpose dam projects on the Columbia River. Writing for *Holiday* magazine in 1949, the future senator expressed his belief that Americans should use their technical expertise and engineering abilities to control flooding, produce cheap electrical power, and, in the case of the arid Columbia plain, build a "new civilization . . . out of the dead desert." The architects of Bonneville Dam, he argued, had proven conclusively that salmon could successfully navigate fish ladders to upstream spawning grounds and that downstream bypasses protected ocean-bound fingerlings from being chewed up in the giant turbines.[52]

Even when he raised tough questions, Neuberger assured readers there were technical solutions to managing the Great River. "What about Grand Coulee Dam," he asked, "a massive roadblock, impassable to salmon headed for some icy rivulet far upstream?" The solution, he wrote, was a "biological experiment conducted on a heroic scale" whereby salmon trapped below Grand Coulee were transferred to enormous hatcheries on tributary streams. Once the fingerlings were released, they enjoyed free passage to the Columbia. When the first generation of salmon returned, Neuberger reported, the fish obediently turned up the Methow, Entiat, Okanogan, and Wenatchee Rivers to their new homes. "Man," the effusive journalist concluded, "had masterminded the salmon's breeding habits."[53] With few exceptions, Neuberger's enthusiasm for imposing human-designed engineering schemes on the Columbia system never abated. Although he had questioned the influence of dams on migrating salmon in some of his earlier writings, with the passing of time he became a true believer in fully developing the great waterway.[54]

When the Columbia's swirling waters inundated much of the lower river, the Corps' Pacific Division was in the midst of a comprehensive update of the old "308 reports" on the Columbia Basin. After the Vanport flood, President Truman directed the Corps to include long-range flood-control planning as part of its survey. The principal feature of the eight-volume final report, published in 1950 as House Document 531, was a broad-scale blueprint to prevent flooding on the Columbia River and its tributaries. Completing the multipurpose dam system, the Corps document concluded, would "create a well-knit system of works for the most efficient utilization of the water resoures in the promotion of a balanced economy." In the River and Harbor and Flood Control Acts of that same year Congress passed legislation authorizing the Corps to begin work on a number of new projects—Albeni Falls, Libby, Priest Rapids, John Day, The Dalles, and smaller dams in the Willamette River Basin. In the interim, the Inter-Agency Committee, Corps of Engineers, Bureau of Reclamation, Bonneville Power Administration (BPA), and regional business groups continued to aggressively promote the big water projects. The committee produced a pamphlet citing overfishing, pollution, logging practices, and unscreened diversion canals as the chief culprits responsible for declining fish runs. While it admitted that dams posed problems, the committee insisted that fish-protection devices—fishways, fish screens, and artificial propagation—would reduce losses. It also praised Bonneville's fish ladders as an unqualified success.[55]

The Bonneville Power Administration was especially adroit in promoting hydropower projects. Shortly after the release of House Document 531, BPA published a pamphlet praising the new dams under construction in the Columbia Basin: McNary, Hungry Horse (South Fork of the Flathead River), Chief Joseph, and Detroit (Santiam River). The agency continued to insist on the need for more electrical power, pointing out that increased economic activity and a growing population after the war had created a shortage of electricity. Delays in construction and building transmission lines and substations had curbed industrial expansion in the Northwest and cost the region 25,000 jobs in large electro-process industries. The "power of the Columbia" had been "America's strong right arm" in the Second World War; the BPA publication observed that the nation once again was looking to the region's power program to strengthen its Cold War defense program. Federal development of the Columbia system held the potential to create 300,000 new jobs and to help reclaim 4 million arid acres of land. Dams on the Great River, the pamphlet concluded, "must be designed and operated as an integrated unit" to permit the full use of the waterway: "This should be the region's goal in developing America's greatest power stream: for all its values and for all its people."[56]

The Corps of Engineers blueprint embodied in House Document 531 would significantly influence the future development of the Columbia River. In submitting the document, the Corps noted that the study was carried out in response to Truman's directive following the disastrous Columbia River floods of 1948, and flood control received the greatest emphasis in the report. The agency also pointed out that the water projects would have "far-reaching significance nationally." Without "artificial regulation," periodic flooding would continue; there would be inadequate water for irrigation; the potential for power generation would go unrealized; and rapids and falls would continue to hamper navigation. Completing the series of dams, levees, and revetments, the Corps argued, would "create a well-knit system of works for the most efficient utilization of the water resources in the promotion of a balanced economy." The report's introduction singled out the "Willamette sub-basin plan" as an urgent project that "should be carried to substantial completion with high priority."[57] Congress responded to the release of the House Document 531 report with passage of the River and Harbor and Flood Control Act in 1950, a measure that authorized the Corps to proceed with several new projects.[58]

The genius of the Corps' plan to rationalize the Columbia system lay in regulating the rivers' flow:

> The potentialities involve . . . a fairly uniform seasonal run-off pattern in the greater part of the basin, whereby the low-water period and the season of high run-off are so timed that regulation of the flow for one purpose often improves conditions for other purposes. Thus, the opportunities are great for controlling destructive spring floods by impounding excess run-off for

release during the summer months, to generate power. Release of stored water in winter to augment the low-water discharge will not only serve to increase power production, but will increase minimum channel depths for navigation, improve industrial and domestic water supplies, and dilute the corresponding wastes that now cause pollution.

A rapidly growing population, the need for more productive farm land, improvements to river transportation, greater hydropower capacity, and protection against destructive flooding, according to the Corps, provided "a compelling argument for expanding the scope of multi-purpose river development."[59] Although Congress dispersed appropriations for the projects in piecemeal fashion, the funding authorizations laundered through Congress in the 1950s continued the dam-building juggernaut in the Columbia Basin.

The Portland District of the Corps of Engineers saw its work as part of an effort to fulfill regional planning objectives initiated during the New Deal years. The projects detailed in the 531 report were "further steps in the overall regional system of improvement." While solid economic analysis justified the Columbia projects, there were less tangible reasons for carrying the work forward, far-reaching purposes that did not lend themselves to "accurate segregation and measurement." The "impressive byproducts" included national security, the protection of human life, improving the nation's standard of living, and stabilizing business and employment. Those were important reasons, the Corps contended, for moving forward with Columbia River development, and they far outweighed "conflicts with resources of lesser significance that cannot be avoided."[60] Even a fully sympathetic reading of the 531 report would suggest that anadromous fishes (and the communities dependent on them for their livelihoods) were among those "resources of lesser significance."

Columbia River scholars err when they conclude that agency-produced documents such as the 531 report ignored salmon. In the years following the Second World War, fishery interests were still important and influential voices in Northwest politics. Their concerns about main-stem dams on the Willamette system and the opposition they expressed in Walla Walla and at other hearings testify to their continuing presence in regional affairs.

The Fish and Wildlife Service contributed a full appendix on fish and wildlife issues to the 531 report, with much of the analysis focusing on anadromous fishes. In the broadest sense, the appendix began, river development "has had a profound influence on fish," especially anadromous species. Furthermore, because of the big projects planned for the Columbia system, the Fish and Wildlife Service insisted that careful consideration be given to "feasible plans which could prevent unnecessary losses." The fisheries narrative candidly admitted that dams were harmful to anadromous fish in several ways—interfering with upstream migration, inundating spawning grounds under deep reservoirs, and creating dangerous downstream passage for fingerlings.[61]

The Fish and Wildlife Service acknowledged that safe downstream passage of smolts through Bonneville Dam "has not yet been solved." For hatchery-reared fish alone, marking studies revealed a mortality rate of 15 percent at Bonneville. Although "such a mortality at one dam would not have a drastic effect, . . . the construction of a series of dams between the ocean and the major spawning areas might seriously reduce the level of abundance." The appendix's authors observed, however, that the problem "would be virtually solved" if some way could be found to safely pass the fish through the dam. While the difficulties posed by single dams could be resolved with only minor fish losses, a succession of dams would present "a combination of problems that cannot be looked upon so optimistically."[62]

The fisheries brief raised the possibility that environmental changes in the Columbia Basin—fish losses during upstream and downstream passage and reduced spawning areas—might create conditions that would make "continued propagation in the headwater tributaries impracticable." This was especially true for the lower Snake River, the agency concluded, where the proposed dams "present the greatest threat to the maintenance of the Columbia River salmon population of any project heretofore constructed or authorized." If the proposed scenario were carried out, the only alternative would be to depend on lower-river tributaries and artificial propagation to maintain anadromous fish runs. But sustaining the lower-river fishery would be no substitute for conserving upper-river spawning runs: "Every effort must be made to maintain the runs in the upper tributaries." The Fish and Wildlife Service viewed the lower-Columbia plan as "insurance to protect the fishery resources" in case other protective measures failed.[63]

The 531 report and subsequent studies carried out by state agencies in Oregon and Washington included abundant evidence that the cumulative effects of dams would harm the downstream passage of smolts. It is also true that federal agencies and private organizations anxious to move ahead with Columbia Basin development conveniently ignored the dangers that multipurpose dams posed to fisheries. The Columbia Basin Inter-Agency Committee, as noted, cited overfishing, sewage and industrial pollution, careless logging practices, and unscreened diversion canals as key factors contributing to declining salmon runs. Although it recognized that dam construction raised additional questions about depleted salmon numbers, the agency noted that fish passage over low-head dams had "long been successfully accomplished by means of simple fishways or ladders." The committee assured the public in one of its periodic news releases that fish ladders, screens, and artificial propagation could partially mitigate those losses and help sustain the runs.[64] In a letter to assistant Interior secretary C. Girard Davidson, David Charlton observed that agencies had a tendency "to 'duck' controversial matters" related to their pet projects.[65]

In the end, the Corps and Bureau of Reclamation did not build all the authorized storage projects. In some instances, Congress refused funding or downsized the plan; in still others private utilities carried out the construction work. In the case of Hells Canyon, the Idaho Power Company constructed the dams (although with a sharply diminished storage capacity). But for the most part, the river developers had their way, setting in motion a series of federal authorizations and appropriations that dramatically altered the Columbia River landscape, including major tributaries such as the Snake, Cowlitz and Willamette. The combined storage capacity of the dams ended the most destructive flooding in the region, especially along the broad floodplain in the Willamette Valley. The construction of McNary, The Dalles, and John Day Dams and the four dams on the lower Snake River turned Lewiston, Idaho, into the Northwest's most distant inland port. Columbia Basin dams also turned the Great River into the most productive hydropower waterway in the United States, its turbines capable of producing surpluses of electricity marketed by the Bonneville Power Administration as far away as California. Each of those authorized structures in the increasingly engineered Columbia system also posed greater difficulties for anadromous fish. As historian Richard White put it, although the river's architects continually expressed their concern for salmon, they "quite consciously made a choice against the conditions that produce salmon."[66]

II

MAKING

AGRICULTURE

MODERN

3

BRINGING PERFECTION

TO THE FIELDS

From sea level to mountains we have a tremendous variety of topography, soils, and climate. This adds spice and variety to our agriculture. We are one of the few states in the nation that has a truly diversified agriculture, with a hundred products going to market each year. . . . We sell a large volume of farm products to other states and nations.—*Agriculture in Oregon* (1965)[1]

If there were a storied beginning to modern Oregon, it would be vested in its agricultural bounty, especially the productive lush green valley of the Willamette. From its first white settlements, Oregon enjoyed a rich tradition of agricultural abundance that gave birth to stories creating an image of a bountiful land filled with promise. The nineteenth-century magazine *West Shore* praised the Willamette Valley for its wheat production and for shipping "its unequaled wealth of breadstuffs to the nations of Europe." The valley's tremendously productive soils had already earned it a "foremost place as a producing State." *Willamette Farmer,* a contemporary weekly newspaper, celebrated the state's agricultural productivity as "the triumph of man over Nature in Oregon" and praised the improvements for promoting the progress of civilization.[2]

"For over a century," the state's *Agricultural Bulletin* declared in 1951, "the Willamette valley has been the magic name inducing a westward migration of people." The majority of those immigrants, state agriculture director E. L. Peterson wrote, came West to establish homes, raise families, and build communities. Because they were "home builders rather than seekers of quick riches," the valley's development "has been steady and conservative." Although some had squandered the wealth of its broad acres and green forests, Peterson observed, forward-looking people had taken conservation measures to assure a promising future. In the first half of the twentieth century, Oregon boosters were equally enthusiastic about Rogue Valley agriculture and reclamation developments in the Klamath Basin, the middle-Deschutes region, and eastern Oregon's Umatilla and Vale-Malheur districts. Although the Great Depression and the Second World War slowed irrigation development, with the end of hostilities newspapers and developers once again boasted about the bright prospects for Oregon agriculture. Willamette Valley residents were restoring forests, rehabilitating rivers, rebuilding soils, and positioning the valley for a future "that could well surpass anything yet envisioned." The big federally sponsored Willamette Valley Project, with its ambition to curtail flooding and provide water for irrigation, promised to increase agricultural productivity when it was completed. This could be accomplished, the agriculture director believed, through "demonstration and education" and leadership "in the administrative arm of state government."[3] Peterson assured readers that the state Department of Agriculture was prepared to lead farmers to the promised land.

With its ideal soil, climate, and productive conditions, a Portland Chamber of Commerce pamphlet declared in 1945, Oregon was positioned "to produce abundantly any field, fruit and vegetable crop native to the temperate zone." In unrestrained and uninhibited prose, *Farming in Oregon* informed readers that no state offered "so many possibilities, has such beauty, and affords so much opportunity for enjoying life." In a section "TO THE AGRICULTURAL HOMESEEKER," it urged potential buyers to consider both developed and partly developed farms and offered practical advice. Settlers planning to acquire "raw land, especially logged-off land," should arrange to have an income from another source until sufficient acreage was available to make a living. Clearing logged-off land was expensive and time-consuming, the pamphlet pointed out, but the effort was worthwhile.[4]

Although California's huge corporate-based agribusiness dwarfed agricultural output in the other western states, Oregon's farm enterprises ranked second only to its logging and lumbering and were important contributors to the state's economic livelihood. Linked together by highways and county roads, Oregon's small towns and rural settlements remained vital to the state's economic and cultural life. But change was coming to the Oregon outback as the Rural Electrification Administration was bringing electric lights, hot-water heaters, electric stoves, and a rapidly increasing number of other electrically powered gadgets to farm families. The parallel development of mechanized operations in the fields and dairies meant an exodus of young people to larger towns and cities. State land-grant universities—Oregon Agricultural College in this case—distributed circulars and advice to agricultural constituents heralding the latest research findings, the newest miracle chemicals, and other information important to profitable agricultural practices. In the long run those initiatives contributed to reduced labor costs, the increased use of petrochemicals, and larger landholdings to achieve economies of scale. Advice to farmers, in Oregon as elsewhere, emphasized "technological efficiency in crop production and harvests" and "advanced designs in tractors, combines, hay balers, and sugar-beet and corn harvesters."[5]

War-induced demographic changes in the rural Northwest had already contributed to increased income, better living conditions, greater production, and more favorable prices for farm products. The U.S. Department of Agriculture (USDA) estimated in 1944 that Oregon's rural farm population had declined by about 20 percent during the Second World War. Approximately 10,000 Oregonians had joined the armed forces and an even larger number (some of them rural non-farm residents) had moved cityward for wartime employment. Oregon's declining rural population mirrored national trends and paralleled that of its neighbors, Washington and Idaho. The USDA report also indicated that the region's agricultural output would be sufficient following the war to meet the public's needs "even if there is no increase in the farm population."[6]

The Readjustment and Development Commission reported in January 1949 that the state's workforce was rapidly shifting from "farmlands to manufacturing and services," with the increased use of "power machinery" explaining the diminishing need for farm workers. Although the rural population declined after 1940, farm income actually increased because of larger units and mechanization. The size of the average Oregon farm increased

from 268 to 387 acres between 1935 and 1954, largely because of technological changes—the substitution of machines for animal and manual labor
power—and the pattern was repeated across the United States. While the
use of heavier mechanized equipment meant larger fields, it also meant
increased soil compaction—and an increasing potential for erosion.

By the end of the war, irrigation was also becoming a more prominent
feature across the arid country east of the Cascade Mountains. The state
Department of Agriculture estimated that more than 1.3 million acres were
under irrigation statewide, with supplemental watering systems in the
Willamette Valley still in their infancy. Counties with large reclamation
projects—Klamath, Deschutes, Umatilla, and Malheur—produced a great
diversity of crops: forage seeds, rye and barley, potatoes, hops, peppermint,
sugar beets, and large fields of alfalfa. But the most dramatic changes were
associated with the use of chemicals to reduce losses from diseases and pests
and to promote plant growth.[7]

Oregon's postwar agricultural literature emphasized efficiency, mechanization, reducing labor as a factor in production, and encouraging farmers
to make use of the most recent information on insecticides and herbicides.
The widespread use of mechanized equipment, especially the tractor,
released land for production that hitherto had been used to graze horses and
oxen. Wheat, the crop with the highest market value during the nineteenth
century, was still the state's leading agricultural product in the early 1950s.
The average yield of thirty-six bushels of wheat in 1958 was an all-time
record, with most of the crop cultivated on the rolling slopes above the
Columbia River in northeastern Oregon, but the Cooperative Extension
Service pamphlet *Oregon's First Century of Farming* reported in the same
year that potatoes, fruits, and vegetables, seed crops, berries, and nuts had
become significant cash producers.[8]

The agricultural systems that took shape in Oregon after 1945 were of a piece
with practices and procedures elsewhere in the United States. Strategies and
approaches were rooted in scientific and technological traditions dating to
1862 with the establishment of the USDA and congressional passage of the
Morrill Act to fund state agricultural ("land-grant") colleges. According
to its enabling legislation, the USDA's mission was "to acquire and to diffuse

among the people of the United States useful information on subjects connected with agriculture in the most general and comprehensive sense of that word, and to procure, propagate, and distribute among the people new and valuable seeds and plants." The Hatch Act, passed in 1887, authorized the establishment of agricultural experiment stations to carry out "researches, investigations, and experiments bearing directly on and contributing to the establishment and maintenance of a permanent and effective agricultural industry." The legislation directed stations to obtain "useful and practical information" and to promote scientific inquiries "respecting the principles and applications of agricultural science." Congress completed modern agriculture's institutional matrix when it passed the Smith-Lever Act in 1914, setting up the Cooperative Extension Service to disseminate research results to farmers. Those legislative initiatives contributed to an explosion in scientific and technical innovation in American agriculture, especially following the Second World War.[9]

With most initiatives originating in Oregon State College in Corvallis, the combined effect of USDA, Experiment Station, and Cooperative Extension Service efforts fostered an ideology of progress and a belief that seemingly endless increases in agricultural productivity were possible. Like other land-grant colleges, Oregon State College had a long tradition of educating agricultural scientists and engineers. By the 1930s its faculty was offering courses in genetics, agronomy, entomology, agricultural engineering, economics, and related subjects, and it was through those disciplines that the most dynamic and spectacular agricultural breakthroughs took place. Writing in *Technology and Culture*, Deborah Fitzgerald observes: "A progressivist fervor that equated progress with science, mechanization, and capital-intensive methods on the farm" guided work in land-grant colleges. Born in a rhetorical tradition of promoting good agricultural practices and advancing the social welfare, land-grant colleges and their affiliated agencies promoted industry-like models of American agriculture.[10]

By 1950 Oregon's Agricultural Experiment Station, the research arm of the School of Agriculture at Oregon State College, employed more than 100 scientists in several college departments and branch stations located in different parts of the state. Funded by state money, gifts and grants, and federal research dollars, the Experiment Station boasted that its work contributed to lower food prices, labor-saving technologies for farmers, and healthier plants and more productive animals. One Experiment Station circular published the exaggerated claim that for every $3.00 investment in agricultural research, consumers received $200 in benefits. Such overblown claims, ostensibly to produce more food for a growing global population, were common to American agriculture during the Cold War. "What can Oregon do to help solve the world's hunger problem?" another Experiment Station publication asked. The agency's response underscored the land-grant college's important role as "an international leader in the battle against world hunger." Promotion of this vital work required stepped-up research efforts to develop more productive crop varieties, more effective ways to combat insects and diseases, and other strategies to boost production.[11]

The Experiment Station was of one mind with industry officials who pointed to agriculture's important role in stabilizing the state's economy. Indeed, department publications touted the industry as the "most important factor in Oregon's general economy." The men and women who tilled the soil and raised animals, a state Department of Agriculture pamphlet declared, were "progressive, alert," and read carefully the latest "experimental and educational work in farming."Agriculture was heroic work because the world would require "all the food that farmers can produce," an effort that would require "everyone to put strong shoulders to the wheel." It is doubtful, the department's *Biennial Report* for 1968 declared, that any business "returns as much to the community." The industry was permanent and clean, created jobs, carried a heavy tax burden, and was still taken for granted.[12]

Because Oregon's diverse agricultural interests were increasingly better organized after 1945, they were able to participate as full partners with state and federal agencies in adopting progressive farming practices. National farm organizations such as the Farm Bureau were well represented across the state, and the legislature and state Department of Agriculture began recognizing major agricultural commodity groups to assist in shaping policy. The commodity organizations vigorously represented their constituencies before the legislature and to the Agriculture Department, and some of the more influential groups began establishing commodity commissions under state law—a relatively new development in the United States.

Oregon's commissions worked closely with the state Department of Agriculture to offer advice and counsel on farming practices, pesticide regulations, tax policy, and a host of other matters. The commissions indicated the growing sophistication of some segments of agriculture and were powerful lobbying groups, with much of their effort directed at fighting off legislative regulation of agricultural practices. The commissions were especially aggressive in defending commodity groups from what they deemed unfair environmental challenges to conventional farming practices.[13]

The Department of Agriculture occupied a central place in Oregon's move towards greater technical efficiency in agricultural production. It was born of the Depression-era need for economy and efficient operations: the Oregon legislature consolidated seventeen government branches into a single agency when it authorized formation of the department in 1931. The enabling legislation mandated the appointment of a state Board of Agriculture representing horticulture, animal husbandry, dairying, poultry, field crops, market gardening, and cooperatives. (The board did not have a "public representative" appointee until 1949.) During its first two decades the department emphasized grades and standards to improve the marketing of Oregon crops, and until the mid-1950s its work involved primarily inspection and regulatory activity. Licenses and fees from the respective commodity groups provided approximately three-fourths of the department's operating expenses, practices that contributed to close working relationships between the agency and its producer clients. From its inception, the Oregon Department of Agriculture emphasized education, voluntarism, and other enticements to encourage farmers to abide by the state's modest rules and regulations.[14]

The Oregon Department of Agriculture, the Experiment Station, and the Cooperative Extension Service were involved in a broad range of activities that had major implications for the physical world and its flora and fauna. The department devoted attention to cloud seeding, controlling ragweed, efforts to contain predators and exotic species, quarantines, crop-pest surveys, and more conventional activities such as promoting agriculture. The state's fast-growing seed-crop industry, which had its commercial beginnings in the 1920s, is a case in point. Oregon's seed industry, especially grass seed, was both planned and serendipitous, subsidized at first by federal conservation programs during the 1930s. The greatest production was in the Willamette Valley, where acres planted to varieties of grass seeds increased. Distant markets brought change to local settings in Oregon's major western valley, repeating historic patterns that had characterized Oregon agriculture since the California Gold Rush.

According to the seed industry's conventional story, Oregon's agriculture was languishing in the early twentieth century when President William Jasper Kerr of Oregon Agriculture College organized a planning conference to seek ways of reinvigorating the lagging industry. From that meeting, Extension Service and Agricultural Experiment Station personnel began urging Willamette Valley farmers to grow seed crops in appropriate locations and to work out production and marketing wrinkles. In a later account, perhaps granting more credit to agricultural planning than it deserved, Leonard Jernstedt of the Oregon Seed Growers League contended that the emergence of the seed industry "was not a chance happening, naturally evolved, but was the result of a plan carefully carried out."[15] Maybe so; but without natural conditions ideally suited to growing forage and grass seed crops, moist and mild winters and relatively dry summers, the planning effort would have been less successful.

Federal subsidies during the 1930s were central to the initial experiments with seed growing in the Willamette Valley, with the product intended to help rehabilitate eroded hillsides in the Tennessee Valley. In its May 1947 report, the PRDC recognized the federal government's pivotal role under the headline "SOUTHERN STATES DEVELOP OREGON." The principal outlet for Oregon's seed industry, according to the commission, was the market for "cover crops in southern states to rebuild exhausted soil." In 1951 Jernstedt also pointed to the importance of New Deal programs such as the Agricultural Adjustment Administration and the Public Works Administration in southern land rehabilitation. Beginning in the 1930s and continuing through the war years, federal payments to cotton farmers for cover crops "supplied the market for our growing industry."[16]

The shortage of nitrogen fertilizers during the war further advanced the need for cover crops in the South, and a modest shift from cash-crop farming to raising beef cattle created a demand for perennial forage-grass seeds. E. R. Jackman, extension crops specialist with Oregon State College, called the seed business "a unique cross-continent cooperative venture." Although the South had poetic names for its rivers—Chattahoochee, Swanee, Rappahannock—"beneath the poetry was mud." Because the region also used heavy quantities of nitrogen fertilizer, Oregon-raised legumes served double duty, returning nitrogen to the soil and preventing further erosion. Jackman estimated that Oregon enjoyed approximately 80 percent of cover-crop sales in the South. At the industry's production end in the Willamette Valley, the cover crop showing the greatest sales value was hairy vetch, a plant virtually unknown before the 1930s. By the early 1940s growers were planting about 90,000 acres to hairy vetch. Common rye grass, which increased from 35,000 acres in 1939 to more than 97,000 acres in 1946, eventually became the Willamette Valley's monopoly grass seed.[17]

Leonard Jernstedt's 1951 declaration that the seed industry had a promising future proved prophetic. Because the business was basically sound, well planned, and favorably adapted to valley conditions, he reasoned, it would provide farmers with a stable income and would require only occasional market adjustments. In a 1951 letter to *Oregon Journal* editor Marshall Dana, Jackman pointed out that Oregon produced more seeds than any other state, growing 90 percent of the nation's bentgrasses, 90 percent of ryegrass, 80 percent of fine-leaved fescues, 90 percent of alta fescue, and 50 percent of Ladino clover. Fifteen percent of its agricultural land grew seed crops, and Jackman claimed that no other state approached Oregon "in the way we have worked seed production into our economy." Of Oregon's 30,000 commercial farmers, 5,000 were seed growers, most of them located in the Willamette Valley. Growing grass seeds was also good for soil conservation, Jackman argued: "We show [the farmer] . . . how to grow seed crops, which nails his soil down tight and improves it, and makes money for him at the same time."[18]

Jackman was Oregon's most prominent agricultural figure during the postwar years, and he believed firmly in the restorative qualities of planting seed crops. In testimony to the Columbia Basin Inter-Agency Committee in 1950, he praised the "erosion prevention and soil improving factors" associated with seed crops. For more than two decades the Extension Service and the Agricultural Experiment Station collaborated to establish grass-seed nurseries around the state in an effort to identify seed crops that would prosper in each of Oregon's diverse climates. "Ladino clover for southern Oregon; alsike for central Oregon; red clover, vegetable seeds, and alfalfa for Malheur; cover crops and seed grass for the Valley; dryland grasses for the Columbia Basin; and bentgrasses for the Coast." Travel through the Willamette Valley, he told the committee, and you will see farms producing more than ever before. "The land is productive again." Growing seed crops, Jackman declared, had effectively stopped erosion and built soil fertility.[19]

Thanks in large part to federal research and marketing support, seed-crop production gradually became dominant over vast expanses of the Willamette Valley. Growers planted approximately 11,000 acres to grass-seed crops in 1940, 100,000 acres in 1950, and about 300,000 acres by the early 1980s when grass-seed production covered more than 32 percent of all valley cropland. In Linn, Lane, and Benton Counties in the southern valley, 56 percent of all crops were planted to grass seeds. In Linn County, the "Grass-Seed Capital of the World," farmers grew more than 40 percent of Oregon's grass seed in 1987 and 75 percent of all the ryegrasses grown in the United States. Growers quickly learned that valley soils varied in their ability to produce certain seed crops. In Polk, Yamhill, Marion, Clackamas, and Washington Counties, farm sizes were generally smaller, and more variable soil conditions permitted a variety of alternative crops. In the southern valley, however, heavy clay-like and poorly drained soils limited options for farmers to annual and perennial ryegrasses.[20] At the production end, the seed-grass business had a striking influence on the Willamette Valley landscape. Ever larger tractors and harvesting machines and an emphasis on economies of scale prompted farmers to eliminate hedgerows and brush separators until some huge fields encompassed more than 1,000 acres.

The grass-seed industry's amazing marketing successes had other environmental consequences. Just as seed production was taking off in the early 1940s, growers encountered problems with blind-seed disease (*gloeotinia*) and ergot (*claviceps purpurea*), infestations that threatened the purity and quality of seed crops. The fungus-like blind-seed disease had spread through valley ryegrass crops in epidemic proportions by the mid-1940s. Following the recommendations of John Hardison, USDA plant pathologist at Oregon State College, ryegrass farmers began experimental burns in 1948 to get rid of the straw residue in their fields. Because open-field burning proved strikingly successful in controlling diseases, growers quickly adopted the practice to sanitize their fields following harvests. This seemingly new approach to the disease problem was reminiscent of Indian burning practices reported by Scottish naturalist David Douglas more than a century earlier. Literally overnight the autumn fires that returned to the Willamette Valley became an inexpensive and magical solution to a variety of problems. Post-harvest burning eliminated blind-seed disease and ergot, and it reduced or controlled a number of other plant diseases. In perennial grass fields where farmers raised the same crop for several years without plowing, burning provided an effective and low-cost technique for getting rid of straw residue, which could reach three to five tons per acre.[21]

Burning achieved even more; it served as an effective tool to control weeds and it cleansed fields of substances that would interfere with efficient herbicide applications when the new chemicals became available in the early 1950s. Burning fields following harvests promoted the vigorous regrowth of perennial grasses, which in turn contributed to higher seed production the next season. Burning destroyed breeding niches for insect pests and helped control plant bugs, and the ash deposits left in the fields recycled nutrients to the soil (potassium, magnesium, calcium, and phosphorus). Torching the post-harvest residue of annual ryegrass crops readied fields for replanting with little or no tillage required, thus reducing labor and production costs. On perennial-ryegrass fields, open burning raised yields from an average of 300 pounds per acre to more than 1,000 pounds. Finally and most important, because burning destroyed shattered seeds following harvests, it enhanced the genetic purity and marketability of grass seed.[22]

Willamette Valley grass-seed farmers had seemingly achieved perfection in their fields. With seed certification procedures firmly in place, markets for Oregon-grown grass seeds increased tenfold. By 1959 the estimated gross income for Oregon's seed-crop farmers was almost $24 million, approximately $11 million of that from the sale of common and perennial ryegrasses.[23] At the center of the Oregon Seed Certification Program, and a key explanation for the state's remarkable marketing successes, was post-harvest burning. The Oregon Seed Council proudly boasted in the mid-

1970s that Oregon grass seeds had 7 percent more "pure live seed" than similar harvests in other climates. The biggest problems, especially in ryegrass production, were volunteer plants that threatened to alter the genetic purity of seed crops. Field burning, an Oregon State University Extension publication reported in 1989, "has historically been the most effective practice used for volunteer control."[24]

There was a downside to field burning, especially in the Willamette Valley, home to more than 70 percent of the state's population. The plumes of smoke that rose above the valley floor each autumn increasingly became a bone of contention between the growing non-farm population in the southern valley and the powerful grass-seed industry, which claimed that open-field burning was absolutely essential to high-quality seed production. Public complaints about smoke escalated when farmers sharply increased the number of acres put to the torch. Those problems worsened when growers burned fields with green regrowth later in the season or under adverse weather conditions. Reflecting in part citizen protests over the right to clean air, anti-burning activism emerged in the mid 1960s in the Eugene-Springfield area, where unregulated burning emissions tended to accumulate. Those conditions were especially severe when atmospheric inversions prevented smoke from dissipating. Field-burning opponents quickly gained the ear of local legislative representatives, and the matter became a contentious political issue that persisted for more than three decades.

The state legislature's initial interest in field burning was cosmetic, focusing on smoke abatement to improve visibility. The seed industry, Oregon State University (OSU), and the U.S. Weather Bureau worked cooperatively to investigate the proper timing for scheduling burns when weather conditions would maximize smoke dispersal; they also sought to prevent concentrated burning in the late fall (which tended to produce more smoke). OSU scientists carried out research in 1965 to determine how environmental variables affected burning and to determine "conditions . . . when significant air pollution reduction can be achieved." Other university and Extension personnel experimented with extensive field plots, investigating the effects of time and the intensity of burning on smoke volume. USDA funds, seed-industry contributions, and the Oregon legislature supported a variety of inquiries into alternative methods of controlling seed infections, marketing straw from the fields, using meteorological conditions to issue burning permits, and reducing the acreage burned. None of those early studies contemplated a ban on field burning.[25]

As opposition continued to mount, the Oregon legislature began to intervene, granting the Oregon Sanitary Authority (renamed the Department of Environmental Quality [DEQ] in 1969) advisory power to determine where and when burning could take place. And then, following a disastrous burning season in 1969, the legislature gave DEQ temporary authority to limit burning on marginal days. The autumn winds that blew into the southern valley created crisis conditions on August 12, 1969, when the Eugene-Springfield area was blanketed with acrid smoke from burning fields. An angry Eugene mayor Les Anderson phoned Governor Tom McCall, who made an emergency late-afternoon visit to the city. Standing atop Skinner's Butte between the Willamette River and the downtown area, McCall dubbed the conditions "the scene that can't be seen." On his return to Salem he immediately issued a ten-day moratorium on field burning.[26]

The appalling smoke conditions of that August day, referred to as Black Tuesday, prompted the 1971 legislature to grant the DEQ permanent authority to prohibit burning on marginal days. More important, state lawmakers mandated that the practice be banned altogether after June 1, 1975. State legislators also established a per-acre field-burning assessment to support research and development, with some of the funds targeted for the smoke-management program. DEQ's actions to prohibit burning on marginal days reduced the volume of smoke blowing into mid- and upper-valley population centers. When protests from the Eugene-Springfield area lessened, the legislature rescinded its permanent ban, allowing field burning to continue under restricted conditions. The growers' political action committee, Farm PAC, was chiefly reponsible for getting the ban lifted.[27]

Before the legislature revoked the 1975 ban, the OSU Agricultural Experiment Station published a brief report suggesting an alternative burning technique, "thermal treatment" with propane burners, to protect against seed diseases. Whether farmers used the open-field method or propane machines, burning provided the necessary "physiological stimulation" to guarantee "economic seed yields." To put it simply, burning worked. By the close of the 1972 season, the Experiment Station reported that Willamette Valley growers had burned more than 270,000 of an estimated 300,000 acres eligible for burning. But with the pending ban on field burning, the Experiment Station emphasized that grass-seed farmers needed practical alternatives in order to survive. Finding effective methods to sanitize their fields was especially important for growers in the upper valley, where heavy and poorly drained soils limited crops to perennial and annual ryegrasses.[28]

When state legislators rescinded the 1975 field-burning ban, they passed an alternative measure that would progressively reduce the acreage to be burned from 250,000 in 1974 to 180,000 acres for 1978. After the election of conservative Republican governor Victor Atiyeh in 1978, the legislature replaced the entire program with a fixed 250,000–acre limit and gave DEQ the authority to regulate burning in accord with federal and state air-quality standards.[29] The federal Environmental Protection Agency (EPA) approved the new program subject to certain smoke-management restrictions. The EPA added other cosmetic restrictions in 1980 through its Visibility Protection Plan, a requirement that states make "reasonable progress" to preserve high-quality visibility in Federal Class I areas (wilderness set-asides and national parks). The Oregon legislature moved again in 1986 to restrict field and slash burning (burning debris in logged-off areas), banning burns upwind of Class I areas on weekends from July 4 through Labor Day.[30]

Open-field burning declined from 315,000 acres in 1968 to an average of about 220,000 acres during the 1980s. Beginning in the late 1970s, some growers began using "propane flamers," a technique that did not provide satisfactory "sanitation" for ryegrass fields. Torching approximately 25 percent of the total burned acreage with propane, however, helped to reduce emissions. Sean O'Connell, the DEQ field-burning coordinator, remarked in 1984 that the political climate stabilized when "we learned how to allow more burning with less impact on Eugene."[31] Beginning in 1982, OSU's Harold Youngberg and William C. Young edited a yearly publication on seed research that included short briefs citing the necessity for continued field burning. In the inaugural issue of *Seed Production Research*, Youngberg praised the close working relationship between the seed industry and the research staff in working through problems.[32]

From its inception in 1982 through the 1990s, *Seed Production Research* scientists repeatedly emphasized that burning was "the only treatment capable of maintaining high quality seed yields over several years in perennial ryegrass."[33] Their message? Annual burning remained the only effective means to guarantee the purity of ryegrass seeds. In a report to the Oregon Public Affairs Counsel, the Willamette Valley seed industry claimed that it contributed about $150 million to Oregon's annual income. Because most grass seed was sold outside the state, the export sales meant millions of "new dollars" for Oregon every year. Those aggregate numbers, the industry indicated, should not obscure the fact that real people depended on a healthy grass-seed industry: store owners, farm machinery businesses, school teachers, truck drivers, and Portland dock workers. Growers also cited an entomologist's report indicating that increases in pesticide use would be required in the absence of burning.[34]

Seed industry officials proved themselves politically astute, requesting in the early 1980s that the state create an Agricultural Research Commission, to be appointed by the Department of Agriculture. Growers proposed that "experts" be appointed to the commission whose concern would be "*research,* not politics." The seed organization recommended that four appointees should be grass seed farmers, two nominees would come from the DEQ, and the final member would be the dean of the School of Agriculture at OSU. The growers believed that the commission would "make intelligent decisions about the direction of further research" and look for "economically feasible alternatives" to burning and for the improvement of smoke management.[35] The seed growers' strategy worked, although politics weighed heavily in every legislative decision involving the industry.

And then tragedy struck on the warm summer morning of August 3, 1988. When seed-crop farmer Paul Stutzman put the torch to his ryegrass fields adjacent to Interstate-5 near Harrisburg, burning embers ignited the dry grasses along the interstate and engulfed the heavily traveled freeway in blinding smoke. By the time visibility had cleared, historian Peter Boag wrote, "23 vehicles lay mangled on the highway in one of Oregon's worst auto accidents: 7 people died and 37 others were injured." Stutzman's field burn once again rekindled the long-simmering controversy. "Protestors," Boag observed, "rallied to the same old colors," and the seed industry once again cranked up its propaganda machine, insisting that field burning contributed only a minor percentage of particulate atmospheric pollutants in the Willamette Valley.[36] In the wake of the freeway tragedy, officials further restricted field burning close to heavily traveled highways, and the 1991 legislative session sharply reduced field burning after 1997. The new restrictions established maximum acreage limits for open burning, emphasizing that farmers should make increased use of alternative sanitation methods. Over the next decade, open burning dropped sharply (83,593 acres in 1995; 56,878 acres in 1997; 49,999 acres in 1999; and 52,934 acres in 2001). Despite weather restrictions and the reduced acreage burned, however, more than 600 citizen complaints were filed with the Oregon Department of Agriculture in the year 2001.[37]

"For Better or Worse, the starling is here," an *Oregonian* editorial proclaimed in January 1950. Although it had taken sixty years for the bird to make its way across the continent, "for weal or woe it is ours now—or perhaps we are its." Descendants of forty pairs of starlings released in New York City's Central Park in 1890, the avian hordes multiplied and swept westward across North America in a relatively brief span of time. The Audubon Society's annual census reported forty-three starlings in the Portland area in 1950, the *Oregonian* announced (although the naturalized birds had been spotted in the vicinity of Scappoose and in the Klamath country for several years). The Oregon legislature had already made the bird an outlaw, classifying it with predatory species—"a rather farfetched indictment," the newspaper declared. Once the birds became established, the editorial concluded: "Woe to that luckless city which the starling chooses as its roosting place. Woe to the cherry orchard." Because of the birds' nuisance factor and the damages they inflicted on cherry orchards, berry fields, and other crops, the Oregon Department of Agriculture became involved in efforts to control the exotic invader.[38]

Beyond the world of insects and weeds common to field and orchard crops, the Oregon Department of Agriculture also devoted attention to a variety of exogenous pests and plants. And for several years the agency made a concerted effort to control starlings that began to appear in increasing numbers in the late 1940s. By the end of the 1950s, the department was employing a wildlife biologist, John Ludeman, to head its starling-eradication program. Through a cooperative agreement between the Oregon Department of Agriculture and the U.S. Fish and Wildlife Service, Ludeman attended a short course in Florida on the use of poisons and gases to control the birds. Because starlings threatened to inflict agricultural damages, the Oregon legislature gave the department an additional "$20,000 eradication appropriation" to carry out its starling control work. To assist with its research program, the agency circulated questionnaires about the starling's movements, feeding, and roosting patterns to determine why the birds flocked by thousands to some areas but bypassed others. In its *Agriculture Bulletin* for December 1959, the department described the starling as "a marauding, unclean bird that no one has patience with." The birds also left "unbelievable collections of droppings" that desecrated the areas where they roosted.[39]

In its "Operation Starling" report, the Department of Agriculture reported the use of an experimental mesh trap supported by forty-foot-high television antennas in a Polk County holly orchard to capture more than 6,000 birds during a nighttime "drive." When the floodlights were turned on, "drivers" banged and shouted their way through the orchard, chasing the starlings through a funnel and into a "catching pen" where they were gassed. In four nights, Ludeman reported netting nearly 18,000 starlings. In central and eastern Oregon feedlots near Ontario and Enterprise where starlings congregated in large numbers, field personnel poisoned french-fried potatoes "with satisfactory results." But when the department experimented with poisoned hamburger in Malheur County, the agency reported, "the starlings would have none of it."[40]

The Oregon Department of Agriculture continued the starling research program with modest success for a few years. The agency took a fourfold approach to its starling control program: trapping, spraying, baiting, and banding. The banding work turned up some interesting information about the starlings' migratory habits: some of the birds were reported as far away as Canada and Montana. The department compiled crude data from its trapping effort, estimating that personnel were able to capture about 10 percent of the roosting population in target areas. Its spraying experiments were more problematic; on one occasion department employees sprayed individual holly trees in western Oregon, reporting 47.4 dead starlings per tree with "no evidence of secondary hazards." "Except for robins," the agency noted, "few *beneficial* [my emphasis] birds were killed in the spray tests." During the winter of 1960–1961 the department conducted a baiting program in the vicinity of eastern Oregon feedlots and reported killing 591,362 starlings. Two years later the Agriculture Department shifted to a different strategy, conducting tests with repellants in holly groves and other starling roosting areas. One researcher reported less than modest success with "Roost No More," a sticky substance applied to the tree's inner branches. The department also admitted in 1963 that its continuing experimentation with light traps "was disappointing."[41]

Creating the proper conditions for a risk-free and profitable agricultural industry included a multifaceted and long-standing effort to eliminate predatory animals. To protect domestic animals and crops, Oregon farmers had participated for more than a century in a concerted campaign to exterminate wolves, bears, cougars, and elk—indeed, any animal that threatened livestock or crops. The state Game Commission paid its last wolf bounty for an animal shot in Lane County in 1946 and ended all bounty payments in 1961. Wildlife agents considered the bounty system ineffective in controlling predators and adopted cooperative government hunting programs to control "nuisance animals." To eliminate losses to predators, the Oregon Department of Agriculture and the Fish and Wildlife Service provided regular funding for extermination campaigns for what it called "the running battle against predatory animals."[42]

The black bear had inhabited most western valley and lowland regions at the time of the first white settlements but had retreated to upland and more remote areas by the 1940s. Although the Forest Service considered the bear a game animal on the national forests, elsewhere bears could be hunted year round until the legislature established seasonal restrictions in the 1970s. Cougars, one of Oregon's most sought-after predators, were not added to the protected list until 1967, when their population numbered only about 2,000 statewide. But agricultural officials insisted that problems remained, especially with "vermin" such as coyotes, cougar, bobcats, and occasional explosions in rodent populations.[43]

State and federal officials waged a relentless and purposeful crusade to

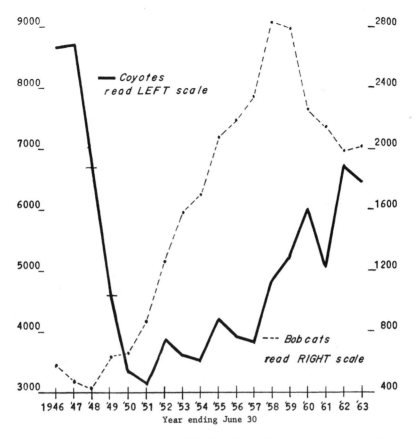

FIGURE 3.1 Coyotes and Bobcats "Take" by Years (Source: *Agriculture Bulletin*, no. 219, Sept. 1963, p. 7)

eliminate coyotes and other predators. The Oregon Department of Agriculture collected annual data on expenditures and personnel involved in each program, the number predator species killed, and a catalog of other efforts to control wild animal "depredations." The agency boasted that the cooperative effort killed more than 9,000 "prime predators" in the fiscal year 1962–1963. Coyotes headed the list, with more than 6,000 taken that year. Agricultural officials praised government hunters for preventing losses to domestic livestock and game birds and eliminating animals that caused damage to industrial forests. Of the total number of predators killed, trapping accounted for most of the animals taken. Figure 3.1 provides a statistical profile of the state's effort to control coyote and bobcat populations.

Only with the passing of time, an increasingly cosmopolitan cast to Oregon's population, and changing cultural values were control activities restricted and some declared illegal altogether.[44]

The wily coyote proved to be the most vexing and controversial of all predatory animals in the postwar era. "The coyote continues to head the list as the worst predator in Oregon," the state's *Agriculture Bulletin* reported in 1963. Better adapted to living amidst human populations than any of the larger predators, coyotes maintained a healthy statewide presence despite concerted federal, state, and county efforts to exterminate them. Historically, livestock interests led the campaign to eliminate the animal, an effort that involved all levels of government and a variety of strategies, from trapping to the use of "coyote-getters," spring-loaded mechanisms containing cyanide. Charged with carrying out the control program, the Oregon Department of Agriculture devoted more attention and budgetary appropriations to exterminating coyotes than to the elimination of any other animal. The agency also took great pride in listing the annual kill for each fiscal cycle.[45] As time passed, the federal government and the state of Oregon placed tighter restrictions on strategies to control the coyote.

"Oregon's cold war against ragweed, bugaboo No. 1 to many who suffer pollen allergies," Oregon's *Agriculture Bulletin* announced in late 1957, "had entered the hot stage." Increasingly solicitous of the state's growing tourism industry, the Oregon legislature appropriated $50,000 in 1957 to carry out a two-year program of ragweed control. The legislative authorization permitted Department of Agriculture personnel to enter private lands in eighteen western counties to spray ragweed infestations, all at state expense. During the first two years of the program, agricultural officials reported good public cooperation with the exception of a few landowners who questioned the herbicides' effect on livestock, especially dairy animals. Using 2,4–D chemical sprays, department employees treated nearly 5,000 acres in the autumn of 1957 and more than twice that acreage the following season. For the first time in many years, the department declared in late 1958, there was no ragweed pollen in the Portland area.[46]

The agriculture department's *Biennial Report* for 1960 contended that the eradication effort had achieved sufficient success to garner Oregon the reputation as "a pollen free state," a fact critically important to the tourist trade. Because ragweed seeds could lie dormant in the soil for several years, agricultural officials suggested that the eradication program would be a slow process. By the mid-1960s, however, the department had added Hood River County to its officially designated Ragweed Control Area. Contract sprayers reported that they were less successful than they might be because of the limited treatment time and the fact that employees had difficulty "'selling' the program to the owners of the high cash orchard crops." But in 1968 the department reported proudly that Oregon had once again earned the reputation for "being free from ragweed pollen," an achievement directly related to western Oregon's spraying program. Two years later, the agency again boasted that the state was "virtually free from ragweed pollen" and was a "haven" for those who suffered allergies from the plant.[47]

Beyond the world of weed infestations, pests, predators, and plant diseases, there always remained the uncertainty of weather, its unpredictability and its seasonal fluctuations in most sections of the state. During the twentieth century, weather conditions became increasingly important to the types of vegetables, fruits, nuts and grains that could be grown commercially in different parts of the country. There was also a Catch-22 to the relationship that developed between modern agricultural specialization and the weather, because it increased the importance of climate to agriculture's success. According to science writer William Meyer, "a new national map of agricultural specialization took shape" as the Extension Service and the Agricultural Experiment Stations directed farmers to more profitable lines of production.[48] The mild marine West Coast climate and soils of the Willamette Valley were ideally suited to certain types of seed crops during most years. The same could be said for the semi-arid plains of the middle Deschutes, which supported an irrigated agriculture of short-season vegetable and seed crops. But introduce floods, drought, windstorms, and other natural disasters, and all bets were off. Capricious weather conditions were unacceptable to modern agricultural systems that required predictability and assurance about the future. Such fickle behavior, agricultural scientists agreed, deserved the attention of specialists.

The Oregon Department of Agriculture initiated a series of "weather modification" efforts in the early 1950s. While rainmakers of various kinds are as old as human interest in the firmament, modern efforts at cloud seeding and other attempts to influence precipitation appealed to the postwar confidence that science could resolve such uncertainties. The Oregon legislature joined this effort in 1953 when it passed an "artificial weather modification and control law" and designated the Department of Agriculture to administer the act. Three of the initial licenses were extended to out-of-state firms, which established offices in northeastern Oregon's wheat districts and in the southern part of the state. The objective of one firm, the Rogue River Traffic Association, was to protect the valley's pear orchards from damaging hail storms. Another southern Oregon company was interested in increasing the snowpack in the mountains, especially the uplands surrounding the Klamath Reclamation Project. The Water Resources Development Corporation of Denver, Colorado, contracted with the Blue Mountain Weather Research Association, representing Umatilla County wheat growers with more than a half million acres of land. The weather-modification law required that all contractors be bonded and that they file notice with the Department of Agriculture before initiating any activities. Companies also were required to publish notices in local newspapers in areas affected by their operations.[49]

Although the Department of Agriculture reported "no particular difficulties" with its licensees in 1956, it mentioned one complaint about the tardy operation of cloud-seeding generators. Ultimately, the postwar fetishism for weather modification proved a short-lived phenomenon. Such efforts brought initial enthusiasm and optimism to a few special-interest groups but then receded quietly from public notice. For the duration of Oregon's law the state never licensed more than six operators, and by 1962 there were only two.[50] The climate-modification venture ran counter to the views of prominent weather scientists, who had argued for some time that altering precipitation patterns had minimal effect and only at the local level. A wise geographer, Isaiah Bowman, wrote in the 1930s that although climatic variations showed broad general trends over long periods of time, "the physical world changes constantly in its *meaning* to man."[51] The commercial desire for predictable precipitation in Oregon's northeastern wheat country spurred growers' interest in rainmaking just as similar profit-oriented objectives encouraged Klamath Basin irrigators to seek "normal" snowpacks to assure ample water for the next growing season. Because such ventures could not deliver on their promises, they disappeared from Oregon's agricultural literature by the mid-1960s.

Nowhere have agricultural practices left a greater footprint on the western landscape than in the expanding irrigation districts that reached across the arid and semi-arid West in the wake of Depression-era water projects. Oregon lacked a single enterprise the size of Washington's Columbia Basin Project. Collectively, however, extensive federal/private reclamation works in the Klamath Basin, the Vale-Owyhee-Malheur area, the Umatilla Project, and the North Unit on the middle Deschutes set in motion water withdrawals that by century's end would lead to extensive litigation involving Indian treaty rights, endangered species, and a host of other problems. Those ongoing water fights center on the over-appropriation of streams, mostly by agricultural interests, and the public's propensity to consume water as if it were unlimited. Although the Oregon Department of Agriculture occasionally acknowledged that farmers would have to share water with urban users, the agency paid virtually no attention to fish and wildlife until it was legally required to do so. At the same time, the department recognized as early as 1968 that increasing demands on the state's water resources would exceed "our visible supply."[52]

Because highly diverse precipitation patterns exist across Oregon and the Pacific Northwest,[53] most of the big irrigation projects were located east of the Cascade Mountain Range within reach of mountain shadows and the annual snowpack that provides water for summer crops. The problem for Oregon's arid regions (and even its western valleys) is that precipitation occurs primarily during the winter months and seldom during the growing season. For winter wheat crops, precipitation in upland areas is usually sufficient to bring the grain to maturity. But the lifeblood for eastern Oregon row crops and fruit orchards is the stored water carried through canals and ditches during the hot summer months. Nowhere are these conditions more obvious than in the productive agricultural district of the lower Umatilla Valley, the site of one of the oldest federal reclamation projects.

First authorized in 1905 by Interior secretary Ethan A. Hitchcock, the Umatilla Project developed slowly over the years, with initial construction activity centered on the "Feed Canal" to carry water from the Umatilla River to arid land in the lower basin. The next step in the expanding project involved the construction of Cold Springs Dam to augment the project's supply. Completed in 1908, the dam stored water for delivery to the Stansfield Irrigation District. In the following years, district engineers designed and built additional canals to bring still more acres into production. McKay Dam, completed in 1927, was an essential feature of that effort. The storage capacity in Cold Springs and McKay Reservoirs and diversions from the Umatilla River theoretically provided water for 22,000 acres. In years of normal precipitation, farmers were able to irrigate an additional 17,000 acres by diverting water from Butter Creek and the upper Umatilla River.

Problems plagued the Umatilla Project virtually from the beginning. Sandy, porous soils in parts of the Hermiston and West Extension Districts made Umatilla the least productive of all the federal reclamation efforts in Oregon. The Feed Canal also experienced seepage problems from the beginning, a flaw that softened the bed of the Oregon Railway and Navigation Company's railroad track and threatened the safety of passing trains.[54] The Feed Canal also soon began to fill with silt from blowing sand along its extensive route, a problem that was partly resolved when Reclamation crews planted brush along the canal's banks. Despite those setbacks, the engineering works "made the desert bloom" as the project area became home to more than 2,000 people by 1920. Although settlement was stable until 1945, the postwar years witnessed an even more significant population increase. The federal project irrigated 12,000 acres in 1920, 13,000 in 1923, and 22,038 by 1950. These water diversions also dried up the lower Umatilla River for much of July, August, and September, turning the stream into a slow-moving, ugly mix of farm runoff wastewater.[55]

Although the Umatilla Project was small and eventually was surpassed by the great groundwater, center-pivot irrigation systems that began to appear in the Hermiston-Umatilla-Boardman area in the 1960s, its great-

Every Oregon community celebrated V-J Day (Victory over Japan) on the evening of August 15, 1945. The announcement of the Japanese surrender set off this merriment at the intersection of Southwest Broadway and Yamhill in downtown Portland. Courtesy Oregon Historical Society, CN006167

Portland's Kaiser shipyards produced large numbers of "Liberty" supply ships during the Second World War. Swan Island on the Willamette River was home to one of the biggest yards. Courtesy OHS, OrHi49686

(Facing page, top) Women, such as the carpenters' helpers in this photo, joined the wage-earning work force in greater numbers during the Second World War. Courtesy OHS, OrHi89798

(Facing page, bottom) In the late 1930s, the Rural Electrification Administration (with the ardent support of the Bonneville Power Administration) proudly displayed photographs showing citizens using labor-saving electrical appliances and equipment. Courtesy OHS, OrHi67024

Labor shortages during the Second World War speeded the development of gasoline-powered chain saws. Courtesy Oregon State University Archives, P61#408

Columbia River floodwaters destroyed the wartime housing project Vanport on May 30, 1948. Located north of Portland near the Columbia River, the project was demolished when a dike broke on a sunny Sunday afternoon. Casualties were remarkably low because the break occurred during mid-day. Moreover, many residents were away from the community. Courtesy OHS, OrHi52427

(Facing page, bottom) The Oregon State College Extension Service encouraged the use of electricity through farm electrification exhibits displaying mechanical milking machines and other devices. Courtesy OSU Archives, P62#5

The Columbia River flood of 1948 served as the great catalyst for dam building on the Columbia system. The Dalles Dam, completed in 1957, flooded the ancient Indian fishery at Celilo Falls, shown here. Courtesy OHS, OrHi92042

An aerial view of The Dalles Dam, March 10, 1957, shortly after Columbia River waters filled the reservoir. Courtesy OHS, CN015349

For at least two decades following the Second World War, the Oregon State College Extension Service encouraged the use of DDT for a wide variety of purposes, including spraying sheep for ticks. Photo by R. G. Fowler Jr. Courtesy OSU Archives, P120:6686, neg. 2434

This Western Pine Association photo shows a forester marking ponderosa pine for selective harvesting. The caption proudly points to the "thick growth of younger pines . . . springing up around the big trees." The consequence of several decades of rigorous fire suppression, the thickly massed second-growth trees became prone to disease and insect infestations, circumstances that have contributed to potentially catastrophic fire conditions. Courtesy OHS, OrHi58459

A nationally renowned journalist and prominent conservationist, Richard Neuberger was elected to the United States Senate in 1954. His wife, Maureen Neuberger, succeeded him when he died early in 1960. Courtesy OHS, OrHi54008

Through his long public career, Governor Tom McCall (1967-1975) touted Oregon's livability, its spectacular mountain backdrops, rushing streams, and coastal vistas. Courtesy OHS, OrHi97348

(Facing page, bottom) A helicopter spraying the chemical herbicide 2,4,5-T over forest land in Oregon's Coast Range in the spring of 1977. Courtesy OHS, OrHi104979

The two big pulp and paper mills at Willamette Falls near Oregon City dumped untreated sulfite liquors directly into the river until Oregon's Department of Environmental Quality forced all mills on the Willamette system to build treatment plants in the late 1960s. Courtesy OHS, Gi12007

57836

This 1958 photo shows a thoroughly industrialized lower Willamette River with log booms and the operations of the Ross Island Sand and Gravel Company in the foreground. Courtesy OHS, OrHi55482

Willamette Valley storage reservoirs, such as Lookout Point Dam on the Middle Fork of the Willamette River, brought an end to most seasonal flooding in the valley. Courtesy OHS, Cn019495

A thriving sawmilling town of about 11,000 people at the end of the Second World War, Bend had a population boom in the 1990s when the community was transformed into a setting for affluent retirees, upscale telecommunications professionals, and tourism and recreational activities. This 1947 photo shows Bend's two big sawmills, Brooks-Scanlon and Shevlin Hixon, in the top left background. Courtesy OHS, OrHi50429

(Facing page, bottom) During discussions about an appropriate statue for former Governor Tom McCall, the *Oregonian's* Jack Ohman drafted this cartoon which plays on McCall's famous 1971 quip on CBS Evening News: *"Come visit us again and again. This is a state of excitement. But for heaven's sake, don't come here to live."* Courtesy Jack Ohman

This aerial photo on Albany's eastern edge (astride Interstate-5) shows the effects of Oregon's pioneering land-use planning system in curbing urban sprawl and protecting valuable agricultural land. The buildings on the right-hand side of the photo predated the land-use planning law. Courtesy USGS est environmental legacy has been the extinction of coho and Chinook salmon in the Umatilla River. It would take threats of tribal lawsuits and several decades of time before a cooperative agreement was reached to leave enough water in the river to support anadromous fish runs. The turnabout in the "management" of the Umatilla River was vested in history and law. Because the Confederated Tribes of the Umatilla Reservation held superior water rights under prior-appropriation doctrine, tribal legal authorities were certain that those rights dated from the time of their enabling treaty in 1855. As the Confederated Tribes and local irrigators traded charges and counter-charges, mediators began seeking a way out of the dilemma.[56]

Tribal officials were aware that the Bureau of Reclamation had given birth to a flourishing irrigation economy which had driven the salmon to extinction. They also knew that honoring their prior-appropriation rights would significantly affect existing water users. The federal government was responsible for both sides of the Umatilla crisis; it had promised water to irrigators—which de-watered the river during the summer—and it also held trust responsibility to protect tribal interests. The route to compromise was not easy. Water Watch of Oregon, an environmental group long critical of wasteful irrigation practices, threatened the potential for a compromise with litigation. For years, Water Watch had urged the Oregon Water Resources Department to enforce state law on the Umatilla Project—in particular, to prohibit "waterspreading," a practice by which irrigators sold "excess" water to farmers outside district boundaries. Water Watch accused the state agency of complicity in this widely acknowledged custom. The *Oregonian* added to the controversy when it published a feature article in October 1991, "The Umatilla Blues," essentially supporting Water Watch's allegations.[57]

With the pending threat of an environmental lawsuit, the Oregon Water Resource Department began to enforce state law banning waterspreading, an action that outraged irrigators. The Confederated Tribes pursued a diplomatic approach, pressuring the Bureau of Reclamation and the irrigation districts to follow the law and to resolve the instream flow problem on the Umatilla River. Through the winter of 1991–92, the tribes, irrigators, the Bureau of Reclamation, state agencies, and Water Watch hammered out an agreement that would provide water for irrigators and restore water to the river.[58]

The extended negotiations led to a cooperative solution when Congress passed the Umatilla Basin Act in 1992. The Bureau of Reclamation began construction on a $48 million federally funded pump-and-canal scheme that would deliver Columbia River water to participating irrigation districts in lieu of water that would remain in the Umatilla River to maintain adequate instream flow for spawning fish. Federal funds also paid for improved fish-passage facilities through the five dams on the Umatilla River. In the end, irrigation-district farmers gave up nothing except illegal water-spreading. The federal government put up the capital to subsidize a technical solution to the over-appropriation of Umatilla River water. As writer Christopher Shelley observes, federal funds "essentially purchased the region's way out of crisis." Although Water Watch of Oregon originally condemned the agreement as "pork wrapped in an Indian blanket," the organization's Web site later boasted that its threatened lawsuit had worked: "For the first time in decades, salmon could migrate upstream to spawn without being trucked through the Umatilla's middle reach." The onset of the twenty-first century finds hatchery-introduced salmon and steelhead returning to the Umatilla River to spawn, through what Shelley calls "a remarkable blend of technology and nature."[59]

By far the most contentious of all reclamation efforts in Oregon—and the most environmentally invasive—is the big Klamath Project straddling the Oregon-California border. One of the earliest federal reclamation enterprises, the Klamath undertaking provided water to approximately 240,000 acres by the close of the twentieth century. Through an extensive series of dams, reservoirs, canals, siphons, pumping plants, tunnels, laterals, and drainage ditches, the project delivers water to irrigate crops such as potatoes, oats, wheat, barley, alfalfa hay, and pastureland. The big engineering works drained marshes and shallow lakes for agricultural use and redirected the water flowing from three lakes and two rivers to irrigate the former lake beds. The project has two principal sources of water, Upper Klamath Lake and the Klamath River, and a series of small reservoirs and streams in a closed basin to the east. Before damming and water diversions dramatically altered the landscape, the Klamath Basin had an estimated 185,000 acres of natural wetlands and provided one of North America's most productive waterfowl breeding areas. Today, there are fewer than 75,000 acres of wetlands, and water transfers seriously compromise even those areas.[60]

It is important to understand the unique natural topography of the basin, especially the relationship between Upper Klamath Lake and Lower Klamath Lake.[61] Water flows through a natural basalt outlet from the upper lake into Link River, a short stream that connects to Lake Ewauna, headwaters to the Klamath River. During the high-water months of spring, water flows from Lake Ewauna into a huge natural marsh known as Lower Klamath Lake, a 94,000–acre area that extends into California. When the Reclamation Service arrived on the scene in 1905, it performed water gymnastics that involved building a dike to prevent the overflow into Lower Klamath and Tule Lakes and then diverting water from Upper Klamath Lake through a canal to irrigate the reclaimed lands. Eventually, federal engineers built a series of dams and canals in the enclosed basin to the east, greatly expanding the area's cropland. At the close of the twentieth century the wetlands that remain from the once extensive Lower Klamath marshlands, now owned or managed by the U. S. Fish and Wildlife Service, depend on returning irrigation flows or water that is available after existing agricultural withdrawals have been satisfied.[62]

And therein rests the dilemma for the Klamath Basin, by any measure the most troubled waterscape in the Pacific Northwest. Cutover timberlands, overgrazed vegetation—especially in riparian zones—and increasing non-point source pollutants in basin tributaries have been at the forefront of public debate. Many Upper Klamath Lake tributaries suffer from high turbidity, low dissolved oxygen, excessive nutrients, pesticide residues, and other problems. A primary tributary, Sprague River, has been diverted and channelized, its riparian vegetation destroyed in many places. The lower section of the Williamson River, another tributary, has very high coliform levels. It is little wonder, therefore, that the 80,000–acre Upper Klamath Lake—a body of water that historically has been eutrophic (highly productive of algae)—has experienced an increase in eutrophication in the last thirty years. The lake's high level of algae has contributed to several large fish kills in the last two decades.[63]

Water conditions downstream from the big lake have become even more problematic, as Link River has exhibited conditions similar to those upstream. Because the Bureau of Reclamation regulates the lake's outlet, the river and tiny Lake Ewauna further downstream developed a severe dissolved-oxygen deficiency during the summer. Below Lake Ewauna in the Klamath River proper, high coliform levels developed in the 1960s— conditions that reflected domestic and industrial sewage discharges. As a result of the Clean Water Act (1972), improved treatment facilities and better regulation of waste discharges reduced the coliform count to legal levels. But the Klamath River's difficulties persist, especially downstream during the summer and early fall when a large volume of water is diverted for irrigation. The runoff of water from agricultural lands, which carries high nutrient levels from farm and grazing land, worsens the situation. Higher summer water temperatures create ideal conditions for algae blooms in areas where the current is sluggish.[64]

Important legal and environmental factors have further complicated the Klamath Basin's water conflicts in recent years: a matter of equity, the prior-appropriation rights of the Klamath Indian Tribe; and the requirements of the Endangered Species Act. Federal courts ruled in the 1990s that the Endangered Species Act required sufficient water in Upper Klamath Lake to protect the endangered Lost River and shortnose suckers; in addition, the courts mandated an adequate flow in the Klamath River to protect threatened coho salmon. Federal District Judge Michael Hogan also determined in 1998 that the Klamath Indian Tribe's senior prior-appropriation rights and the Endangered Species Act held precedence over more than 2,000 Basin farmers who use Bureau of Reclamation water to irrigate crops. The water issue proved volatile, especially in drought years when fields went dry, and there were threats of violence. And the Bureau of Reclamation, long the farmers' partner in expanding the Klamath Reclamation Project, was suddenly cast in the role of upholding federal law (the Endangered Species Act) and carrying out the government's trust responsibility to treaty tribes.[65]

Environmentalists joined the Klamath Basin water fight when several organizations, including the Oregon Natural Resources Council (ONRC), challenged the federal government's management of the Tule Lake and

Lower Klamath Lake National Wildlife Refuges. Environmentalists accused the U.S. Fish and Wildlife Service, managers of the refuges, of allowing too many acres to be turned over to agricultural production. ONRC spokespersons pointed to the area's importance to waterfowl and other birds following the great Pacific flyway, the westernmost migratory route in the United States. Studies indicated that 80 percent of all migrating birds stopped over at Tule Lake, Lower Klamath, and the three other basin refuges. Although habitat and water quality had deteriorated, the extensive Klamath water system was still critically important to migrating waterfowl and other wild animals. The refuges confronted inadequate water supplies, pesticide pollutants highly toxic to birds, and the virtual disappearance of tules, a reed that once dominated the marshlands. While environmentalists viewed the Klamath refuges as an opportunity for the prudent management of fish and wildlife, farmers underscored agriculture's $200 million business in the Basin, an industry heavily dependent on reclamation project water.[66]

The Klamath Basin water crisis led to the formation in 1999 of the Hatfield Upper Klamath Basin Working Group, a broad-based citizens' organization seeking ways to resolve the legal warfare. Among the ideas floated before the group were proposals to return the former 680,000–acre Indian reservation, currently the Winema National Forest, to the Klamath tribe. Others included converting farmland back to marshes, providing adequate water for the refuges, and protecting another 22,000 acres leased from one of the refuges. The thirty-one-member Hatfield Working Group included representatives from agriculture and timber firms, state and federal agencies, conservation and recreation organizations, and the Klamath Tribe. Although news reports were vague on the issue, restoring the old Klamath reservation to the tribe would be linked to guaranteed water deliveries to the reclamation project. One non-Indian, Earl Miller of Bonanza, spoke the truth at a local meeting when he remarked that the government's termination of the Klamath tribe in the 1950s "wasn't a situation of a willing seller. It was a condemnation."[67]

The environmental conflict in the Klamath country has taken other unusual twists, sometimes pitting agency against agency and with conservation groups sometimes on opposing sides of issues. The Bureau of Reclamation limited flows to Lower Klamath refuge in the fall of 2000 to

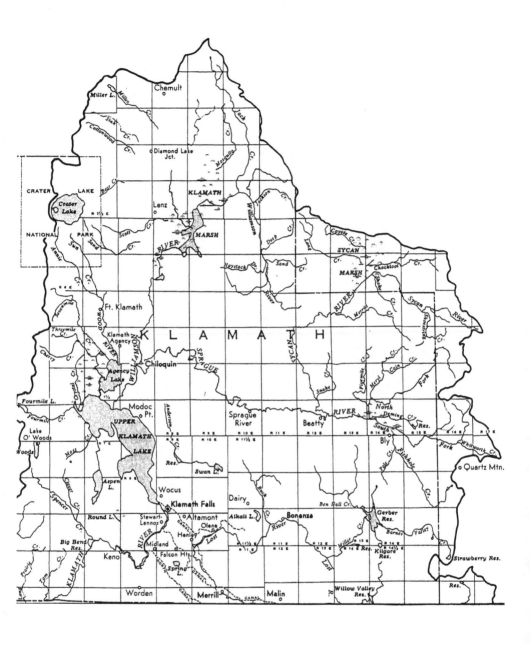

Klamath Basin. Courtesy State Water Resources Board.

maintain an adequate water flow in the Klamath River to safeguard endangered salmon and to raise the level of Klamath Lake to protect the endangered suckers. Conservationists argued that Reclamation had more than enough water to carry out all its environmental responsibilities if it restricted deliveries to irrigators. The Wilderness Society's Don Barry charged that the failure to provide adequate water to the refuges violated the spirit and intent of federal law. While farmers were irrigating crops on the refuge, he pointed out, "refuge marshes are allowed to go dry." Because of the limited water supply in 2000, some conservation groups began to call for an end to commercial agriculture on land leased from the refuges.[68]

As the prolonged drought in the Klamath Basin entered its second year, local farmers and the Klamath Water Users Association met to learn more about predicted shortages for the summer of 2001. While the Bureau of Reclamation was asking farmers to leave a portion of their fields fallow to make water available for other purposes, some in the audience threatened to open their irrigation headgates despite a ban on watering their fields. Paul Simmons, an attorney for the Klamath Water Users Association, thought the shortages were likely to be annual events. Farmers and the wildlife refuges, the Klamath *Herald and News* contended, were casualties of the need for more water in Upper Klamath Lake to protect suckers and increased water downstream for salmon. With snowpack in the uplands around the Klamath Basin less than 50 percent, some farmers saw a conspiracy in the emerging crisis—environmentalists were out to close down agriculture completely. A Bureau of Reclamation official told the audience that the agency had lost its flexibility to provide more water.[69]

The most unusual environmental battle in the spring of 2001 involved a Williamson River algae-harvesting company and its lawsuit against the Natural Resources Conservation Service (NRCS—formerly the Soil Conservation Service) and the Corps of Engineers. In an effort to restore the lower Williamson River to its historic stream-flow pattern, The Nature Conservancy had planted willows and other streamside vegetation and was ready to remove a dike to return the river to its original channel. Trouble emerged when Klamath Valley Botanicals, a company harvesting algae from

Upper Klamath Lake, filed a lawsuit to protect its use of the river to move harvesting barges to its property near the restoration work. "All we're after," company president Shannon Hamilton remarked, "is the quiet enjoyment of our property." The NRCS and the Corps had planned to return more than 165,000 cubic yards of dredged fill back into the river as part of its restoration work, a project that would prevent Klamath Valley Botanicals from using the waterway. The lawsuit charged that the restoration project's permit failed to consider the economic effects on the company. Hamilton asked that the restoration work be undone and the river returned to its straightened, channelized condition.[70]

The Klamath Basin made national news headlines in April 2001 when the Bureau of Reclamation announced that Upper Klamath Lake water would not be available for the Klamath Reclamation Project during the current crop season. Based on scientific findings by the U.S. Fish and Wildlife Service and the National Marine Fisheries Service, the Bureau cited the need to provide sufficient water for the endangered short-nosed sucker in Upper Klamath Lake and for threatened coho salmon in the lower Klamath River. Don Russell, chairperson of the Klamath Water Users Association, told the *Klamath Herald and News* that the project could not continue to operate from year to year under crisis situations: "We can get a wet water year and still be in trouble." Speaking for the Bureau of Reclamation, Jeff McCracken called the situation tragic: "We've been delivering water in this community for 94 years in a row," but for the present "our hands were tied by Mother Nature and the law." During the hot summer months that followed, farmers charged federal officials with putting fish ahead of human survival. The water cutoff to the project affected about 1,400 farmers and ruined crops on about 200,000 acres. A *New York Times* editorial in July defined the problem as an overtaxed ecosystem: "too many claimants for too little water— farmers, fish and other wildlife, towns, downstream salmon fishermen whose business has long been declining and Indian tribes whose water claims predate those of the farmers."[71]

Writing for *High Country News* in August 2001, Rebecca Clarren chronicled the social and economic chaos taking hold in the Klamath Basin. While incidents of domestic violence increased, Klamath Indians who supported the water shutoff were refused restaurant service, and federal employees feared for their own safety. In early July, as local law-enforcement personnel stood by, a large group of farmers and their supporters leveraged open the headgates to one of the main irrigation canals. The Bureau responded by bringing in the FBI and National Park Service Police to guard the headgate. Clarren reported that the long-neglected Klamath Basin had suddenly become a negative symbol for the Endangered Species Act. Although the media tended to reduce the conflict to farmers vs. fish, the issue was actually much more complicated. The Klamath Basin, she observed, had "a long history of trying to stretch a finite water supply to meet the needs of farmers, Indians, and wildlife, and for most of the last century, farmers have had priority."[72]

In the midst of the Klamath Basin tempest, Interior Secretary Gale Norton asked the National Academy of Sciences to appoint a scientific panel to review the decision to cutoff water to the Klamath Project. The panel's preliminary findings, released in February 2002, caused another firestorm when the scientists faulted the Fish and Wildlife Service and the National Marine Fisheries Service for basing their decision on "inadequate" science. In the midst of angry finger-pointing, charges of "junk science," and frenzied radio talk-show hosts out to discredit federal agencies, the *Oregonian's* Michael Milstein provided a more rational explanation: "The National Academy of Sciences panel did not fault the federal biologists as much as it found that they had been walking a legal tightrope through a deteriorating ecosystem without a safety net." The real culprit, Milstein argued, was the Bureau of Reclamation's long history of providing water to project farmers as if the supply were unlimited.[73]

While farmers, environmentalists, and tribal people continued to point fingers of blame, the Bureau of Reclamation quietly found reason to supply the Klamath Reclamation Project with its full quota of water for the 2002 crop season. Later that summer another train wreck unfolded, this one of a different order, when the water flow in the lower Klamath River slowed to a trickle just as salmon were entering the stream to begin their journey upriver. Because of the low water, the salmon stacked up in the lower river and naturally recurring diseases spread among the crowded salmonids, killing approximately 33,000 fish, perhaps the nation's greatest die-off on record. A California Department of Fish and Game report released early in 2003 found that low water had impeded the upriver movement of fish and caused them to bunch together in shallow, warm water. The higher temperatures lowered the water's dissolved oxygen content and created ideal conditions for the rapid spread of deadly parasites and disease. The department also reported that water diversions from upstream reservoirs— all operated by the Bureau of Reclamation—were responsible for the low water flow in the lower river. Interior Department officials, however, continued to search for alternative explanations for the 33,000 dead salmon. It should also be noted that during the summer of 2002 the Bureau ignored tribal officials on the lower river who were asking for more water. Donald Worster's 1986 essay on the dilemmas of water diversion put the case bluntly: "What once had seemed so simple and beneficial—the capturing and diverting of water to make the dry lands over into a ripe, blooming garden—came to be seen as a Pandora's box of ills opened and loosed on the West."[74]

Agriculture in Oregon and elsewhere has remained a risky business, subject to unpredictable market forces, changing consumer tastes, the rising costs of pesticides, herbicides, and fertilizers, environmental restrictions, and increasing struggles over the allocation of scarce resources such as water. As the Klamath Basin disaster indicates, environmental conflicts have emerged everywhere across Oregon's agricultural landscapes in the last few decades. Intensive farming practices—especially through the use of pesticides—expanding urban boundaries, and the public's heightened interest in health and food safety have put farmers in the daily headlines. Through all of those controversies, state agricultural officials and their federal associates have promoted the notion that scientific expertise and greater technical efficiency can resolve agricultural and environmental problems.

For most of the years since the Second World War, agricultural science has given privileged status to production, to overcoming limits rather than setting boundaries to human practices. Believing too strongly in the efficacy of scientific, technical, and engineering solutions has sometimes led to a false optimism (and in the case of the Klamath Basin to failed expectations). Policymakers, scientists, and industry leaders developed parallel strategies in urging farmers to make use of a growing variety of new chemicals that appeared on the market after the war. At the urging of agricultural officials, farmers sprayed an increasing volume of pesticides, herbicides, and fertilizers on their fields and orchards, practices that would pose some of the more controversial and persisting human health issues in all of modern agriculture.

4

THE WONDER WORLD

OF PESTICIDES

All over the West the best Panzer divisions of the weed army are at present
in hopeless confusion. They conquered state after state, and there seemed to
be no stopping them, but lately the weapons of defense have caught up with
their offensive tactics, and the conquered peoples are using new weapons
with spirit and high morale.—E. R. JACKMAN[1]

Because agriculture is important to Oregon's economic well-being, the
industry has been a powerful institutional player in establishing state
regulatory policies, including the use of chemicals to treat a host of
agricultural problems. Oregon Department of Agriculture administrators
and land-grant college scientists established collegial working relations with
farmers and chemical companies, thereby creating an environment receptive to scientific and technological solutions to problems affecting agricultural producers. Similar cooperative relationships existed in the world
of forestry, where the Oregon Department of Forestry, land-grant college
foresters, timberland owners, and chemical companies cooperated to seek
scientific solutions to a host of forestry issues. What emerged from those
mutual relationships in agriculture was an overweening confidence in the
human ability to control nature, the widespread assumption that the
world could be shaped to suit human purposes. By eliminating pests in
orchards and fields and applying chemical fertilizers to soil, farmers could
increase production and play their part in feeding a hungry world. That
scenario, played out in a series of United States Department of Agriculture (USDA) memoranda and circulars, found a receptive home in Ore-

gon where agriculture was the state's second most important industry. Federal and state administrators, Corvallis-based scientists, and their chemical company allies convinced farmers that there were scientific solutions to a broad range of agricultural problems.

For a time, the modern chemical revolution—especially the introduction of wondrous synthetics such as DDT (dichlorodiphenyl-trichloro-ethane)—was an amazing success. Without clearly understanding their ecological consequences for animal and human health, newspapers and agricultural agencies touted the benefits of the new miracle chemicals. The new compounds, especially DDT, provided stunning evidence of the ability of humans to reduce environmental risks and to make the world a safer place. DDT killed bugs, eliminated diseases, provided security to hardworking farmers, and suggested a problem-free future. State agricultural agencies were part of the USDA-Extension Service network that celebrated the promise of the new chemical. Oregon State College and Extension Service entomologists worked closely with the Oregon Department of Agriculture to sell the virtues of the new pesticides to the public. At this early date federal and state officials were concerned primarily with the efficiency of pesticides—whether they performed as well as manufacturers claimed. But even early in their use, there were concerns. Ecologist and writer Rachel Carson worried as early as the summer of 1945 about the broader effects of DDT on waterfowl and birds that depended on insect food.[2]

With the surrender of Germany in April 1945 and with the Japanese in serious retreat in the Pacific, the Oregon legislature authorized the Department of Agriculture to carry out systematic surveys "for the discovery, control and eradication of dangerous insect pests and plant diseases." Agency employees traveled the state between August 1945 and June 1946, visiting every county and putting more than 21,000 miles on department vehicles, to conduct the first of the department's annual surveys of extensive insect and disease inventories. The department established a laboratory, collected and classified specimens, shared information with experts in Washington, D.C., and worked in cooperation with the U.S. Plant and Disease Control. In its 1946 report, the department's Division of Plant Industry observed that the quickened pace of transportation, especially increased air travel, would "make it easier for insect enemies to settle here." The state should ready itself, therefore, for new insect and plant disease problems, including such "unwelcome visitors" as Japanese beetles, a variety of moths and maggots, European corn borers, gypsy moths, and Dutch elm disease. If infestations could be detected before they became serious, pests could be controlled and losses to growers prevented.[3]

Agricultural officials urged farmers to use chemicals to bring greater efficiency to their operations, to increase productivity—and thus to increase profits. The state Department of Agriculture report for 1946 observed that farmers were "rapidly becoming more weed conscious" and were trying "new methods of weed control." The new chemical sprays and dusts, many of them developed for military purposes, were being shifted to civilian use with the end of hostilities. These new insecticides and sprays required the Oregon Department of Agriculture to be "continually on the alert" to make certain that the labels on "economic poisons"—a trade term for chemical compounds such as DDT and chlordane—were not misleading. Herbicide, insecticide, and pesticide use increased dramatically with the passing years, but of all the new chemical agents that attracted early attention DDT ranked at the top.[4] Everyone agreed that the insecticide was highly toxic, but it appeared to pose no harm to humans.

Writing for the *Oregonian* in late August 1945, Stewart Holbrook referred to DDT as the "Atomic Bomb for Parasites of the Forests." Holbrook recounted the recent history of a hemlock looper infestation in Oregon's coastal forests south of the Columbia River and the amazing successes DDT achieved in combating the outbreak. Using the metaphors of war, the Portland writer reported that state forester Nelson Rogers was in "overall command of the war against the loopers." State and federal entomologists used a mixture of DDT, "which looks like a high grade of white flour," at the rate of one pound to two gallons of fuel oil. Two small aircraft flown in from Yakima, Washington, loaded the liquid aboard at a makeshift landing strip and applied the spray in late June. The results of this first effort to use DDT on a West Coast forest were astonishing, according to Holbrook: "The death of the loopers happened so quickly, so silently, yet so manifestly, that I could do little but wonder."[5]

In an accompanying editorial, "Science Licks the Looper," the *Oregon-*

ian reported that DDT "brought millions of dead loopers to the needle carpet of the earth twenty minutes after a plane's passage." This was a real-life miracle, the newspaper declared, because the mixture did no harm to "bees, birds, fish—even mosquitoes." Although scientists and loggers would have to wait until the following spring to determine DDT's effectiveness in destroying the looper, the *Oregonian* was certain that millions of board feet of valuable timber had been saved. The editorial suggested the bright promise that DDT held for the future: "With production controls released for the manufacture of DDT, the people of the world may expect great benefits from the miracle powder—in health and sanitation, in agriculture, in the comfort of living. . . . Soon DDT will be available to householders and dwellings may be freed of flies, mosquitoes and other pests." The federal government released the insecticide for civilian use on August 1, 1945 (on the advice of its manufacturer), so the DDT sprayed over Oregon's northern coastal forest preceded that decision by at least one month. Manufacturers were turning out more than 2 million pounds of DDT per month by 1944, and by the time the chemical compound hit civilian markets, it had an unmatched reputation for safety and success in combating insects. By the end of the war, the public also had access to a complete range of instructional manuals.[6]

Oregon's agricultural agencies began spreading word about the effectiveness of DDT as early as May 1945. State extension entomologist and plant pathologist Robert Rieder sent circulars to county agents recommending the use of DDT to control an outbreak of tuber flea beetles in potato plantings. Because the chemical effectively controlled the flea beetle, the circular informed growers, the War Production Board had released limited quantities for use on potatoes. Although companies in potato-growing areas were ready to distribute DDT, the circular added a warning:

Caution. DDT is a powerful new insecticide about which we need much more information. *Its use is restricted to potatoes* and growers obtaining it will be asked to register when purchasing it. There just recently has come to our attention a case of very severe consequences resulting from the eating of radishes which were dipped in a DDT containing solution. The case was hospitalized, and came close to dying.[7]

Although Oregon's Cooperative Extension Service continued to advise caution in the use of DDT, there were few sustained arguments against the chemical for at least two decades.

During the summer of 1945 an Oregon State College Agricultural Experiment Station entomologist carried out experiments using DDT to control mosquitoes at the Washington County Flax Growers' plant in the small town of Cornelius, west of Portland. The mosquito problem was confined to a ditch that held the factory's wastewater until the late-winter months, when the company drained the ditch into a nearby stream. Each year when the mosquito season approached, Flax Growers officials attempted to control the pests—without success. Finally, a nearby Shell Oil Company official agreed to supply the plant with DDT on an experimental basis, providing that an Experiment Station entomologist supervised the spraying and collected data. Flax company personnel made two applications, with striking results. While the chemical's residual action lasted only about two weeks, entomologist Ernest Anderson called Shell's DDT mixture "very effective in controlling the mosquito."[8]

DDT was first used to control mosquitoes in the Portland area in 1947. To protect the public from the pest, Multnomah County and City of Portland officials formed a mosquito-abatement authority and used aircraft to spray more than 100,000 acres with a DDT-diesel oil mix. A. W. Lindquist, an entomologist with the Experiment Station in Corvallis, reported great success during the 1947 spraying season: "a 96 percent degree of control." After the destructive Columbia River flood of 1948, which promised to worsen the seasonal bout with the insects because standing water in lowland and swampy areas created ideal breeding habitat, Lindquist assured the public that the percentage of DDT used for the 1948 spraying program would not be harmful to plant or animal life. Spray officials also pointed out that the effort to control the mosquito had the full cooperation of the U.S. Public Health Service, the Oregon Board of Health, and the state Department of Agriculture's Bureau of Entomology and Plant Quarantine.[9]

Oregon State College (OSC), the Extension Service, and the Agricultural Experiment Station were involved in using DDT and other pesticides from the beginning. The OSC entomology department initiated small-scale experiments with DDT in the summer of 1944, a full year before the government released the insecticide for civilian use. The early distribution of the chemical to OSC entomologists is striking, because the federal government did not begin its massive distribution program until 1945. The department used DDT with amazing success in an experimental spraying program to control housefly infestations around the college's dairy barns. A later department memorandum noted: "The housefly in 1948 was nonexistent on the campus." But DDT created the most optimism among growers of fruits and nuts, vegetables and grain crops, and livestock producers. The OSC Extension office sent memoranda to county agents around the state offering advice on the use of DDT to control filbert worms, the hairy vetch weevil, onion maggots, other plant-feeding insects, and ticks and lice on livestock. The results were, again, dramatic. Filbert orchards required only one application of DDT as long as the ground under the trees was thoroughly treated; extension entomologists also pointed to striking successes in treating the hairy vetch weevil with aerial dustings of DDT.[10]

Extension Service memoranda provided detailed information about proper mixtures and application methods and praised the insecticide's successes. Field personnel provided newspapers with the latest evidence praising "the new wonder insecticide" and its effectiveness in controlling pests. In one of Oregon's first large-scale commercial uses of the compound, the *Oregonian* reported that Gresham-area raspberry growers had used DDT to combat an outbreak of the banded leaf roller, an insect that had ruined much of the previous year's crop. Because the raspberries were destined for military use, a local experiment station official obtained the release of sufficient quantities of DDT to gain "complete control" of the banded roller. While other insecticides achieved 85 percent control, entomologist Joe Schuh found that figure unacceptable, because a single worm in fruit destined for the cannery would condemn the entire lot. But spraying DDT on 250 acres of raspberries proved 99 percent effective and saved the crop. Don Mote, head of the department of entomology at OSC, reported similar tests on other Oregon agricultural pests with highly promising results.[11]

Extension Service circulars, Oregon Department of Agriculture memoranda, and OSC entomologists all emphasized cost savings in the use of DDT. Whether the intended uses were for food crops or livestock, officials customarily referred to "costs per head" or "costs per acre," suggesting that the chemical was both readily available and cheap. Extension entomologist Robert Every told a gathering of Marion County livestock breeders in 1947 that DDT would effectively control ticks and lice in livestock at a "materials" cost for sheep of about two cents per head and five cents for cattle and horses. In a memorandum to county agents (circa 1947), entomologist Don Mote stressed the chemical's effectiveness in controlling sheep ticks. "DDT is readily available," he wrote, "safe to use, gives good control, and is relatively inexpensive." At current prices, he estimated that the cost of the insecticide would be about three cents per head. Another OSC entomologist, S. C. Jones, tested DDT's effectiveness in controlling prune thrips and found the chemical "less costly and more pleasant to use" than older insecticides. Historian Thomas Dunlap explained the new wonder insecticide's popularity: "DDT's combination of high toxicity to many insects, low mammalian toxicity, low cost, and suitability for aerial spraying invited its use in areas that had been, before World War II, free of insecticides."[12]

Swiss chemist Paul Muller's discovery of DDT and its toxic effects on insects in 1939 marks the onset of the modern industrial use of chemicals. Muller, who won the Nobel prize for chemistry in 1948, told a reporter that when the results of his research became obvious, "it became our duty to bring this chemical revolution to the knowledge of the Allies." The United States Army's chief of preventive medicine announced in January 1945 that DDT would prove to be "the War's greatest contribution to the future health of the world."[13] With the return of peacetime, the increased use of the chemical signaled an important divide in the evolution of modern agriculture and its relation to public-health issues; but in the immediate postwar period, no federal agency had the authority to prevent civilian sales of DDT.[14]

Because civilian use of DDT began immediately after the Second World War, the literature associated with its use is rife with military metaphors. The same language had been used a generation earlier; historian Mark Fiege found similar references to metaphors of war in his study of irrigated agriculture on southern Idaho's broad Snake River Plain. Farmers whose point of reference was World War I described their struggle against crickets, gophers, weeds, jackrabbits, and other pests in terms of social conflict. They viewed the bugs and animals invading their fields as enemy hordes that must be destroyed. Fiege contends that the seemingly Edenic setting of Idaho's Snake River Valley was "less a garden than a place of trouble and conflict: a battleground." Weeds were similar to invading armies, and citizens must engage in war to protect their crops and livelihoods.[15]

In the U.S. Department of Agriculture *Yearbook* for 1947, F. C. Bishop, a Bureau of Entomology and Plant Quarantine research scientist, published a short article about humanity's "constant battle against insects." Because they were "our rivals here on earth," Bishop held high hopes that scientific investigation would enable man to "overcome his insect foes and reduce the extent to which he provides food for the insect hordes." We should never forget, he cautioned, "that the battle is still on and that insects have the advantage because they can multiply with incredible speed." Bishop distinguished between good and bad insects and urged entomologists to work diligently "to control our insect enemies and to protect our insect friends." Insecticides such as DDT, however, represented "only one phase of our armament in the war against insects." In words that would come to haunt the USDA within two decades, Bishop warned that promiscuous and improper use of chemicals such as DDT had the potential to endanger "beneficial forms of life, including insect predators, fish, frogs, or even birds."[16]

The most notable incident of Oregon's "war" against insects involved a May 1947 "Mormon Cricket invasion" in the northeastern part of the state.[17] The Corvallis *Gazette-Times* headlined its May 23 story: "OLD OREGON TRAIL RED WITH DEAD CRICKETS OVER SIX-MILE STRETCH." The newspaper reported that the cricket hordes had "'broken' the bait line" near Hermiston and were advancing on the small military outpost at Ordnance. Seagulls, virtually unknown in the arid wheat country, had arrived in large numbers and were "attacking the left flank of the crickets." Machines were brought to the aid of the seagulls, blowing "poisoned mash into their 12–mile front." On the same day, the *Oregonian* printed bold front-page headlines: "MACHINES, CHEMICALS, MACHINES BATTLE CRICKET HORDE." Governor Earl Snell told reporters that men and equipment were "already on the battle lines" and that airplanes were being brought in and "will try strafing the advancing horde with chlordane." The *Oregonian* reported that the USDA was using a dozen "blower-spreader outfits" mounted on old army weapons carriers to spray the insects. And in an accompanying editorial—"MORMON CRICKETS MARCH"—the newspaper reminded its readers, "although we have shattered the atom, we aren't really so smart as we thought we were."[18]

The *Oregonian* noted that the mysterious account of "hungry seabirds . . . summoned from the coast, winging far inland, to combat the marks of the crickets, is a spontaneous theory that pleases the fancy." But the story of the seagulls' providential arrival does not rest in sound biology. The account of seagulls arriving in the Salt Lake Basin in great numbers and saving early Mormon crops is authentic enough, but it deserves fuller explanation. There were and continue to be large colonies of gulls living around the great Salt Lake, just as there are large colonies of gulls living along the Columbia River and in the Malheur Basin. An abundance of food in a certain areas—such as a "banquet of pot-bellied crickets"—became common knowledge to gulls and other birds, and they flocked to the appetizing bounty in great numbers. Biology, rather than divine intervention, explained the seagulls' presence in nineteenth-century Utah and in postwar northeastern Oregon.[19]

The Portland newspaper continued to use military references in spinning out the rest of the Mormon cricket tale. On May 27 it reported that "the battle . . . was in the 'mop-up stage,'" and the airplane used to spray chlordane over "the insect army" had been released from duty. Seagulls, however, "rejoined the battle," advancing from Morrow County into Umatilla County for the first time. But the grasshoppers were not to be easily deterred: "CRICKETS SWITCH ADVANCE, MOVE ON ORDNANCE DEPOT," the *Oregonian* announced in its May 28 issue. "Stopped in their march toward the farm fields of the Umatilla Valley," the crickets moved towards the depot as ground crews continued to spread poison "along the 12–mile front of the insect army." But the worst of the cricket threat was over. A low-flying plane, an *Oregonian* editorial announced, had "stopped the advance of the gluttonous insects by use of a new chemical called chlordane."[20]

A recently developed chlorinated hydrocarbon like DDT, chlordane proved highly toxic to the insects; it killed more than 95 percent of the grasshoppers in treated areas, with residual effects that lasted more than two weeks. The *Oregonian* was pleased that DDT had proved to be "of tremendous importance to agriculture." Moreover, the spraying program in northeastern Oregon appeared to resolve the question of its effect on humans and domestic animals. Chickens fed with poisoned grasshoppers for several days "showed no ill effects," and DDT appeared to be nontoxic to pest control personnel as well.[21]

Several years after the Mormon cricket incident, *Oregonian* writer John Denny celebrated the hundredth anniversary of economic entomology in an article titled "Bug Warfare in the Northwest." Although humans had been waging war on bugs for nearly a century, they were only holding their own, despite the impressive array of "new chemical weapons" available. If humans relaxed their effort, Denny warned, bugs would multiply, "leaving famine, pestilence and death in their wake." Insect pests were "smart and resourceful," he wrote, and pointed to DDT-resistant flies who "drink the once-lethal dose like orange juice and spit it in your eye." Denny praised the new chemical weapons in the human arsenal, especially chlorinated hydrocarbons such as DDT and its relatives: methoxychlor, TDE, lindane, chlordane, aldrin, dieldrin, endrin, toxaphene, and heptachlor. He also urged people to use the chemical insecticides with caution; some of them were "as lethal to man as they are to bugs, . . . [and] should be handled like a loaded gun."[22]

For several years after the war, state and federal officials paid little attention to spray residues and the persistence of chemicals in the environment, issues that would eventually become matters of public concern. Oregon Department of Agriculture reports regularly emphasized its "sound, cordial, and cooperative" relationship with OSC, the Agricultural Experiment Station, and the Extension Service. The department's ties to constituents and to industries that serviced agriculture remained close and intimate. When Oregon and Washington fertilizer manufacturers offered to advise state agricultural officials about fertilizer issues, the Oregon department announced the appointment of an advisory committee in the two states. The department also continued its systematic surveys to discover, control, and eradicate "dangerous insect pests and plant diseases." The agency's 1948 report referred to "new problems" and "increasing demands on the department for survey work"; requests for additional funding usually followed such statements.[23]

Department of Agriculture personnel carried out their survey work with missionary zeal, spending the months of April through October "tramping through fields and orchards looking for insect and plant diseases." With the onset of winter, the team was back in its Salem laboratory studying the summer's take and preparing specimens for its collection. In cataloging its seasonal work, the survey team was constantly on the watch for potential pest invaders, including worms, weevils, maggots, the gypsy moth, and the Japanese beetle. Among threatening plant diseases, the department listed the Dutch elm disease in its 1952 report. Surveyors also were on the watch for native insects that traditionally fed on "wild and unimportant hosts but which might transfer their attention to cultivated crops."[24]

Oregon agricultural officials also devoted increasing attention to "economic poisons." The department reported in 1948 that it was becoming "increasingly difficult" to regulate the industry. The agency also faced problems with the chemical poisoning of bees, especially from materials that were extremely toxic such as chlordane, cryolite, parathion, benzine hexachloride, and DDT. Department entomologists devoted more time offering advice about the proper use and care of new chemicals. There were also difficulties with the increased use of herbicides such as 2,4–D, especially when sprays "drifted" and killed non-target plants. The department's annual reports included extensive appendixes listing the licensing of new pesticides, laboratory tests on fertilizers, herbicides, and insecticides, petitions asking for damages from drifting sprays, and the number of predators killed.[25]

To enforce its modest regulatory policies, the Department of Agriculture emphasized cooperation, education, and voluntarism to achieve compliance with state laws and mandates. Operators were "encouraged" to follow proper instructions in mixing chemicals, to determine the frequency of application, and to observe safe "handling" procedures. The department urged employees to stick to the letter of the law and stated that not *"one iota more of compliance with the provisions of any law . . . than the provisions of the law itself provide shall be required"* (my emphasis). The agency's leadership warned against the tendency of employees "to become confused between what [they believe] the law should provide and what it actually does provide." Department personnel needed "to understand the limitations of the authority which they exercise" and to confine the rules and regulations firmly "within the scope of the authority in the legislative enactment."[26] In brief, the state agriculture agency's primary responsibilities were to serve its agricultural constituents and the industries that provided technical advice and material assistance.

By mid-century, chemical manufacturers and agricultural officials were praising the new chemicals' stunning successes in controlling insects and diseases. In the words of science writer John Wargo, the new synthetic pesticides were symbols of postwar progress, promising control over environmental risks "never before experienced." Agricultural interests also tended to see new weed-killing herbicides as symbols of responsible land stewardship. Because herbicides made plowing unnecessary for certain crops, severe weather conditions were less likely to erode the undisturbed soil. "No-till" agriculture became the new buzzword, a symbol of progressive farming practices. Representations of success were everywhere, especially in quantum increases in crop yield and decreases in erosion. Moreover, once farmers applied sprays, their very invisibility created the perception that risks to human health and the environment were nonexistent.[27]

What we now know about the new wonder chemicals—insect resistance, persistence of residues and their accumulation in plant and animal tissue, age-related variance among humans to the pesticides, and the multiple ways that humans are exposed to them—surfaced only in piecemeal fashion and over several decades of time. As early as 1945, tests with DDT alarmed biologists at Maryland's Patuxent Research Refuge, who worried about its effects on fish and wildlife. Although the military had used DDT with great success to control troublesome insects such as lice, chemical screening to determine its toxicity to humans was in its infancy, and its release to the pubic in the summer of 1945 came despite the alarm of some scientists who wanted more information on the chemical's long-term effects. As late as 1957, Rachel Carson biographer Linda Lear declares, the American public still knew little about the long-term effects of DDT and other chlorinated hydrocarbons such as chlordane, dieldrin, and aldrin.[28]

Despite the good press given the pesticides and herbicides in Oregon, there were early words of caution, suggestions that not all was well with the new world of environmental control. In a lengthy memorandum in 1949,

Leroy Childs, an entomologist with the Hood River Agricultural Experiment Station, reported on the annual meeting of the American Association of Economic Entomologists. Based on sessions he attended, Childs thought it prudent to warn that many of the new sprays were "extremely poisonous" and presented potential health hazards to workers and consumers. It was important, he wrote, to obtain more information about spray residues. The ongoing USDA Pure Food and Drug Administration hearings, he cautioned, might establish residue tolerances so low that substitute sprays would have to be identified.[29]

Other sessions at the entomologists' meeting addressed the accumulation of DDT in soil and insect species that were apparently resistant to the chemical. Soil samples from apple orchards that had been sprayed for three or four years showed remarkable accumulations of DDT. Moreover, investigations taking place across the country indicated that DDT and its close relatives such as chlordane persisted in the soil for a long time. Sensitive plants were especially prone to injury in those instances where DDT accumulated in the top layer of soil. Because people continued to use the chemical, Childs advised investigators to look for soil contamination and symptoms of DDT poisoning in legumes and grasses. Because of the chemical's efficiency, he feared that entomologists had been "lulled into a sense of security . . . that is not entirely justified." The increasing resistance of houseflies, mosquitoes, ticks, and greenhouse mites to DDT poisoning, Childs pointed out, suggested that the miracle agent might not be the final answer in controlling insects.[30]

Childs also reported that scientists had been hard at work developing analogues to DDT and related hydrocarbons, chemicals "more selective in their performance" that could be used with reduced risks. University and commercial researchers were proving that there was "no limit to the chemical combinations that can be devised possessing insecticidal properties." It was only a matter of time, the Hood River entomologist believed, before insecticides would be developed with potencies "equal to or surpassing those of DDT and at the same time less hazardous." Parathion, a highly toxic chemical that required great care in handling, was among the new chemicals discussed at the entomologists' meeting. Childs reported that carelessness in manufacturing and mixing sprays in the field had caused several deaths. Like most others in his profession, he believed that if applicators took proper precautions, parathion could be sprayed with little danger. Childs concluded his lengthy report on an optimistic note: "As a result of chemical research now underway, it is evident that the entomologist will not be lacking new insecticides for his consideration. At this time, the job ahead seems almost endless."[31]

In the meantime, the hitherto "nonexistent" housefly problem on the OSC campus had resurfaced. Two OSC entomologists released a study in late 1952 indicating that control with DDT was no longer possible because resistant strains of houseflies had developed. Workers at the college dairy barns had been using chlorinated hydrocarbon insecticides since 1944; until 1950, they reported, DDT and similar compounds had provided excellent control of flies. The entomologists reported that they first suspected the development of resistant strains of houseflies in 1949. In the following years, workers increased the number of sprays and used higher concentrations to gain some measure of control. "The summer of 1952," they observed, "has been the most difficult thus far."[32]

A sound argument can be made that Oregon researchers were at the cutting edge in exchanging scientific information regarding the use of DDT and other insecticides.[33] They recognized very early the development of resistant strains of insects, the highly toxic effects of certain chemical compounds, the persistence of certain sprays, and the accumulation of trace chemical residues on fruits and vegetables. The question that most entomologists raised at the time, however, was the level of risk. Leroy Childs and other scientists believed that proper handling, using the correct mixing formulas, and practicing safe spraying procedures would reduce risk to acceptable levels. There were other problems, especially an overworked Department of Agriculture laboratory staff charged with testing the toxicity and persistence of new pesticides and herbicides. Beyond the lab, powerful chemical companies and agricultural groups were constantly pressing for the approval of new chemical compounds.[34]

The housefly problem at the OSC barns provides an interesting case study in the use of the new chemical compounds. In their first applications of DDT in 1945, entomologists used 1.125 pounds to effectively control flies for the summer season. Four years later, the same buildings required five separate sprayings of "68 pounds of actual insecticide to give good, but not excellent control." During the summer of 1952, researchers placed sticky fly strips in the barns to estimate the number of flies. As the season progressed, the number of flies increased, the researchers reported, and "at no time was there a decrease due to the application of insecticides." At that point, the entomologists determined that it was necessary to revise the "entire control program." The scientists concluded that it was futile to control houseflies "by the chemical means known at present." Although DDT was effective at first, "this was a short lived hope," because flies resistant to one insecticide were often resistant to others.[35]

The increasing volume of pesticides and herbicides being sprayed on Oregon's forests and fields finally prompted the state legislature to pass a chemical-control law in 1951. The new measure reflected the spirit of the Federal Insecticide, Fungicide, and Rodenticide Act of 1947, requiring chemical manufacturers to label products to explain proper usage. Oregon's new law, however, provided little in the way of enforcement. The legislation also required licenses for airplane pilots who were spraying agricultural chemicals and for ground operators applying herbicides. In cooperation with OSC, the state Board of Aeronautics, and the state Board of Health, the Department of Agriculture set up an "operators' school," in effect a short course in applying chemicals; it also published an operator's manual available for the price of printing and mailing. Responding to increasing public concern about some spraying programs, the legislature acted again in 1953 and 1955 when it passed and then strengthened measures to regulate herbicide use. The new laws required air and ground applicators to register their herbicide spraying equipment with the state Department of Agriculture. To obtain a license, operators were required to pass an examination demonstrating a knowledge of the herbicide's characteristics, proper procedures for spraying, and precautions under certain weather conditions and an understanding of the state's rules and regulations.[36]

The hazards associated with the new chemical compounds were not limited to commercial agriculture. The state Board of Health warned in 1952 that the increasing number of new garden insecticides on the market threatened the public's health. Because some of the products were "extremely dangerous," users should wear protective clothing, respirators, and rubber gloves. According to state health officer Harold Erickson, the failure to read labels carefully was the chief problem: "Since the advent of DDT, insecticides have been so widely used that it seldom occurs to us that some of them may be dangerous." Erickson singled out the organic phosphorous group as especially risky, because of several fatalities in some sections of the country. He urged home gardeners to be on the watch for parathion, tetraethyl pyrophosphate, and similar compounds and advised merchants to urge buyers to wear protective clothing.[37]

Newspapers reported occasional tragedies with the chemicals. One was a young Hood River child who spilled a bottle of orchard-spray concentrate on himself in the spring of 1954. The poison critically burned the youth's legs and upset his respiratory system. The *Oregonian* took a personal interest in the story, addressing the accident in an editorial and then publishing an article by Loren Milliman, one of its former editors. Although "economic poisons" were important "in producing more food for more people," Milliman warned, they were also very dangerous. TEPP (tetraethyl pyrophosphate), the insecticide that harmed the Hood River boy, could be fatal if taken internally or absorbed through the skin. Milliman explained why farmers used such deadly poisons on their crops: "Because, left untreated, the crops would provide a banquet for bugs"; left uncontrolled, the bugs "would soon swarm over the earth and men would starve." In an accompanying editorial, the *Oregonian* admonished those "who treat the 'poison' label lightly" and urged citizens to use good judgment and to be more vigilant in keeping dangerous chemicals out of the reach of children.[38]

Because of negative publicity and the threat of more stringent legislation regulating pesticide use, state agriculture officials met with public-health personnel and chemical manufacturers in Portland to develop a course of action. The state Agriculture Department's Walter Upshaw told the group that his agency would exercise its full authority and monitor pesticide sales more carefully. Such an effort, he suggested, might head off "ill-advised legislation." The department's chief chemist recommended following USDA policy in permitting sales of certain chemicals for commercial use only. Representatives of two companies producing TEPP announced that they had discontinued small-container sales and withdrawn such items from store shelves. Ralph Sullivan, a state Board of Health administrator, assured the Portland gathering that his department would not propose "unrealistic" legislation but would seek ways to prevent home accidents. Because home gardeners failed to read labels, he said, the potential for accidents was high. Sullivan cautioned against singling out organic phosphates for special regulation, because "every medicine cabinet is loaded with toxic material." In what turned out to be a strategy session to avoid more restrictive legislation, the Agriculture Department's Upshaw concluded that the issue was one of adult education. The meeting closed with chemical-industry representatives pledging to do their part in an educational campaign.[39]

Education, cooperation, and voluntary compliance became the watchwords of the Oregon Department of Agriculture. Because pesticide and herbicide use was relatively new, the department's annual reports declared that much of its work related to chemicals was educational. As required by the herbicide law, state agriculture officials published an operator's manual that attracted the attention of other states and several foreign countries. The department also sponsored short courses on the OSC campus and supplemented those with field workshops around the state. During the 1950s Oregon Department of Agriculture officials also began investigating an increasing number of damage claims for losses related to herbicide spraying.[40]

Although the chemical control of weeds was still a relatively new practice in the 1950s, the number of acres treated with herbicides was striking. Licensed operators sprayed or dusted more than 1 million acres in 1957 and an even larger acreage the following year. Because the number of complaints about weed-killing sprays had risen sharply, the 1957 legislature again amended the state's herbicide control law, delegating to the Department of Agriculture responsibility for receiving and checking complaints and serving as arbiter in determining the extent of injury. The department reported a fivefold increase in alleged or verified complaints for the 1958 crop season, with most of the complaints involving injuries to crops from aerial spraying with 2,4–D and 2,4,5–T. Agency scientists proposed a technical solution to the problem of drifting sprays, recommending the use of amine formulations of the weed killers in lieu of the ester version. The latter was excessively volatile, especially when sudden temperature increases caused sprays to drift to nearby orchards and crops.[41] Although state officials devoted many hours to investigating such complaints and attempting to resolve citizen disputes, through the 1950s there is no evidence that the department was concerned about the long-range toxic effects of pesticides and herbicides.

Chemical sprays posed still other problems. Robert Every, the Extension Service entomologist in Corvallis, sent a circular to county agents in October 1962 warning about the dangers of treating cattle with CO-RAL spray. In a memorandum marked "Confidential," Every reported that several animals had died in eastern Oregon's Wheeler County even though the spray had been applied "in accordance with recommendations." Similar incidents, he noted, had taken place in other parts of the country. The Chemagro Corporation, manufacturers of CO-RAL, had "acted promptly. . . . in a manner that reflects the integrity of the company" and had withdrawn certain batch numbers from the market. Because CO-RAL had a "long record of safe and effective use as a livestock spray," Every asked county agents to treat the information about the spray's toxic affects confidentially.[42] Incidents of this kind did not make it into the Oregon Department of Agriculture's annual reports nor its more widely circulated *Agriculture Bulletin*. Extension Service and OSC entomologists obviously wanted to protect their close ties with agricultural constituents and the chemical manufacturers who funded much of their research.

Although widespread public opposition to broadcast herbicide spraying did not emerge until the 1970s—and then most of the focus was cutover timberland—there was considerable concern much earlier about the effects of drifting sprays, especially the most popular weed killers, 2,4–D and 2,4,5–T. Some landowners expressed concern about the toxic effects of herbicides on livestock, especially dairy cattle. Although the Department of Agriculture found no evidence of harm on treated pastures, the agency acknowledged that more careful research would be helpful. While the department's annual reports do not indicate a growing public restiveness, an increasing number of legislative restrictions suggests that not all was well. The 1961 legislative session amended the state's food laws "to cope with new problems in the field of residues and additives." There were also increasing difficulties with new fertilizers containing insecticides and growth regulators. Agricultural officials complained about the department's work load and admitted that its personnel could effectively examine only those pesticides "that have significance to our state's agricultural economy."[43]

By the 1960s, inspectors were regularly testing raw agricultural products for spray residues, especially 2,4–D, the most widely used herbicide. Because most of the complaints filed with the department concerned 2,4–D, the state required a special assessment on sales of 2,4–D products to support research at Oregon State University (the institution became a university in 1961). The Agriculture Department occasionally complained that purchases outside the state and the federal government's exemption from the tax lowered its potential revenue. In a "residue project of special interest," field personnel tested dairy farms, fish and wildlife, and city water supplies. The department's laboratory scientists determined that "no actionable residues were found" in the milk samples it tested.[44]

The vexing problem for entomologists, chemical manufacturers, and agriculturalists was the continued development of pesticide-resistant insect strains. In an extended study at OSC of insects affecting onions, entomologist H. H. Crowell reviewed early efforts to control onion maggots, some of the research dating to the 1920s. When researchers first used DDT, the chemical "looked promising," but chlordane proved to be more effective. However, the insects quickly built up resistance to the chlorinated hydrocarbons, and by 1953 and 1954 chlordane no longer effectively controlled the maggots. In the following years, Crowell wrote, "resistance to chlordane and other chlorinated hydrocarbon insecticides appeared in almost every onion growing area." Beginning in 1958, researchers turned to organic phosphate insecticides, especially Trithion, ethion, and Guthion. In his 1961 report, Crowell recommended new test runs with furrow treatments of fungicide-insecticide combinations and "continued studies on onion maggot biology and control in order *to be better prepared for the eventual selection of strains resistant to the phosphate insecticides*" (my emphasis).[45]

As federal inquiries into persistent chemicals broadened, OSC entomologists continued to support the use of controversial pesticides. Paul Ritcher, head of the department of entomology, wrote to the Food and Drug Administration in 1960 requesting clarification on a "food additives amendment" published in the *Federal Register*. The issue involved pesticide residue in mint oil and whether it was subject to the food-and-drug law dealing with food additives. Ritcher noted that mint growers had first used DDT as foliage treatment to control mint flea beetles in 1946 and then switched to aldrin and dieldrin in 1953 to effectively control insects. Although the entomology department had never found traces of DDT in mint oil, it assumed that use of the chemical was permitted because it enjoyed federal sanction. Even if slight residues were found, Richter reasoned, they "would be below the currently established tolerances for insecticides, including DDT." On behalf of Oregon mint growers, Ritcher urged the Food and Drug Administration to permit the use of DDT and related compounds through the 1960 season.[46]

Oregon State College agricultural scientists were in the midst of the growing public restiveness over the use of chemical agents on orchards and other food crops. Because of the public's heightened sensitivity about the dangers that chemical sprays posed to human health, Oregon farmers and chemical producers grew increasingly worried about liability when insecticides and herbicides were misused. When the federal Food and Drug Administration seized Oregon-grown cranberries for suspected pesticide contamination in 1959, the case became a national story with serious industry repercussions. In response, the Oregon Department of Agriculture and OSC sponsored a one-day conference in Corvallis on the proper use of pesticides. Growers who failed to properly follow directions, officials warned, "place in jeopardy use of these chemicals in the whole agricultural industry." To avoid such restrictions, speakers recommended education as the "key point in putting agriculture's house in order." Virgil Freed, an OSC chemist, told the audience that chemicals had brought great benefits to the American people and that both farmers and consumers had a stake in their continued use. Pesticides and herbicides helped farmers produce quality crops; they held down consumer costs; they enabled farmers to produce more food with fewer workers; and insecticides such as DDT aided in "the control of human disease and suffering." Freed assured the extension agents and growers that authorized chemicals were safe if used properly.[47]

When the Food and Drug Administration called for "zero tolerance" in chemical residues in dairy feed, J. D. Patterson, the Oregon Department of Agriculture's chief chemist, charged that achieving such a standard was "scientific fantasy." Writing for the department's monthly *Agriculture Bulletin*, Patterson reported that the Agriculture Department followed proven methods—reliable techniques that established the degree of probability that chemical residues would turn up in dairy products. Because the department lacked funding to carry out inspections in "all cases," the agency used "publicity and education" to help producers eliminate potential contamination. When Patterson spoke a few months later to a "Dow Dealer's Short Course" in Salem, he applauded commercial sellers for their educational role in the use of agricultural chemicals. Oregon's law, he noted, already required the sale of highly toxic material in "unbroken packages." That requirement, Patterson told the audience, "resulted from the loud clamor which followed the death of a child in Hood River county as a result of the careless handling of TEPP." To retain such chemicals for agricultural use, the Department of Agriculture convinced the state legislature to restrict dealers from selling certain products to "just anyone."[48]

"For the first time in the history of the world," Rachel Carson wrote in her 1962 best-seller, *Silent Spring*, "every human being is now subjected to contact with dangerous chemicals, from the moment of conception until death." Like the proverbial fire bell in the night, Carson's classic study of synthetic pesticides—in particular, chlorinated hydrocarbons such as DDT—reverberated across the United States and beyond. The reactions to *Silent Spring* included angry denunciations and threats of lawsuits from pesticide manufacturers and large agricultural organizations, some of whom went into siege mentality.[49] The controversy surrounding the book also revealed the friendly ties between the chemical industry and land-grant-college scientists who depended on the companies' largesse for research dollars. According to historian Thomas Dunlap, "*Silent Spring* marked a watershed, as the private, scientific debate became a public, political issue."[50] Oregon's part in this story gained momentum in the mid-1960s with the growing public awareness of the dangers posed by persistent pesticides. In most instances, however, federal actions to limit and ban certain pesticides took place in advance of similar moves in Oregon, where agriculture and lumbering powered the state's economy. Although no precise data are available, because Oregon's economy was centered in growing foodstuffs and trees, one may assume that chemical applications across the state were considerable.

As the storm of public discussion swirled around *Silent Spring,* the Oregon legislature amended the state's food laws to cope with what the Department of Agriculture called "new problems in the field of residues and additives." Laboratory personnel also were investigating fertilizers containing insecticides, growth regulators, and trace elements of other chemicals, and they were looking into new animal feeds containing drugs and antibiotics to control diseases. The department used its expanding laboratory facilities to assure that chemical compounds in such materials were "accurate and within the legal range." The state laboratory also operated at the edge of overwork, complaining in its 1964 report that the pressure to test new pesticide chemicals limited the samples that could be examined. Because personnel and laboratory facilities were excessively busy, sampling was restricted to "materials *in most common use*" (my emphasis). The Department of Agriculture also licensed an increasing number of "chemical applicators" to operate in the field: 389 commercial sprayers (or dusters) in 1964, 420 in 1965, and 435 in 1966. At the national level, by the late 1960s the total use of DDT had surpassed 640,000 tons. And alarming stories continued to multiply. *Time* reported spectacular fish and wildlife kills in the Canadian province of New Brunswick, directly attributable to DDT. *Time* cited the use of the spray to control spruce budworm and the way the chemical compound worked its way up the food chain.[51]

Two years after the publication of *Silent Spring,* the USDA circulated Memorandum No. 1565, *U.S.D.A. Policy on Pesticides,* signed by Secretary of Agriculture Orville Freeman. The publication underscored federal responsibilities in "the methods and materials for the control of pests." The document outlined USDA's education and regulatory program "in the never-ending struggle to protect man, his food and fiber supplies, and his forests from the ravages of pests." The Freeman memo expressed "vital concern" for those who used pesticides and the effects of "pesticide pollution" on wildlife, air, and water. It was USDA policy to use methods of pest control that "provide the least potential hazard to man and animals." The directive promised that the department would "exercise constant vigilance" and urged pesticide users to protect humans and to avoid unnecessarily exposing crops, animals, and fish and wildlife to chemical contamination.[52]

With the increasing criticisms of pesticide use, OSU scientists became more assertive in defending chemicals used to kill insects and weeds. Virgil Freed told an *Oregonian* reporter in 1966 that pesticides had saved more lives than penicillin, as DDT had wiped out malaria and typhus in many areas of the world. Freed, who was participating in a large U.S. Public Health Service study to understand how chemicals affect insects and weeds and their side effects on humans, told the *Oregonian* reporter that scientists were also studying why insects built up immunities to chemicals such as DDT. Laboratory houseflies at the university, Freed reported, were now able to survive 100 times the normal dose of DDT.[53]

As the move to restrict and ban targeted pesticides, especially DDT, gained momentum, Oregon Department of Agriculture officials and OSU scientists became increasingly defensive. Ben Allen, the Agriculture Department's assistant director, wrote an editorial entitled "We Are Fortunate" for the March 1968 issue of *Agri-Record,* the agency's new monthly publication. Because city dwellers did not understand recent scientific and technological advances in agriculture, Allen reasoned, they were unaware of the critical role that chemical pesticides played in food and fiber production: "Consequently the 'residue' of 'Silent Spring' lingers on with the general public." The public's failure to understand the key role that pesticides played in the state's agricultural abundance, Allen warned, might create a "consensus obstructing continued progress."[54]

The 1969 session of the Oregon legislature reacted again to the public's growing restiveness about chemical sprays, updating its pesticide law and creating an oversight body, the Committee on Synthetic Chemicals in the Environment. The committee was another Department of Agriculture recommendation to the legislature; committee members were to be "knowledgeable and capable persons within Oregon who are experienced . . . with all the problems relating to pesticides." The measure specified that committee members would represent chemical manufacturers, the trucking and railroad industries, the Department of Geology and Mineral Industry, the Public Utility Commission, and the Department of Forestry. The state Board of Health, the state Pollution Control Center, OSU's Environmental Health Center, and two "public members" would also serve. The state chemist was permanent chair and a non-voting member.[55] The makeup of the committee and its advisory status insured that it would not challenge the department's pesticide authority but would serve as a public-relations buffer for the agency.

The same legislative session also gave the Department of Agriculture exclusive authority to control the sale and use of pesticides. The amended law clarified the agency's prerogatives and responsibilities:

> The Department is authorized to take all measures or to make any seizure or embargo necessary and proper to protect property, or the health or life of animals or humans, from the incorrect sale, storage, handling, transporting, disposal or use of pesticides, residues from the manufacture of pesticides or other dangerous chemicals.[56]

Although the legislation streamlined the department's enforcement mechanism in transporting and selling chemicals, it did little to redirect the agency's attention to the long-term threat that pesticides posed to ecosystems and human health.

Although some of his contemporaries referred to Governor Tom McCall as a "cosmetic environmentalist," the governor was in the forefront of those who questioned the continued use of persistent pesticides such as DDT. When he signed the amended pesticide bill into law in July 1969, McCall announced that he wanted Oregon to ban controversial chemicals such as DDT. The governor urged state agricultural interests to look for acceptable substitutes, and he used an August pesticide symposium at Oregon State University to call for a phaseout of DDT. McCall told the 450 assembled scientists that the newly appointed Committee on Synthetic Chemicals would establish rules for pesticide policy in the state. One year later McCall acted, banning the use of DDT except where it was required to protect "the health, food sources, natural resources and general welfare of our citizens." Under no circumstances, however, could citizens use DDT in aquatic environments or places "where there is any possible hazard to health or environment."[57]

From wonder chemical to the target of citizens, lawmakers, and a growing number of scientists, persistent pesticides such as DDT were the subject of public debate across the United States. "At first there was no doubt about DDT, World War II miracle pesticide," Richard Floyd told the OSU symposium. It had saved millions of lives and helped control malaria, cholera, typhus, Rocky Mountain spotted fever, and encephalitis. But, Floyd noted, traces of DDT could now be found in plants and animals around the globe and in the oceans. Organizers of the OSU gathering distributed a World Health Organization (WHO) statement underscoring DDT's importance in fighting diseases in developing countries. Although the WHO wanted to move away from using DDT, the organization worried that alternative measures would be too expensive.[58]

"Is DDT a chemist's Frankenstein monster?" Floyd asked at the Corvallis gathering. Denzel Ferguson, a Mississippi State College zoologist soon to become director of Oregon's Malheur Field Station, warned scientists that pesticides represented an unexamined "selective force" that might contribute in subtle ways to evolutionary influences that would lead to long-term biological changes. Ferguson asked why some insects built up resistance to chemicals but others did not. Ohio State University environmental biologist Tony Peterle thought persistent pesticides should be banned immediately. "Time is a commodity we no longer have," he told the gathering; it was "unwise, if not sheer folly, to continue loading the environment with persistent toxic compounds." OSU's Virgil Freed told the scientists that Oregon had been lucky, because the state's Department of Agriculture guidelines on the manufacture and use of pesticides had minimized potential problems. Freed thought stricter controls might be needed, but he opposed a ban of DDT. He argued that technical adjustments—better application techniques, buffer zones, and rotating chemicals—would minimize the environmental dangers.[59]

In the months leading up to the Environmental Protection Agency's 1972 ban on DDT, Freed, state Department of Agriculture chemist Virgil Hiatt, and newspaper editorials continued to argue for the restricted use of the chemical. Freed was especially aggressive, writing in an *Oregonian* op-ed piece that there was no evidence "to point to any particular hazard of DDT to higher animals—and particularly man." He pointed to the "enormous benefits" humans had enjoyed from its use and argued that certain pest problems could be controlled only with DDT. Freed praised the Extension Service and the Oregon Department of Agriculture for their vigorous educational programs in the safe handling and careful application of sprays. Chemical pollution in Oregon, he believed, had been minimal. Appearing before an Oregon House of Representatives task force on pollution, Hiatt told the committee that banning DDT would force housewives to live with wormy apples, flies, and mosquitoes: "I can't believe in prohibition that would lead to worldwide epidemics of malaria and other diseases." Hiatt told the legislators that persistent pesticides were being blamed for problems that should be attributed to other sources.[60]

To support the continued use of DDT, the *Oregonian* solicited an essay from J. Gordon Edwards, a professor of entomology at San Jose State University. Edwards was a leader among a group of scientists who opposed a complete ban on DDT, and he was sharply critical of those who wanted to prohibit its use altogether. The "barrage of anti-DDT propaganda," he asserted in his 1969 essay, was "untruthful and misleading" and bereft of facts. There were some things in life, such as "motherhood, patriotism, and DDT that simply do not need defending," he reasoned, "at least, not until recently." Too many uninformed critics were shouting "wolf," he wrote, including "sensationalist" news media that had convinced the public that anti-chemical arguments were vested in scientific fact. Experiments with mice and birds were misleading, because they involved concentrations of DDT many times stronger than the amounts ingested by humans. "Many sincere citizens [were] becoming concerned enough to listen to the scientific evidence that refutes the charges against DDT," Edwards concluded.[61]

The *Oregonian* regularly pointed to sharp disagreements among "experts" and praised the move toward "control" rather than a total ban. That Oregon had already drastically reduced the use of DDT "should quiet some of the semi-hysterical, doomsday wailing about DDT," the newspaper declared in January 1970. The *Oregonian* urged that "emotionalism . . . be replaced by scientific evidence." That fall—with a complete federal ban looming—

the paper praised the state director of agriculture, Walter Leth, for insisting that objective and realistic evaluations be used in policy decisions. Farmers were, according to Leth, being forced to increase production on fewer acres because land was being lost to housing, retail, and industrial development. Pesticides enabled farmers to produce more food and fiber with less labor and on fewer acres. The *Oregonian* also cited Freed's work calling for reasonable restrictions rather than a total ban.[62]

When Environmental Protection Agency director William Ruckleshaus banned the domestic use of DDT in June 1972, to take effect at the end of the year, the announcement followed three years of contentious debate within President Richard Nixon's administration. In a memorandum circulated to agency personnel in November 1969, Secretary of Agriculture Clifford Hardin directed the department to give greater attention to persistent pesticides such as DDT. He prohibited use of the chemical except where it was essential to protect "human health and welfare and for which there are no effective and safe substitutes." Hardin also banned the use of DDT to spray shade trees, house, garden, and tobacco insects, and to treat swamps and wetlands. The agriculture secretary's order also reflected the department's powerful agricultural constituency: "The scientific evidence now available does not establish that the use of DDT constitutes an imminent hazard to human health." Although there were "some adverse effects" on select fish and wildlife species, those influences did not "constitute an imminent hazard to fish and wildlife or the environment."[63]

In the order banning the pesticide, Ruckleshaus found the "long-range risks of continued use of DDT" unacceptable and outweighing any of its benefits. He based his decision on a federal report authored by Dr. Emil Mrak which concluded that persistent pesticides adversely affected ecological relationships. When Ruckleshaus made his announcement, most states, including Oregon, had already drastically limited DDT to only a few crops. The Oregon Environmental Council's executive director, Larry Williams, praised Ruckleshaus for his "considerable courage . . . in the face of strong opposition from industry." No one, Williams wrote the EPA director, "had the fortitude to withstand buffeting from industry until now."[64]

Despite the ban, Knud Swenson, head of OSU's entomology department, argued that DDT was still important to control cutworms in cabbage and cauliflower and root weevils in berries and ornaments. Swenson speculated that it might be a year or two before an adequate substitute would be available. Dave Nelson, a vice president with the Oregon Farm Bureau, was "very disappointed" in the Ruckleshaus announcement, because he said there was no scientific evidence that DDT harmed human health or posed a threat to fish and wildlife. In a grudging editorial on the ban, the *Oregonian* observed that DDT had been subject to heated dispute "since the late Rachel Carson published 'Silent Spring,' even though there has been no silent spring."[65]

Although the public debate over the use of DDT and other pesticides would never reach a consensus, the ban on its use in the United States established the precedent that scientific findings and citizen pressure could lead to prohibitions against targeted chemicals. Oregon's pesticide story mirrored the larger debate being waged at the national level. When the ban on DDT went into effect, the major federal law regulating chemical sprays was the outdated Federal Insecticide, Fungicide and Rodenticide Act of 1947, a measure drafted when there were few environmental concerns. In Oregon, the Agricultural Chemical Control Act of 1951 (periodically revised) still served as the state's principal legislation regulating pesticides. At the federal and state levels pesticide regulation had been left to the USDA and the Oregon Department of Agriculture, two agencies closely associated with agricultural producers and the companies providing them with services. Moreover, in Congress and in the Oregon legislature, influential individuals chaired agriculture committees and controlled funding for pesticide legislation.[66]

Agriculture's exclusive control over pesticide policy began to change when Congress passed the National Environmental Policy Act in 1969 and established the Environmental Protection Agency in 1970. Congress further strengthened federal pesticide legislation when it enacted the Environmental Pesticide Control Act in 1972. Although those were compromise measures, they reflected an increasingly influential environmental movement. The Oregon Environmental Council, founded in 1970, had also been actively involved in pressing federal and state agencies to ban DDT. At the federal level, the Nixon administration provided environmentalists with direct access to policymaking through the EPA, an agency that historian J. Brooks Flippen calls "a staunch ally to environmentalists." By the mid-1970s EPA guidelines shifted the burden of proof to chemical companies, requiring the firms to provide evidence that a pesticide was benign before authorizing use of new pesticides.[67]

The DDT ban was limited to use in the United States only. Domestic production for foreign sales continued unabated; as recently as 1996, United States ports shipped the pesticide at the rate of one ton per day. Of the nearly 6 billion pounds of all pesticides sold in the global marketplace in the last fifty years, nearly half of that volume was manufactured in the U.S. Despite the fact that traces of DDT were discovered in the fatty tissue of arctic fish and Antarctic penguins during the 1970s, the chemical industry was successful in promoting DDT for use in the developing world. The fact that the chemical showed up hundreds of miles from where it was used indicated the global nature of the pesticide problem.[68] The ban on the use of DDT and its analogues in the United States would not be the last struggle waged against a toxic pesticide.

Farm workers were also involved in the move to ban DDT. In California the United Farm Workers Union was successful in including health and safety clauses in contracts with growers, and some contracts explicitly banned pesticides such as DDT, aldrin, and dieldrin before the federal ban. Robert Gottlieb points out, however, that the pesticides used in lieu of DDT were invariably more toxic to humans even though they broke down more quickly in the environment. Cesar Chavez of the United Farm Workers' Union led the effort to protect California's agricultural laborers from pesticide poisoning, at one point calling the issue "more important . . . than even wages." When Chavez asked leading environmental organizations for help in protecting farm workers from hazardous pesticides in the early 1970s, he met with mixed results. In a study of the United Farm Workers and environmental politics in California, historian Robert Gordon attributes the unwillingness of the mainstream environmental groups to provide support to indifference, style, and class and ethnic divisions.[69]

Conditions were no different in Oregon, where the grower-dominated Board of Agriculture establishes policy guidelines and advises on legislative policy. Strongly opposed to union organization, the Oregon Farm

Bureau and other grower organizations have fought tenaciously to dominate agricultural politics and especially pesticide policy in the state. With the emergence of PUCN (Northwest Treeplanters and Farmworkers United) in 1985, Mexican-American farm workers have fought for union recognition, higher wages, and protection from pesticide exposure. Often working without adequate training and protective gear and laboring in hand-harvested operations, agriculture employees are especially vulnerable, because current Oregon law provides little protection for farm workers. Although the 1999 legislature passed a pesticide-use reporting bill, lack of funds and enforcement has made the legislation a dead letter. In a special series published in December 1999, the *Oregonian* revealed the health problems agricultural workers confronted with organophosphates, especially Guthion, the long-standing spray of choice to control codling moths in fruit orchards. Agricultural workers have been on the front lines—the proverbial canary in the coal mine—in studies of the effects of organophosphates on humans. Even low-level exposure to Guthion can cause nausea, dizziness, and headaches.[70] Although the controversial spray was the subject of intensive politicking at the national level in the late 1990s, the EPA still sanctioned its use.

The quest for efficiency, the attempt to achieve economies of scale, and cooperation between the private sector and public agencies to advance agricultural production also extended to modern forest management and the efforts of Oregon's most powerful industry to increase output from public and private timberlands. Although there are parallels in technique, appeals to science, and friendly working relations between government agencies and the private sector, policymakers and the public clearly treated forestry and agricultural issues as separate from one another. Planners and planning strategies played a much greater role in forestry, involving much discussion and several legislative attempts to achieve a sustainable forest industry and stable communities. There were no comparable efforts in agriculture.

III

INDUSTRIAL

FORESTRY

MANAGEMENT

5

PLANNING AND TECHNICAL

EFFICIENCY IN THE FORESTS

The sweetest smell to the Oregonian is that of sawdust. Roughly 65 cents
of every dollar in incomes derives from wood and wood products.
    The lumber industry has made spectacular progress in recent years.
With smart leadership, the industry has introduced conservation methods, both in its mills and in the woods.—*Newsweek,* November 9, 1953

S ocial innovation and social experiments characterized many of the
    reform strategies of Franklin Roosevelt's New Deal administration.
    The Resettlement Administration made a halfhearted effort to estab-
lish model communities for refugees from failed farming enterprises and
disintegrating industrial centers; the Farm Security Administration provided labor camps for agricultural migrants in search of work; the Tennessee Valley Authority produced electricity and attracted industry to the
distressed upper South; and the Bonneville Power Administration attempted
to accomplish the same for the Columbia Basin.[1] New Deal policymakers
sought other planning solutions to the nation's acute economic and social
distress. With the encouragement of the president's much favored National
Resources Planning Board and the Public Works Administration, federal
officials in the Northwest formed the Pacific Northwest Regional Planning
Commission in 1934.[2] Its proponents argued that planning would stabilize
industries and communities, soften boom-and-bust cycles, alleviate social
distress, rationalize the exploitation of resources, and contribute to sustainable

communities. Because of the region's heavy dependence on timber harvesting, planning approaches appeared to hold special promise for the
Northwest, especially among progressive-minded individuals. When the
National Resources Planning Board reported in 1937 that the region had
the potential to be the country's permanent woodlot, powerhouse, and
source of food, the *Oregonian* boasted that the Northwest still had sufficient
timber reserves to make it "the continent's most important lumber yard."
But experts also contended, the newspaper declared, that sound conservation principles were "more important in the Pacific northwest" because
of its tremendous supply of raw material.[3]

Throughout the 1930s there were voices urging caution, greater care,
and stewardship in managing the region's timber wealth. As early as 1934,
C. J. Buck, regional forester of the Forest Service's Region 6, suggested that
Northwestern states adopt sustained-yield management policies to avoid
"a day of social and economic reckoning." The regional chief singled out
Oregon's Clatsop County as a living example of the effects "of our present
'cut-out-and-get-out' policy upon families and communities." Sustained-
yield timber production, he believed, would stabilize the region's economy
and its cultural and civic life. An assistant Region 6 forester cautioned in
1935 that Coos County would face "an appalling tragedy" unless corrective measures were taken. And toward the end of the Depression decade,
Forest Service official Edward I. Kotok joined those who were concerned
about the long-term viability of lumber communities. Although the Northwest's forest problems were "yet in the making," Kotok observed "adverse
symptoms" already apparent in certain districts.[4] With the onset of the Second World War and an economy operating at full employment, however,
social planning lost its appeal as Congress moved aggressively to dismantle certain New Deal programs, including the National Resources Planning
Board.[5]

In the Pacific Northwest, however, the planning process lived on in occasional studies and surveys. Charles McKinley, a consultant to the Pacific
Northwest Regional Planning Commission, carried out one of the better-
known surveys of natural resources shortly after the war. War-related prosperity had erased all doubts about the permanence of the Northwest
lumber industry, McKinley concluded. In its place, "patriotism and profit"

and the turn to private enterprise had created an atmosphere "inhospitable to sober reflection" about conserving timber resources for future generations. "The demand blew the ceiling off manpower and prices," McKinley declared, and unprecedented postwar profits brought a mad scramble for marketable timber and much overcutting. Unless the region adopted a plan to reduce annual harvests "to the total perpetual yield from forests," he predicted a period of intense industrial activity for a few decades, "then suddenly dropping to the stillness of dead communities." McKinley added that the frenetic pace and huge scale of logging activity were also disrupting streams, destroying spawning grounds, and raising water temperatures. Illusions born of the postwar boom would, McKinley feared, "obscure the relations between basic resources and a healthy economy."[6]

Despite much breast-beating about the region's timber reserves, occasional voices urged restraint during the postwar years. "National welfare, in war and peace," an *Oregonian* editorial argued in April 1945, "is dependent upon a permanent and adequate supply of wood." Although Oregon entered the 1940s as the leading lumber-producing state, the newspaper's editors worried that demand vastly exceeded growth in many sections; moreover, labor shortages and increased harvests during the war years had delayed planting. The editorial pointed to the need for federal and state governments and private timber owners both to develop long-range plans for stabilizing the industry and to prepare for a heavy postwar demand for lumber. On Oregon's southern coast, the Coos Bay *Times* warned that the area's production capacity far outstripped annual timber growth. Local mills could not operate at their current rate of cutting "without depleting the supply of green timber." To "rationalize" timber inventories, the *Times* recommended larger single ownerships and firm commitments to establish diminished but steady harvesting schedules. "Even for Oregon with its top-rank timber resources," the paper warned, "it is now later than we like to think."[7]

Ideas about implementing sustained-yield management policies on Northwest forests did not originate with planning agencies during the Great Depression. In an effort to limit the volume of timber on a glutted market during the 1920s, industrial-forestry representatives in the Northwest were the first to develop the concept. David Mason, a Portland forester with close ties to lumber trade associations, developed the most articulate sustained-yield proposals in an attempt to rationalize chaotic harvesting rates that were perpetually driving down timber prices. Cooperatively managed units of public and private timberland would, he believed, slow harvesting rates and bring stability to communities. Limiting harvests was especially attractive to the largest private timberland owners, who had a vested economic interest in restricting the volume of marketable saw logs. Regulating the supply, they believed, would stabilize prices, and small mills without access to timber would be left without logs to cut. For more than five decades this fuzzy but grand scheme to control the market attracted the interest of foresters and legislators alike.[8]

Sustained-yield proposals gained special credibility during the Great Depression, a time of glutted markets and depressed prices for virtually all commodities. Until at least the Second World War, industry leaders in the Pacific Northwest joined with the Forest Service in giving general support to sustained-yield proposals. The region was also at the center of the only two pieces of federal legislation authorizing cooperative sustained-yield units: the Oregon and California (O & C) Revested Lands Act (1937) and the Sustained-Yield Forest Management Act (1944). Passed into law at the end of a lengthy period of depressed prices, neither piece of legislation achieved significant long-range success. Dramatically different market circumstances after 1945 and a booming construction industry made the objectives of the legislation—controlling harvests—a moot issue to its industrial supporters.[9] Despite changing market conditions, policymakers continued to praise the virtues of sustained-yield management. Oregon's Postwar Readjustment and Development Commission believed that sustained-yield programs would prevent ghost towns and "abandoned sawmill communities." The commission was optimistic that the federal policies would stabilize employment in timber-dependent areas and assure a dependable supply of wood products for the future.[10]

Despite the ideological battles that ensued, the Forest Service and BLM persisted in their efforts to implement cooperative sustained-yield units. Given the competitive nature of postwar American capitalism and the bullish lumber market, it is not surprising that blocking-up public and private timber into large sustained-yield units would stir controversy. The loudest public protests involved recommendations to implement cooperative agreements involving BLM's extensive O & C Revested Lands in western Oregon. When the agency promoted a cooperative unit involving Fischer Lumber Company property and BLM land north of Springfield along the Marcola River, the initiative brought a storm of dissent from small-mill owners and from Ellery Foster of the International Woodworkers of America (IWA). At the Eugene hearings in January 1948, mill owners and the IWA objected to the Marcola unit, charging that such cooperative units were "undemocratic and monopolistic" and "wrong in principle." The problem, Foster later told an audience in Quincy, California, was how to "achieve sustained yield without strengthening monopoly."[11]

While firms with large timberlands remained quiet about potential cooperative agreements, associations representing small-mill owners and the IWA kept up a sustained attack on such proposals. Delegates representing the Western Forest Industries Association, Northern California Lumbermen's Association, and Western Montana Lumbermen's Association met with the IWA in Eugene for two days in June 1948 and adopted a resolution "opposing monopolistic features in the management of natural resources, including timber, by the departments of agriculture and interior." R. T. Titus of the Western Forest Industries Association argued that cooperative agreements were the wrong approach; if private timber was being cut faster than it grew, the proper solution would be to make more federal timber available. Titus also called for two kinds of federal action: land exchanges that would give holders of "immature timber" cutting rights to "mature public timber" of equal value, and payments to private landowners for practicing good forestry. Before adjourning, delegates approved a four-point policy statement: "equality of opportunity" in forest management decisions; construction of a network of access roads into public timber; making timber sales small enough for small operators; and requiring more than a single bidder on federal sales.[12]

Crow's Pacific Coast Lumber Digest published the most disparaging criticism of the cooperative sustained-yield policy, calling it a "sustained grab," one more big step "in the direction of the Russian plan, wherein crackpot, night-school loggers sent out of Washington will . . . swarm over our forests" and tell lumbermen how to cut down their trees. Filled with Cold

War rhetoric, the *Crow's Digest* editorial lambasted companies "newly sprung from bankruptcy to affluence as a result of war prosperity," who were "joining hands with power-hungry federal bureaus" to convert Oregon's public forest lands into a vast private estate. "No more un-American plan was ever concocted." The backbone of timber-dependent communities, according to the *Digest,* were small loggers and mill operators, people who purchased local supplies: "Kill the little operators and you kill the little communities. Kill the little communities and you kill the Eugenes and the Roseburgs and the Lebanons and you kill the state."[13]

In the midst of those raucous public controversies, the *New York Times* reported that "the lumber capital of the Pacific Northwest" had shifted from western Washington to western Oregon in the vicinity of Eugene. This section of Oregon, the *Times* observed, held "the nation's greatest remaining stands of first growth Douglas fir" as well as a sizable volume of western hemlock, Sitka spruce, and western red cedar. Because of previous heavy cutting on private land, the percentage of public timber on the market had been steadily increasing. "More and more," the newspaper pointed out, "lumber operators are looking to national and state-owned forests to piece out their operations." The *Times* also reviewed federal cooperative agreements in which public and private holdings were intermingled and noted that to date only one such arrangement, involving the Forest Service, had been successfully negotiated. The BLM had been unsuccessful thus far in setting up similar programs for its O & C lands. BLM official Daniel Goldy told the *Times* that future prospects for establishing cooperative sustained-yield units were "somewhat dim and remote."[14]

Federal sustained-yield proposals remained contentious issues in western Oregon through the 1950s and beyond. Despite Forest Service and BLM efforts to promote cooperative programs, all of them failed. The Senate Interior Committee held hearings in Eugene and other Oregon communities in 1948, and newspaper editorials and letters to the editor—equally divided on the issue—poured into BLM's Portland office. Recognizing that they had stirred a hornet's nest of opposition to cooperative units, BLM officials backtracked and looked for compromise. When assistant Interior secretary C. Girard Davidson proposed changes to make cooperative agreements more acceptable to small-mill owners, R. T. Titus offered faint praise for the revisions: "The fenders have been straightened out [on the O & C] jalopy . . . the body painted and the upholstery renewed, but underneath it is the same old chassis of long-term agreement for a very small percentage of operators."

Responding to the heavy market demand for more logs, BLM increased the allowable cut on O & C lands in 1949.[15] The initiative included an authorization for building access roads to reach otherwise inaccessible timber. The increased harvests that followed were immensely popular in O & C counties which received half of the revenue from timber sales. For more than two decades they thrived on huge timber revenues, building good schools, courthouses, and road systems. That prosperity proved fleeting, however, as timber supplies diminished and environmental restrictions during the 1970s began to place additional curbs on harvesting.

Urged on by production-oriented politics at home, BLM proved itself receptive to the booming market for lumber products. In a January 1956 public hearing in Eugene, "log-starved mills" in Clatsop County urged the BLM to modify its "marketing area" policy, mechanisms built into the O & C Sustained-Yield Law of 1937 to limit timber sales to local mills. Sawmill, union, civic-group, and other representatives urged the federal agency to rescind its sales policy. But public timber stands were contested ground, and Polk, Benton, Marion, and Linn Counties supported the retention of the O & C marketing areas. Removing the restrictions limiting timber sales to nearby mills, opponents argued, would be disastrous for communities in Willamette Valley market areas.[16]

In the Smith and Siuslaw river drainages southwest of Eugene, a proposed salvage operation further strained the marketing-area policy for O & C lands. To facilitate the sale of an estimated 500 million board feet of salvage timber, felled by a fierce windstorm in 1951, BLM officials asked the Secretary of the Interior to waive its marketing restrictions for sales to local mills. The downed trees were infested with insects and threatened the area "with the hazard of additional epidemics of insect attacks and fire." Agency officials argued that Eugene-Springfield mills should be allowed to compete for the sales, because the geographical center of the timber was approximately equidistant between the coastal mills and those in the upper Willamette Valley. Although the downed timber was within the Gardner-Reedsport-Coos Bay marketing circle, the area's proximity to the Willamette Valley, the availability of good access roads, and the deteriorating condition of the timber called for a stepped-up harvesting schedule. There was an additional caveat: Because the rough terrain made it impossible to salvage only downed timber, BLM officials insisted that "intermingled green timber must also be removed." Adding standing trees to the sales would create rational logging operations and provide a sufficient volume of timber to make contracts profitable.[17] With the growing influence of environmental organizations in shaping harvest policy by the late 1980s, including green timber with salvage sales would prove increasingly controversial.

Interior Secretary Douglas McKay removed salvage-timber harvests from marketing-area restrictions before he left office in late 1955 to run for the U.S. Senate against Wayne Morse. The decision brought a deluge of letters from both opponents and proponents of the marketing-area requirement and, according to historian Elmo Richardson, likely contributed to his defeat in the 1956 election. Shortly after he succeeded McKay, Fred Seaton commissioned Paul Graves, a Syracuse University forestry professor, to review the effectiveness of the marketing-area concept. Graves reported that BLM's marketing-area restrictions were obsolete, because O & C timberlands were too small to provide adequate timber to stabilize communities. With Seaton's support, BLM director Edward Woozley removed the marketing-area requirement for all O & C sales in April 1957. The Eugene *Register-Guard* praised the decision for rescinding an anachronistic feature of the O & C law, the naïve belief that public timber sales could be used to stabilize communities.[18]

Public Law 273, the much acclaimed Sustained-Yield Forest Management Act of 1944, has an even stranger story than the O & C legislation. The Forest Service and the Simpson Lumber Company of Shelton, Washington, in 1946 signed the only public-private cooperative agreement under the law—to cooperatively manage more than 200,000 acres of timberland for the next century. In the end, a burgeoning lumber market rearranged the economic conditions that originally propelled both the O & C Sustained-

Yield Law and the Sustained-Yield Forest Management Act. As historian Paul Hirt indicates, the five sustained-yield units on national forest land and the one cooperative unit "were mild to dismal failures." Market conditions also derailed BLM efforts to forge public-private cooperative agreements in western Oregon. Rather than bringing economic stability to dependent communities, Hirt argues, the legislation "contributed to an increasing vulnerability of the national forests to the fluctuating demands of the private market."[19]

Montana's well-known progressive journalist Joseph Kinsey Howard outlined the most significant ideological problems with sustained-yield proposals in the spring 1950 issue of *The Pacific Spectator*. The essence of the controversy, he wrote, was centered in a good-faith effort at community planning and assuring timber supplies for future generations. The problem, he contended, was that "almost alone among civilized nations," the United States had permitted timber owners to harvest their forests at will.[20] The whole idea of sustained yield was to protect designated mills from destructive competition. "And there's the rub," Howard remarked. "Sustained-yield management is *impossible* [emphasis mine] if competition is unrestrained." The newspaperman argued: "For better or worse it's a planned economy," an effort to restrain free enterprise and protect the interests of "the greatest number in the long run." Howard closed his essay with an admonition from Gifford Pinchot: "Times change and the public needs to change with them."[21]

Joe Howard was right in at least one respect: times had changed. An ardent New Dealer and close friend of writer Bernard DeVoto, Howard had spent his career fighting railroad interests and the Anaconda Copper Mining Company and its control of Montana's economy, politics, and newspapers. Howard, whom novelist A. B. "Bud" Guthrie, Jr., referred to as "Montana's conscience," was arguing for planning in an age of anti-planning, conservation and prudent stewardship of resources in an era of unrestricted development. Cold War rhetoric and appeals to national defense tainted ideas about planning, suggesting that they were socialist or worse. In the future, market forces would play a much greater role in determining what happened on public and private forests in the Pacific Northwest. Despite claims to the contrary, the Forest Service, like the Bureau of Land Management, was moving towards increasing the allowable cut on the national forests. At the same time, industry leaders continued to request more access roads, better funding for firefighting and insect research, and increased harvests of dead and dying timber.

Central Oregon's Deschutes-Klamath district, with perhaps the most extensive ponderosa pine stands in the world, presented an impressive sight to travelers before railroads reached Klamath Falls and Bend in 1909 and 1911 respectively. Urling Coe, an early Bend physician, remembered "an open park-like forest, without any underbrush," where lightning-caused fires periodically burned "the dead pine needles, cones and twigs that had been blown to the ground by the wind." The result was a forest floor clear of debris and destructive pine beetles. Coe noted that over centuries of time, annual fires "had produced the beautiful open forests free from dangerous underbrush, and killed so many pine beetles that they were held in check."[22] Fire shaped the pine forests that Coe so admired, and, according to historian-ecologist Nancy Langston, "without fires those forests changed to something utterly different."[23] When two large mills—Brooks-Scanlon and Shevlin-Hixon—began cutting into their extensive timber holdings in 1916, the elimination of fire and rapid timber harvests began to create a very different forested landscape.

In the midst of the Great Depression, Forest Service chief Ferdinand Silcox worried that few lumber companies had "any real intention of practicing sustained yield." He cited the situation in Bend, where Brooks-Scanlon and Shevlin-Hixon—with a huge milling capacity—had "no plan for sustained yield or for the permanence of the community." Moreover, some of the land the companies had "cut over in the past 18 years has been devastated." At the present harvesting rate, Silcox thought the companies might have a fourteen-year supply of private timber remaining. In response to a request from the companies to increase Forest Service sales, Silcox remarked in a 1934 letter that national-forest timber should not be "used to support any such program of ultimate community disintegration." Ten years later, Forest Service chief Lyle Watts warned again about heavy logging on the Deschutes plateau, where cutting was taking place at three times the sustained-yield capacity. If operations continued at the present pace in the ponderosa district, he predicted, retrenchment in operations would be required in both the Deschutes and Klamath Counties.[24]

In truth, those counties still held impressive ponderosa timber stands at the close of the Second World War. Forest Service surveys from the mid-1930s reported "extensive, accessible and valuable" ponderosa pine in Deschutes County, a large block of about 140,000 acres extending "north and west of Bend on the lower slopes of the Cascade Range"—nearly half of it privately owned. The surveys indicated that ponderosa pine made up nearly all of the lumber shipped out of the region; and for the ten-year period ending in 1935, Deschutes County mills produced 25 percent of central and eastern Oregon's pine lumber. The Deschutes report also underscored the relationship between the county's forest abundance and its dependent communities: "The history of the economic development of Deschutes County is largely the history of its lumber industry." The survey concluded that when the private timber was gone, the allowable harvests from the national forests "will mean a drastic reduction of the average annual lumber production of the county."[25]

Similar conditions prevailed in Klamath County, where the Forest Service survey described ponderosa pine as the most widespread and economically important species. Occurring in "pure or nearly pure stands" from the lower slopes of the Cascades to the sage desert to the east, ponderosa pine covered about four-fifths of the county's forest land. Because the topography was favorable to railroad and road construction, the Forest Service observed, pine harvests were "progressing very rapidly." The Klamath survey indicated that loggers were taking only ponderosa pine, and on private lands only "unmerchantable trees" were left. With the big Weyerhaeuser plant producing an immense volume of pine timber, Klamath Falls mills turned out nearly half the total output east of the Cascade Mountains. From the time the Union Pacific Railroad extended its line from Ontario to Burns in 1925, loggers had also been cutting their way through nearly "pure" ponderosa stands in the southern Blue Mountains rimming the northern Harney Basin. The Edward Hines Lumber Company followed soon after, building a state-of-the-art mill in Burns and extending a railroad fifty miles north to Seneca. The Forest Service survey reported that Harney County mills were also cutting almost exclusively in ponderosa pine.[26]

Cultural practices—keeping fire out of the woods and cutting only large diameter ponderosa pine ("highgrading")—had consequences for the forest environment. Eliminating all fires from the woods became an article of faith shortly after the turn of the century when lumbermen from the upper Midwest began purchasing huge blocks of timber across central and eastern Oregon.[27] As technology and fire-protection strategies advanced during the coming years, timberland owners pressured state and federal governments to subsidize their fire-protection efforts. Both the Forest Service and the Oregon Department of Forestry would eventually build substantial bureaucracies to satisfy those demands.[28] In the Bend area, Shevlin-Hixon and Brooks-Scanlon, the Oregon Department of Forestry, and the Forest Service eventually became full partners in the Smokey Bear and Keep Oregon Green public relations campaigns. The Bend *Bulletin* joined the effort, repeatedly praising the success of Keep Oregon Green as a "novel and revolutionary" approach to fire prevention.[29]

Since "Smokey Bear and his aides have taken up the guardianship of the woods," the fire prevention effort in central Oregon had been a striking success, the *Bulletin* declared. The annual toll from "the red devil, fire," had mounted each year until the mid-1940s, when "losses from fires . . . dropped to the near minimum, despite dry years and increased use of the woods for recreation purposes." Acknowledging the newspaper's long record of service in forest-fire prevention, the Forest Service conferred on the *Bulletin* its annual Smokey Bear Award in 1960. In bestowing the honor, the Deschutes National Forest supervisor praised the newspaper for its editorials and news stories "in presenting the fire prevention message to the public."[30] The close filial relationship between the big mills, the Forest Service, and the *Bulletin* should come as no surprise. Bend was heavily dependent on its forest bounty, and as one of my former students, Barbara Arnold, has written, "both the timber industry and fire prevention fit largely into the culture and economy of the area." Keeping the forest from going up in flames fostered a sense of community cooperation, provided additional employment during the fire season, and brought revenue into the local treasury.[31] To understand the common reference points binding the larger community is to recognize civic bonds that crossed class lines and linked the population with its forest bounty.

As it did elsewhere in Oregon, timber production soared in the greater Bend area after the Second World War. Brooks-Scanlon shifted much of its cutting activity to its properties west and north of Bend to capitalize on its purchase in 1943 of 30,000 acres of prime timberland near Sisters. The company also purchased 10 million board feet of national-forest timber in a burned-over area at Minto Pass west of Sisters in early 1946. And then as the spring snows were melting from the eastern slopes of the Cascades, the company extended its logging railroad across the old McKenzie highway west of Sisters to facilitate the movement of logs to Bend. In May Brooks-Scanlon workers began to move equipment into its timberlands west of the Santiam highway near Black Butte. The *Bulletin* reported that the first train to arrive in Sisters presented an impressive sight "as the engine, jammer, four caterpillar tractors, two 'grease shacks' on flat cars and some 25 flats crossed over the McKenzie highway on the big red grade separation trestle." In April the company also moved its big portable logging camp to a new location west of Sisters. Although the mobile railroad town had moved nearly a dozen times since 1916, this was the first time it was located near an established community.[32]

The Shevlin-Hixon Company was also on the move, relocating workers and families and 350 assorted buildings in the spring of 1947 to a new location ten miles southeast of Chemult. From a site southeast of LaPine, workers skidded the portable buildings to the company's railroad tracks and hoisted them aboard flatcars for shipment to the new location. "Preparation of the townsite," the *Bulletin* reported, "is now underway with streets being graded and a water system being installed." Crews were also at work building new logging railroads to gain access to the surrounding timber.[33] Those industrial advances, ever more distant from Bend and deeper into the magnificent stands of ponderosa pine, suggest the two big mills' insatiable appetite for pine logs.

Lumber production soared, central Oregon's economy boomed, and the region saw an inrush of people—many of them to Jefferson County's North Unit irrigation project. (Agricultural production alone had more than dou-

bled Jefferson County's population during the 1940s.) In the early stages
of recreating itself as a tourist center and with its thriving riverfront industrial area, postwar Bend (in Deschutes County) looked very different to
Clarence Goodwillie when he returned to the town in August 1948 after a
forty-year absence. The visitor from New York fished in newly completed
Wickiup Reservoir, drove the paved road to the top of Pilot Butte, dined
at Bend's Pilot Butte Inn, and walked along scenic Mirror Pond. Goodwillie
would have noticed, too, signs of construction in the greater Bend area and
in nearby Jefferson and Crook Counties, both in the midst of spectacular
growth during the 1940s. To the northeast, Crook County's sawmilling activity, centered in fast-growing Prineville, drew upon private and public timber in the nearby Ochoco Mountains and helped boost the city's population
by 62 percent during the 1940s. By comparison, Deschutes County's modest 17 percent growth rate reflected Bend's more mature presence as a lum-
ber-manufacturing center and the services it provided to central Oregon.[34]

Central Oregon's hot wood-products market was also showing signs of
moving beyond ponderosa pine to commercially less valuable lodgepole
pine. A Portland-area buyer appeared in central Oregon in late 1945 seeking 100 carloads of "jack" pine poles for the Rural Electrification Administration (REA) to build power lines through the rural West. The REA
approved lodgepole, when soaked in a solution of creosote or pentchlor-
ophenol, for use between 100 and 120 degrees north latitude. "The shortage of cedar and the vast supply of this type of pine,"the *Bulletin* reported,
explained the emerging lodgepole market. New Zealand's director of
forestry, A. R. Entrican, who visited central Oregon in the late 1940s, told
the *Bulletin* that lumbermen should "turn their attention to the lowly lodgepole pine" because it would become a valuable commodity. Using lodgepole and other species, the New Zealand forester suggested, would slow "the
depletion of the present commercial woods."[35] Just as markets and cultural
beliefs were remaking the ecology of ponderosa districts, the same forces
were now reaching into more distant places dominated by lodgepole pine,
a species that thrived in disturbance regimes.

Oregon was easily the most timber-dependent state in the nation after the
Second World War. In 1947 the state had more than 2,000 large and small

lumbering and logging operations in the state with a combined payroll exceeding all other employment. Although production west of the Cascades dwarfed central Oregon's output, the greater Bend area was reaching record levels just three summer seasons removed from the end of the war. Operations were "moving into high gear," the *Bulletin* enthused in August 1948, and building materials were in heavy demand. Although loggers expected to work into the winter months, Brooks-Scanlon and Shevlin-Hixon were already filling their Deschutes log ponds with an abundance of logs in preparation for the snowy season. The *Bulletin* proudly announced that, except for the annual two-week holiday in early July, the big mills had operated without interruption during the previous year. Brooks-Scanlon loggers in the Sisters area were operating full bore, the newspaper declared, with the big virgin timbers headed by rail for the Bend plant.[36] As the 1940s drew to a close, however, rumors began to circulate that Shevlin-Hixon would close, because central Oregon's timber supply could no longer support the combined cutting capacity of the big mills.

"SHEVLIN-HIXON SELLS TO BROOKS-SCANLON," read the Bend *Bulletin* headline on November 21, 1950. The announcement detailing the sale of the firm's timberlands, logging equipment, and mills also reported that Shevlin-Hixon loggers were expected to cut trees until mid-December. When workers finished cutting the log inventory, saws would cease running and the mill would close for good. Seeking to place a positive spin on the business agreement, the companies issued a joint statement. After professing concern for the community's economic welfare, the two firms blamed an overactive lumber market for central Oregon's diminishing ponderosa pine. The heavy demand for lumber during the war and the need to relieve the critical housing shortage afterwards "forced us to deplete our timber reserves at a rate far in excess of our own desires." Calling the situation "unfortunate," the statement claimed that the community would be better served through consolidating operations.[37]

The Shevlin-Hixon sale had far-reaching implications for central Oregon. The company opened its first logging operation in 1916 on the west side of the Deschutes River about six miles south of Bend. In the intervening years, the firm steadily extended its logging activities to the south, eventually reaching through parts of Deschutes, Lake, and Klamath Counties. For thirty-four years, Shevlin-Hixon followed the timeworn practice in the ponderosa region of cutting marketable old-growth timber and leaving snags and smaller trees to serve as reseeding stock. When the company felled its last tree in the snowy vicinity of Chemult in December 1950, loggers were working about seventy-five miles from Bend.

In addition to the cutover landscape, there were immediate social costs associated with Shevlin-Hixon's sale to Brooks-Scanlon: the loss of 850 jobs, 225 in the woods operation near Chemult and the remaining 625 in the vicinity of Bend. At the time of the sale, the Shevlin-Hixon plant was operating two shifts using four electrically powered band mills. When the last log moved through the big Deschutes mill just before Christmas, an overly sentimental and mawkish *Bulletin* writer reported that Jack Mahoney, who had cut the mill's first log in March 1916, "slabbed the final log and joined with other old timers in shedding tears as the final boards fell from the saw."[38]

It all had to come to an end, of course. Although the story is a familiar one, the brevity of the Shevlin-Hixon operation is still striking. "Since the end was inevitable," the *Bulletin* declared rather sanctimoniously, it was better "for it to come as it has rather than to have the operation linger along for a few uncertain years." The newspaper then catalogued the company's contributions to the community in payrolls and purchases of supplies, and as a good citizen in supporting local causes. Shevlin-Hixon's passing, the *Bulletin* declared, would mean a period of readjustment, because it would be unrealistic to expect all former workers to find new employment immediately. The moment was ripe, the paper suggested, for Bend to "take leave of a past in which The Shevlin-Hixon Company had an important part and look to other fruitful years ahead."[39]

Shevlin-Hixon's withdrawal from central Oregon did not slow the harvest of old-growth ponderosa pine. Within six months after the company closed its logging operation near Chemult, Brooks-Scanlon divided its crews in the area beyond Sisters and moved half the workers to the old Shevlin site to continue cutting the remaining stands of pine. The *Bulletin* announced the arrival of the first log train at the Brooks-Scanlon mill pond in August 1951, a seventy-car load "dispatched to Bend over the Great Northern line." The movement of the big ponderosa timbers was a considerable operation. Diesel-powered caterpillar tractors snaked the logs from where they were felled to a loading area, sometimes up to a half-mile distant. The timbers were then hoisted onto trucks and hauled another three miles to the company's railroad loading deck. At that point, steam-operated machines placed the logs on flat cars; "as many as forty railroad cars are loaded and ready to go in a single day." Readers should not worry about the woods being laid waste, the *Bulletin* observed, because Brooks-Scanlon was practicing selective cutting, leaving "a large percentage of the mature trees for reseeding." Hans Milius, the company's forester, told the *Bulletin* reporter that operations had to be spread over larger areas to properly conserve standing timber. Despite the efficiency of modern technology, Milius assured the community, large companies with tree farms were practicing good conservation.[40] The future would show that the Brooks-Scanlon forester was putting an overly positive spin on harvest rates that were clearly unsustainable.

Brooks-Scanlon's sprawling mill complex along the Deschutes River continued to turn out a huge volume of pine lumber. During the summer of 1950, company officials reported that loggers were cutting an average of 350,000 board feet a day in its two major woods operations. The old Shevlin enterprise south of Bend was shipping about 200,000 board feet of timber to the mills every day. At the manufacturing end of this far-flung production system, two eight-hour shifts turned the pine logs into finished lumber. And—perhaps indicating that the huge ponderosa harvests were taking a toll—Brooks-Scanlon began cutting commercially less valuable lodgepole pine in the vicinity of Camp Abbot to augment its log supply. The Bend *Bulletin* also reported increased operating expenses because loggers were "forced to go farther and farther away from their mills to obtain timber." The paper told readers that despite the heavy cut, "timber is assured for the future because of federal sales."[41]

In truth, the heavy take of ponderosa pine from private timberlands was exacting its costs. Historically, Brooks-Scanlon and Shevlin-Hixon had first cut larger trees on lower slopes, with the earliest cutover areas producing a huge volume of exceptional lumber. Geographer Sheldon Erickson's 1953 study of the upper Deschutes Basin predicted that, except for timber exchanged in agreements with the Forest Service, most of the private lands would be exhausted in the near future. Mills would be able to continue operating at their present capacity only by relying on federal-forest and Indian-reservation timber. Because most of the remaining commercial timber was on federal land, Erickson believed the time was ripe to implement sustained-yield forestry to place the industry on a more permanent basis. The change would require building more access roads into higher reaches of the national forests.[42]

In response to the increased demand for harvestable timber, the Deschutes National Forest began authorizing the construction of permanent access roads into its extensive back country in 1948. The road-building costs were subtracted from the sales price of the successful bidders on timber tracts. This rapidly expanding network of new roads would, according to the *Bulletin,* improve fire protection and provide recreational entries "into lake country, mountain streams, alpine peaks and camp grounds." Growing automobile use also spelled the end to the wanderlust town of Shevlin in the late summer of 1952, when the portable buildings were moved for the last time to a permanent location south of Bend on the east side of The Dalles-California Highway. Brooks-Scanlon gave the houses "a complete face-lifting" and wired them for electricity. Four years later, the logging-railroad era itself closed when the company train delivered twenty carloads of ponderosa logs from the Sisters area to the mill pond in Bend. The steel rails were then sold on the international market to India and fleets of logging trucks replaced the daily trains.[43]

The modern era of logging and milling had come to central Oregon. While the elimination of fire and harvesting only the most marketable ponderosa pine brought major changes to central Oregon's forests, the technology and strategies used to harvest timber had also changed. Logging with ponderous Lidgerwood skidding machines operating from railroad tracks became more efficient in the 1920s when crews began using horses and "big wheels" to haul timber to rail sidings. In the 1930s gasoline-powered tractors replaced horses and the modern logging arch replaced the cumbersome big wheels. By the Great Depression logging camps housed fewer horses but required additional buildings to shelter Caterpillar tractors, trucks, and electric generators. Finally, in the most spectacular technological shift of all, gasoline-driven power saws replaced the timeworn practice of felling and bucking trees by hand. The new technologies were faster and highly productive—and more environmentally intrusive than the use of hand, animal, and steam power. By the early 1950s, logging operations required one-third the number of workers to produce a similar volume of timber.[44]

If technological advances made woods operations more productive, the same could be said for the Brooks-Scanlon manufacturing plants along the Deschutes River. The company periodically modernized its facilities, including an extensive overhaul of its sawmill in 1955—the first major change since its installation in 1923. A new steel framework of I-beams and girders was put in place to support heavier saw carriages and conveyors, the first step in renovating the plant to handle both large and small logs and to increase the mill's productive capacity. Brooks-Scanlon upgraded its operations again in 1957, adding a new green chain (for cutting fresh logs), and planing mill and five dry kilns. The company was also putting in new logging roads, especially in the area beyond Sisters, to provide access to more timber. And then, anticipating Oregon's ban on wigwam burners (for burning log slabs and bark), Brooks-Scanlon installed a chipping machine and a half-million-dollar "barker" designed to strip logs of bark. The *Bulletin's* Phil Brogan praised the new equipment for making use of otherwise wasted material. Five railcars of chips a day were being shipped to pulp mills in Longview, Washington, by the late 1950s, and the debris from the barker was "hogged" and then burned to generate electricity for the big mill. The smokeless burner meant an end to soot and cinder blowing across the city.[45] Although Bend was still a sawmill town at the onset of the 1960s, there were signs of change that would eventually transform the community from a predominantly working-class sawmill town to a very different kind of economy and cultural milieu.

Writing in 1949 in the midst of the great production boom, Michael Bigley, a Eugene consulting forester, reflected for readers of *The Timberman* that little had changed in the lumber industry's long migratory movement across the continent. "With the exhaustion of the open market timber supply in Washington and northern Oregon," he remarked, "the past several years have witnessed a shift of mills to southern Oregon and northern California with a consequent disturbance of local communities." Bigley doubted that anything could be done about the industry's migratory habit: "it is the tailend of the movement which started in the Northeast many years ago, moved to the Lake States, the South, and then shifted to the West." He was more optimistic, however, about the Pacific Northwest's prospects, because foresters "had lessened the . . . impact of this cut and get out philosophy."[46] Bigley's optimism about the influence of foresters was misplaced, because markets, not stewardship, dictated harvesting practices on both public and private timberlands.

The postwar years also witnessed a mad scramble for access to more timber and the emergence of a new corporate giant in the Douglas-fir region, the Georgia-Pacific Company. Beginning from a small financial base in Georgia, the company developed into an industrial giant with its largest holdings on the northern Pacific Coast. Between 1956 and 1959 the firm purchased the Coos Bay Lumber Company in southwestern Oregon and 120,000 acres of timberland; the Hammond Lumber Company and 127,000 acres on California's redwood coast; and elsewhere in Oregon the lush 200,000 acres, manufacturing plants, and logging railroads of the Booth-Kelly Lumber Company in Springfield and Toledo's C. D. Johnson Lumber Company and 120,000 acres.[47] Georgia-Pacific's huge acquisitions of standing timber, its sharply increased rate of cutting, and its sales to create a "cash flow" to help pay for its purchases, revolutionized the West Coast lumber industry.

From northern California's Hammond timberlands north through Coos Bay to the Booth-Kelly and C. D. Johnson forests, Georgia-Pacific's old-growth sales sustained for a decade mills that otherwise had limited access to timber. Like most private timber harvests at the time, Georgia-Pacific's were determined by a strong lumber market, good prices, and easy access to ocean transportation. The Georgia-Pacific buyout was a momentous development for the southern Oregon coast where the Coos Bay Lumber Company's large old-growth forest was the primary attraction of the purchase. Estimated at 6 billion board feet, a "timber-rich jewel," according to the *Oregonian,* the forest represented "one of the largest private holdings in Oregon." To create a cash flow to finance the purchase, Georgia-Pacific sold large stands of old-growth Douglas fir to the Coos Head Lumber Company, the Moore Mill and Lumber Company, the U.S. Plywood Corporation, and smaller mills in the Coquille Valley. Those combined sales contributed to what one local forester called the "rapid liquidation phase" of the old Coos Bay Lumber Company lands. Further north, on the former C. D. Johnson holdings, Georgia-Pacific sold large old-growth timber stands along the steep slopes of the lower Siletz River gorge to Willamette Industries, again to help defray the costs of the original purchase.[48]

After moving its corporate headquarters to Portland in 1954, Georgia-Pacific continued to sell timber to southwestern Oregon buyers; it also increased its own timber cutting to supply its expanding manufacturing facilities on Coos Bay. According to one employee, the company's sales "boomed the work in Coos County," and when the firm "immediately jumped" its own cutting quota, it had to hire additional loggers.[49] The rapid liquidation of Georgia-Pacific's timberlands eventually would bring social and economic disaster to communities along the Coquille River and around Coos Bay. The company's extensive road system and tardy reforestation program also increased the potential for landslides and the siltation of streams.

To the *Coos Bay Times,* however, Georgia-Pacific's corporate presence meant bright prospects for the future. When the company announced plans to build a new plywood plant on the bay, the newspaper's prediction seemed to be confirmed. The 260-person workforce in the new facility would provide a large number of jobs and allow the company to use its large timber holdings better. The move would further stabilize Georgia-Pacific's operation, according to the *Times,* and "go far to guarantee a longer span of job creating in the area." The newspaper's executive editor, Forest W. Amsden, thought the company's announcement "served notice that it is here to stay—an active, wealth-producing, valued member of the Southwestern Oregon community." Georgia-Pacific followed with an advertisement in the *Times* assuring readers that its "forestry program" would ensure "that we will grow a volume of timber at least equal to the volume harvested."[50]

Georgia-Pacific's market-driven timber-management practices were not the only large corporate interests affecting the Coos Bay area. As its large timber holdings on the southern Oregon coast reached marketable age, the Weyerhaeuser Company built an immense sawmill facility in the town of North Bend along the Coos Bay waterfront. A lively California lumber market and, as the *Coos Bay Times* put it, "an ocean highway . . . a rifle-shot away" moved the company to action. Weyerhaeuser began building maintenance shops and an extended road network into the Coos back country even before the mill began sawing logs in 1951. As early as the summer of 1950 loggers were at work falling timber that would be trucked to water and then rafted across the bay to the big mill. At the same time, company foresters were carrying out extensive surveys of Weyerhaeuser's timber holdings to determine which sections to harvest first. The firm began dredging operations to create space for a log dump on the east side of the bay, an activity that contributed to the further industrialization of the Coos waterway. To facilitate shipping, the Corps of Engineers began dredging the estuary's channel and the Coos Bay bar to a depth of forty feet to provide passage for fully loaded ocean-going ships.[51] In the midst of those frenzied developments, Weyerhaeuser's public relations releases continued to praise the permanent nature of the Coos Bay operation and the benefits it would bring to the community.

With the postwar building boom in full form, Stuart Moir, an executive and counsel to several Northwest lumber trade associations, offered a striking view of the industry's perspective. Writing for the *Oregon Business Review* in 1957, he pointed out that the region's forest-resources economy was firmly linked to world affairs. The demand for improved living standards around the globe, especially the need for better housing in the underdeveloped world, would directly affect the Pacific Northwest because it was "a timber-wealthy region." Oregon and Washington, with 39 percent of the standing timber in the United States, produced more than 30 percent of the nation's lumber supply. But Moir also cautioned that sawlog sizes were decreasing, harvests were far in excess of annual growth, and transportation costs were increasing because logging was taking place farther from centers of manufacture.[52]

The best hope for the Pacific Northwest, Moir reasoned, was in "becoming the nation's woodlot." The region's soil conditions, climate, and rapidly growing trees "make the Pacific forests the most magnificent in the temperate zone, if not in the world." Moir then outlined the precepts for industrial forestry: a felled tree was not a tragic loss, but "an opportunity for another tree to grow"; trees had to be thinned and harvested to realize the greatest good from forests; and timber needed protection from fire, insects, and disease. The focus of Moir's argument was the need to increase timber harvests on publicly owned land, "which must bear the burden of bridging a national shortage until an adequate new crop is ripe." With tree farmers providing a helping hand, nature would be "able and willing to produce this new crop" to assure timber supplies for future generations. Moir cautioned readers about his one great fear: "Our economic thinking is too frequently clouded by sentiment." In order to make Northwest forests secure, Moir urged others to join with the forest industries to keep the region's timberlands producing "forest crops on a sustained-yield basis."[53]

As the 1950s advanced, industry officials and many public-sector foresters began promoting the idea of "conversion," turning "decadent" old-growth forests into vigorous, fast-growing new stands with projected harvesting cycles of between 80 and 100 years. Although the idea originated earlier in the century, the virtue of converting old-growth forests to fast-growing new stands became an article of faith—especially in Oregon, with its still sizable stands of virgin timber. There was a neat symmetry to the argument posited in an industrialized science in which forests were turned into production-driven "factory regimes" or plantation models. As historian Richard Rajala points out in his study of Northwest forestry policy (including British Columbia), forest research informed public policy only if corporate and government officials approved. The industry's emphasis on profits, Rajala writes, precluded any "meaningful integration of science and forest law." In order to achieve legal status, science needed the sanction of dominant interests in the political economy.[54]

Oregon's booming lumber industry functioned in a favorable political and economic environment for most of the 1950s and 1960s, exercising exclusive control over the state's forest policy regulations and dominating the Oregon State Board of Forestry. The threat of federal regulation, which had surfaced during the 1930s, was reduced to the whimpers of a few disgruntled foresters and outdoor enthusiasts. The state's heralded Forest Conservation Act of 1941, the first of its kind in the United States, served primarily to buy off the threat of federal regulation rather than to function as a tool to conserve natural resources. The primary weakness of the act, Rajala argues, rested in "trade association control over its silvicultural content." Forest Service officials involved in drafting the legislation thought the bill would do little more than legalize clearcutting, practices already widespread in logging operations.[55]

In a major address to the Society of American Foresters in Portland in 1955, Forest Service chief Richard McArdle expressed concern that America might not be able to meet its lumber needs by the year 2000. Referring to the much-anticipated Timber Resources Review (TRR), the chief noted that forest growth in the United States was now close to the harvest rate, a worrisome development given the projected sharp increase in wood-products consumption by century's end. To add to those difficulties, timber quality was declining primarily because of dwindling stands of old-growth timber and a diminishing volume of "preferable" species. Although technological advances had offset some of the shortcomings in quality, the problem looming in the near future was reforestation. Growth figures for the Pacific Northwest, McArdle argued, were lagging because of the huge volume of mature timber. As old-growth harvests increased and second-growth trees reached measurable size, the region's total volume of wood fiber should increase.[56]

In another important talk at the Society of American Foresters' gathering, Crown Zellerbach vice president Edward Stamm charged that lack of access roads was threatening "maximum sustained-yield output in the Douglas fir region." More than one billion board feet of mature timber was lost each year because of the government's paltry appropriations for building access roads. "Old growth is not money in the bank," the Crown Zellerbach executive observed, but rather was deteriorating, rotting, wasting away. Young growth would eventually take over, because "conversion of old growth is inevitable." Stamm thought there was sufficient virgin timber remaining to keep three generations of foresters busy "in converting it to new forests."[57] Indeed, building more miles of access roads in the national forests—an industry objective since the end of the Second World War—gathered momentum as the building boom proceeded apace. The PRDC, heavily sprinkled with corporate representatives, also urged more access roads in its monthly reports to the governor.[58]

The Portland business community as well joined in seeking more funds to build access roads. With the assistance of twenty business executives, Lewis and Clark and Reed Colleges released a wide-ranging study of Oregon's forest problems in 1956. The report emphasized the importance of federal timber sales to the state's economic health—20 percent of Oregon's labor force was employed directly in the wood-products industry—and recommended "a virtual revolution" in the management and sale of government timber. Because the national forests held the greatest volume of standing timber, mature trees should be cut as rapidly as sustained-yield limitations would permit. The report called for an additional 14,478 miles of access roads so that the state could realize the full allowable cut. Those national forests requiring the greatest increase in roads included Mt. Hood (2,400 miles), Siskiyou (2,462), Umpqua (2,442), and Willamette (3,800). The study also recommended that harvest rotations be shortened to further increase the allowable cut.[59]

In an article published in the *Oregon Business Review* in 1954, Region 6 assistant forester Walter Lund predicted that the state's lumber industry would enjoy a bright future because it enjoyed "a tremendous resource in its forest lands." Writing for the same journal three years later, William Hagenstein, executive vice president of the Industrial Forestry Association (IFA), was even more boastful about the industry's prowess. Citing near-record timber harvests and tree planting, an "outstanding" record in protecting against fire, and sharp reductions in insect infestations, Hagenstein called 1956 the "busiest forestry year" in the industry's history. He praised the "tremendous progress in private forestry" and applauded forest landowners for their contributions to the state's economy. Because the forest industry was critical to Oregon's well-being, he argued, the public had a responsibility to insist that government forests increase their annual cut. Hagenstein also warned about political forces seeking to set aside "vast areas of public forests to single-purpose use . . . as 'wilderness areas.'" If adopted, such programs threatened to "destroy a portion of our economy."[60]

Public and private foresters remained optimistic about Oregon's timber-based economy through the 1950s, confident that better reforestation practices, more efficient use of wastes, and reduced harvest cycles would provide ample wood fiber for the future. George Schroeder, a Crown Zellerbach forester, wrote in the *Oregon Business Review* that the state's present volume of standing timber could sustain annual harvests of 8 to 10 billion board feet for fifty years. By restocking old-growth stands with fast-growing young trees and practicing intensive forest management, he believed, the annual cut could be increased to 14 billion board feet. "Intensive management," however, presupposed controlling forest fires, pests, and disease; salvaging dying timber; and building an extensive network of access roads. Schroeder praised the tree-farm movement, the industry's initiative that had "revolutionized thinking in forestry." With "prevention aids" such as the Smokey Bear and Keep Green programs, cooperative efforts in insect control, and measures to reduce losses from diseases, Oregon had made great progress in forest conservation.[61]

Widespread public concern about wilderness areas and the ecological and environmental implications of massive timber harvests, however, still lay somewhere in the future. As the 1950s drew to a close, the lumber industry, the Forest Service, leading political figures, and the most widely read newspapers believed that all was well with the region's forests. The most important public issues were reforestation, protecting against fire and insects, making greater use of wood wastes, assuring adequate timber supplies, and providing economic stability. As the nation's leader in wood products, the *Oregonian* boasted, the state was "husbanding its timber resource as a permanent asset." A Stanford Research Institute study indicated that Oregon's timber resource would continue to be the strength of its industrial growth. Although the state should broaden its economic base, it could benefit from expanding pulp and paper manufacture and making greater use of forest and mill residue.[62]

The Forest Service's extensive Timber Resources Review (TRR), published in 1958, represented a capstone for the postwar era; it looked proudly backward at impressive production records and forward to a future full of promise. In the foreword, McArdle assured the public that although there might be occasional local shortages, no "'timber famine'" loomed on the horizon. It was equally clear, he warned, that increasing demands for lumber products would require "an intensity of forestry practices that will startle many of us." The TRR summary was full of optimism, asserting that the national forestry outlook "could hardly be other than favorable." Indeed, it can be said that a cult of optimism underlay the TRR report: a belief that improved forestry practices and more efficient use would increase the nation's yield of wood fiber.[63]

With the benefit of hindsight, it is easy to read into the TRR report certain management predispositions that would have wide-ranging implications for Northwest forests. The summary includes repeated references to the "necessary intensification of forestry" practices to enable the nation to meet the needs of a growing population. The extent to which forest management must be intensified, it reads, "is much larger and far greater than the general public or most experts" believe. Buried in the 713–page report are details of advanced forestry practices, with most of the discussion centering on reducing timber destruction from "catastrophes by fire, insects, and disease." Curtailing those damages, it concludes, would significantly influence future timber supplies. The battle against insects and disease was already an achievable objective with congressional passage of the Forest Pest Control Act of 1947. Preventing human-caused fires, intensifying fire control, directly attacking forest pests and diseases, and developing new insecticides and methods of using them had the potential to double the timber supply.[64] In effect, the TRR document embodied an ideology of forest management on a grand scale, one that would require an enormous financial and technical investment. "Foresters in this age of technological optimism," historian Paul Hirt observes, "did not shrink from the task."[65]

When the Georgia-Pacific Corporation decided in 1981 to move its headquarters from Portland to Atlanta, Georgia, company chairman and chief executive officer Bob Flowerree had lived in Oregon with his wife and family since the end of the Second World War. Born and raised in New Orleans and educated at Tulane University, Flowerree went to work for the C. D. Johnson Lumber Company in Toledo, Oregon, in 1946. He quickly rose to a management position; when Georgia-Pacific purchased the Johnson firm in 1951, Flowerree moved laterally into executive positions with the rapidly expanding company. Although Georgia-Pacific owned little timberland in the late 1940s when it made its first purchases along the Pacific Coast, by the time Flowerree headed the firm, it owned more than 5 million acres in the region. The corporate decision to move to Atlanta involved strategic considerations. Many of the company's building-products and chemical operations were located in the Southeast; its major markets had shifted to the eastern United States; and the firm's corporate leadership was optimistic about the region's business prospects. A company biography published in 1980 points out that the "New South" appreciated industrialization and was willing to "protect and preserve the competitive enterprise system." And unlike the American West, in the South national forests did not dominate timber ownership.[66]

When *Oregon Business* reporter Robert Hill asked Flowerree for his personal thoughts about moving to Atlanta after thirty years in Oregon, the executive's response was blunt: "You have to go where the job is. It's just as simple as that." He was equally candid about his corporate peers: "The question is, do you put your living before your work? That's what I've told people around here. If they want to play, they'd better stay in Oregon. If they want to work, then they'd better get where the job is." Although he regretted leaving the state, Flowerree told the reporter, "you just have to do what you have to do." The Georgia-Pacific executive also believed that Oregon and the Northwest would no longer dominate the national lumber market. High transportation costs, sharp production increases in the Southeast, and substitute building materials suggested that future markets for Northwest producers would be limited to areas west of the Rocky Mountains. When Hill asked Flowerree about differing patterns of land ownership in the South and the Northwest, he replied: "You have to get back to basic philosophy on that. . . . in a Communist state, like Russia, they own everything and it's the most inefficient operation in the world." In contrast, Flowerree insisted, private timberlands were better managed and far more efficient.[67]

As business head of the nation's leading plywood producer and the industry's largest distributor of building materials, Flowerree directed a business empire that was not only national but international in scope. When it made the move to Atlanta, Georgia-Pacific was Oregon's largest "publically held corporation" and in the midst of a significant expansion of its chemical-manufacturing capacity, including the purchase of oil and natural-gas facilities in Louisiana to feed its large production facilities. Of the company's more than 5 million acres of timberland, 450,000 acres were located in Oregon, much of it the highly productive land purchased in the

1950s and 1960s. Those large "buyouts" placed Georgia-Pacific at the forefront of industrial activity in Oregon for more than thirty years, practicing unprecedented harvest rates and the rapid liquidation of old-growth timber from the company's forest lands.[68] A booming California construction industry and good sales contributed to strong local economies and an influx of people who took jobs in the woods, mills, and service sectors. Through the 1960s and early 1970s, with occasional minor seasonal downturns, times were good for most working-class people.

As log trucks rolled down the slopes of the Coast Range and loggers splashed timbers into nearby waterways, policymakers paid little attention to the effects of road construction and cutting practices on stream quality.[69] The most important thing to understand about forestry, University of Oregon business professor Louis Hamill wrote in 1960, was the process of converting "an unmanaged forest containing much old-growth timber into a managed forest containing only second-growth timber." Because heavy cutting had diminished supplies on private land, the time had come, Hamill insisted, for the Forest Service to change the way it calculated the allowable cut. In that formula, "the thing to change is the rotation age: the length of time which is considered necessary to grow the kinds of trees the Federal agencies want." Hamill offered even more aggressive measures to increase the cut on the national-forest system: put recreation areas into timber production; reduce the rotation age from 100 to 80 years; extend road networks into virgin forests; expand salvage and thinning operations; and undertake more intensive timber inventories. There was "good reason," he believed, for western Oregon's forest products industry "to demand a greater cut from the Federal forests."[70] For nearly thirty years after 1960 the Forest Service repeatedly raised the allowable cut after successive inventories revealed a greater volume of timber than previous calculations indicated. The dramatic production figures from the national forests during those years reflected more a computer generated fantasy than the actual volume of federal timber.[71]

With few restraints to limit harvests, the market was the primary arbiter of timber cutting between 1960 and the mid-1980s. Oregon's production figures led the nation, nearly equalling the production of Washington and

TABLE 5.1

LUMBER PRODUCTION, 1945–1979

(IN MILLION BOARD FEET)[72]

	1945	1950	1955	1960	1965	1970	1975	1979
Oregon	5,003	5,239	9,181	7,401	8,206	6,680	6,342	7,312
Washington	3,257	3,606	3,118	3,377	3,958	3,189	3,104	3,841
California	2,260	4,262	5,319	5,160	5,032	4,979*	4,153*	4,639*

* Includes Nevada

California together—two states with relatively healthy forest products economies. Southwestern Oregon newspapers applauded monthly production records and made frequent references to the permanence of the industry. As table 5.1 indicates, Oregon's production reached a postwar high of more than 9.1 billion board feet in 1955, 8.2 billion board feet in 1965, and 7.3 billion in 1979. California's output, in contrast, reached a high of 5.3 billion board feet in 1955. Washington, where the heaviest cutting had taken place before the Second World War, reached a postwar high of 3.9 billion board feet in 1965. Those striking production figures obscured public costs in degraded landscapes and damages to fish and wildlife populations on both federal and private lands. Conservation interests made some headway during this period in convincing Congress and the Forest Service to expand recreation programs, but only if such activities did not interfere with timber production.

Until Oregon's economy gradually became more diversified during the late 1980s, Oregon's employment indicators reflected the fortunes of the wood products industry. A severe slump in forest-related employment in 1957–1958 contributed to an outmigration of approximately 60,000 people. A technological revolution in the agricultural sector, cutover districts in the lumber industry, and increases in worker productivity in the mills and woods operations prompted an internal migratory movement of considerable magnitude. While the greater Portland area grew by more than 34,000 people in the 1950s, every eastern Oregon county except one (Wasco) lost population. On Oregon's sparsely populated south coast, the center of booming logging and lumbering activity, Coos and Curry Counties gained more than 9,000 immigrants. In the mid-1960s, University of Oregon economist James Tattersall argued that a healthy Oregon economy required a vibrant national construction industry, an observation that explains population growth in the two decades after 1945.[73]

As private timber supplies diminished during the 1960s, the Forest Service bent to those realities and raised the allowable cut on federal lands. Forest researchers and scientists began to frame other technical and scientific solutions to the supply problem: refining clear-cutting as a scientifically sound harvest practice, increasing the use of chemicals as a forest-management tool, and experimenting with genetically selected "super trees." Throughout this period, the logic of the marketplace and a production-oriented science, not the requirements of healthy ecosystems and streams, directed forest practices. Although those policies confronted a rising environmental movement during the 1970s and court challenges invoking the Endangered Species Act in the next decade, the politics and economics of national-forest timber management remained constant. "A commodities management approach to the national forests coupled with a commitment to 'full utilization,'" historian Paul Hirt points out, continued to enjoy ideological hegemony in the forestry profession and the agency until the late 1980s."[74]

6

INTENSIVE FORESTRY

AND CITIZEN ACTIVISM

First they said 300 acres of old growth timber was necessary for each
pair of spotted owls; then they went to 1,000, now they're saying 1,200.
People can propagate on a 120 by 90 foot lot, so do you need 1,200 acres
for a spotted owl?—JOHN BALL, Secretary-Treasurer, International
Woodworkers of America (1982)[1]

Before environmental impact assessments, court directives, and public access to information began to slow federal timber harvests,
Northwest loggers cut a huge swath through the region's national
forests between 1960 and the late 1980s. Bending to pressure from politicians and the forest products industry, the Forest Service developed
intensive management programs to provide a rationale for the high harvest levels on national forests. The heavy cutting of Northwest forests
extended to private lands, where few regulatory measures existed to protect against disturbances to forest and stream environments. Although the
Oregon legislature enacted one of the nation's first forest-practices acts in
1941, the measure did little to alter harvesting practices on private lands or
to protect stream habitat. Drafted by trade association leaders and adopted
by the legislature without change, the Oregon Forest Conservation Act did
little more than to demonstrate the lumber industry's hegemony in state
politics.[2] Except for minor rules changes, the 1941 law was not revised until
1971, when the Oregon legislature passed another industry-sponsored
measure.

The board then circulated a proposed set of rules appropriate to conditions in each of Oregon's three forest regions—northwestern, southwestern, and eastern. The act directed the State Board of Forestry to draft regulations establishing minimum standards for a variety of activities on private and state forest lands. State Forester J. E. Schroeder called the 1971 revision "significant legislation" that would assure future timber supplies and protect "soil, air, and water resources," fish, and wildlife. A casual reading of the proposed rules shows them to be vague, lacking specificity, subject to individual interpretation, and without effective policing and enforcement mechanisms. In building roads, logging operators were directed to "minimize" the amount of material entering streams "whenever it is physically and economically possible." And "wherever possible," loggers should avoid steep canyons and areas prone to landslides. In locating new roads, operators should "minimize . . . stream crossings" and, where it was practical, they should cross streams at right angles. Finally, the rules stated, roads should be "no wider than necessary."[3]

The review draft of the rules drew a storm of critical letters from individuals and environmental groups around the state. Charles Coate, a professor of history at Eastern Oregon College in La Grande, thought the new rules offered some improvement, but he criticized the absence of enforcement provisions and the failure of the guidelines to establish firm "*minimum* standards" for forest practices. He pointed to deficiencies in qualifying phrases such as "minimal," "minimize," and "wherever possible." If there were to be requirements for logging, they should be "standards that are followed without exception." Coate urged the forestry board to rewrite the rules with greater precision. Larry Williams, executive director of the Oregon Environmental Council (OEC), offered a similar assessment, calling the regulations impossible to enforce because of "the ambiguous language and the latitude of judgment left to the timber operator." Williams urged the forestry board to draft meaningful regulations.[4]

In a letter to J. E. Schroeder, Anne Squier, of the Oregon Shores Conservation Coalition, expressed her displeasure "with the disregard by your staff and by the committee as a whole of all comment from individuals or groups outside the industry." Squier, who attended a meeting of the Forest Practices Act Implementation Committee (composed of Board of Forestry members), complained that conservation groups were not given draft documents circulated by Associated Oregon Industries. She reminded Schroeder that the Department of Forestry was a public agency, charged with protecting and conserving forest resources. It was "regrettable but evident," she wrote, that the department was "acting like the arm of the timber industry" rather than an objective regulatory unit of state government. In a follow-up letter, Schroeder said that Squier's letter was disturbing, because the committee's work had extended over several sessions and included input from many groups. It would be better, he wrote, "if you could give credit to how far environmental considerations have been brought" through the new regulatory requirements.[5]

Squier had good reason to question the Implementation Committee's work, because its recorded deliberations confirmed her suspicions and those of others who questioned the new rules. At its first meeting, the committee agreed to preface the new guidelines with the statement: "The following practices shall be complied with where applicable or necessary to accomplish." The committee dropped the phrase "are recommended." Throughout its deliberations, the committee substituted "shall" for "must," "should" and "shall" for "will." In its discussions of a rule protecting stream drainages, members chose "minimize" in place of "prevent." And in the section on logging in steep, narrow canyons, the committee reworded the directive to read: "Where practical alternatives exist, avoid steep, narrow canyons." Members also expressed concern that the Oregon Department of Environmental Quality's stream regulations would be more stringent than the new forestry rules.[6]

Schroeder also engaged in a war of words with Eric Allen, publisher of the *Medford Mail Tribune,* when the latter printed a critical editorial entitled "HOW LONG WILL OUR TIMBER LAST." Allen took the Department of Forestry and its appointed board to task for failing to consider the public's interest in its deliberations. But Allen's larger concern was the absence of leadership, someone who would ensure that the state's economy would "not suffer, unduly and all at once" from overcutting on any class of forest ownership. The cut-out-and-get-out process in the Great Lakes states, he reminded Schroeder, was still fresh in everyone's memory. Oregon's lumber industry was "too myopic and too concerned with today's dollar to take serious cognizance of day-after-tomorrow." Given that absence of foresight, Allen thought the State Board of Forestry should step in. If "its present industry-dominated membership" was unable to provide leadership, the board should be reconstituted with "more environmentally concerned citizens" capable of taking the long view. The board's big problem, Allen wrote, was its fixation on "the exact letter of the laws under which it operates" rather than being in the vanguard of conserving the state's forests.

> I think it is a case, literally, of the Board being too close to the trees to see the forest, and thereby utterly failing to act to protect the long-range interests of the state—its economy, its scenery, its environment, and its livability to which we all give such lip-service.[7]

Since its inception, the state Forestry Department's working principle had been "flexibility" and cooperation with forest owners, premises that it continued to practice after the Second World War. Although the state legislature did not adopt another major revision of the Forest Practices Act until 1987, the Department of Forestry periodically updated its rules, often responding to federal initiatives such as the Clean Water Act. When the agency drafted new reforestation guidelines in 1974, operators were asked to "*minimize* [my emphasis] compaction and movement of topsoil" to prevent it from getting into streams; when forest maintenance crews applied oil to roads, the work was to be carried out "in such a manner as to *minimize* materials from entering streams." In the department's Northwest Region, workers were expected to take "adequate precautions" to prevent leaks and accidental spills. In another rule change involving the protection of streams and lakes in eastern Oregon, logging operators were asked to "*consider* leaving buffer strips."[8] Through the postwar decades, the agency's operative words were always hedged in terms that gave logging operators considerable license in carrying out their activities. While the department insisted that it had moved fully into the environmental age, critics persistently pointed to shortcomings in harvest practices on private and state forest lands.

A growing public condemnation of clear-cutting, an industrial harvest practice especially popular in the Douglas-fir region, began in the early 1970s.

Environmental groups, congressional committees, and a few foresters equated the practice with destruction and ruin and urged that it be stopped. Region 6 forester Charles Connaughton recognized the threat to the industry in a *Journal of Forestry* article published in 1970. Why, he asked, did environmentalists associate clear-cutting with devastation? The likely answer, he reasoned, was that in some instances clear-cuts were ugly, and in any setting their impact was sudden and severe. But as malpractice? "Nothing could be further from the facts." Although the public outcry over clear-cutting had been adverse and negative, Connaughton thought it was being "debated with more heat than light." Clear-cutting was important to foresters as a practical silvicultural tool, and the profession "must take steps to defend it strongly, intelligently, and aggressively." Foresters had their work cut out, the Region 6 chief argued; it was important, therefore, to impress upon the public that clear-cutting was "a sound ecological practice" and that foresters were committed to managing forests for both appearance and production.[9]

Forest-industry groups and their public-relations offices were quick to answer Connaughton's call for an aggressive defense of clear-cutting. "Think of the forest as a crop," one industry pamphlet declared, and consider the harvest practice similar to one of nature's great disturbance events such as fire, landslides, and windstorms. With species such as Douglas fir that require full sunlight for quick growth, clear-cutting was the most efficient way to regenerate forests for future production. It was unfortunate, the pamphlet contended, that a method so "beneficial to man . . . has received the most severe criticism." A Weyerhaeuser Company in-house document circulated to employees in 1971 expressed a similar argument, declaring that the objective of good forest management was the production of wood fiber. To achieve that end, foresters used a variety of silvicultural practices, including one of the most essential, clear-cutting. Its benefits included production and harvest efficiency and cleaner logging, which means "the next crop can be planted or seeded faster and more efficiently—with genetically superior trees."[10] The in-house circular cited the importance of clear-cutting to Weyerhaeuser's High Yield Forestry program. Foresters who were doing conscientious and professional work, the circular declared, were concerned about attacks on the Forest Service and "the misunderstandings on which much of the criticism is based." Weyerhaeuser's ability to clear-cut, "with the understanding and trust of the public," was central to the company's economic productivity. Even-aged management practices were necessary to quickly regenerate cut-over lands, to attain optimum forest growth, and to meet the nation's housing needs.[11]

William D. Hagenstein, head of the influential Industrial Forestry Association and an outspoken proponent of clear-cutting, defended the practice in a widely read trade-journal article in 1970. "Mention clear-cutting today," he wrote, "and emotions run riot. Editorial writers foam. Politicos decry it. Students parade against it. Some foresters wonder about it." But for the Douglas-fir region, Hagenstein contended, clear-cutting was an ideal silvicultural practice. Early foresters quickly learned a basic fact about Douglas fir: the tree's intolerance to shade required clear-cutting to speed the reproduction process. Hagenstein then listed its benefits: clear-cutting replicates fire, enhances multiple use, "transforms a biological desert" into a lush new forest, and helps watersheds by replacing old growth "with higher water yielding young trees." Clear-cutting, Hagenstein concluded, was honest and workmanlike.[12]

The Senate Subcommittee on Public Lands joined the clear-cutting debate in early 1972, issuing a report highly critical of Forest Service policy and citing the need for congressional oversight. Although the committee did not believe Congress should legislate forestry practices, its members agreed that the public had an interest because clear-cutting involved environmental questions. The subcommittee report attributed the national-forest controversy to several developments: the introduction of clear-cutting in eastern hardwood forests, the Forest Service's decision to increase the allowable cut, the public's increasing environmental awareness and its desire to participate in decision-making, and the Forest Service's failure to respond to environmental concerns. The Senate committee cited two major problems with the agency's practices: allowing harvests in areas with fragile to soil or scenic value or that posed problems with regeneration, and permitting harvests that harmed the environment. The committee's report advised the Forest Service to exercise greater care in regulating clear-cutting on the national forests.[13]

The Portland-based Industrial Forestry Association's William Hagenstein authored the association's response to the Senate subcommittee doc-

ument on clear-cutting. Hagenstein, who established his credentials by citing his two graduate forestry degrees and more than forty years' experience "on the ground in the woods," praised the report for recognizing the
difficulty of legislating professional forestry practices and for accurately portraying the major problems in national-forest timber harvests. He also
applauded the committee for recognizing the federal forests' important contributions to meeting the nation's future timber needs. Hagenstein identified a few "objectionable conclusions," however, in which the committee
accused the Forest Service of making timber production "a priority activity in federal forest management." The opposite was actually the case, he
insisted, because recreation, aesthetics, and other non-timber values prevailed in forest management. Moreover, Forest Service sales were always
below the total allowable harvest. "Instead of giving timber management
the priority it deserves," Hagenstein reasoned, "many forests are left very
unhealthy and much valuable wood is dying and going to waste."[14]

As head of the region's most powerful forest-products trade association,
Hagenstein was busy on several fronts, mailing a circular to IFA colleagues
on clear-cutting that included his forthcoming testimony before the House
Committee on Agriculture in June 1972. In the cover letter to the IFA, he
wrote that the twenty-six congressmen opposed to clear-cutting were
"mostly big city slickers from the Midwest and East and the usual bunch
from California." His presentation to the House committee would repeat
the standard industrial litany—"the ecological necessity" for clear-cutting.
Fifty years of accumulated scientific knowledge, he argued, clearly showed
which trees "must be harvested by clearcutting in order to reproduce their
kind and keep them in the inventory of useful trees for human benefit."
In a unique twist, Hagenstein's testimony also asserted that if foresters were
to perpetuate desirable species, "it is nothing short of ridiculous for proposals to be made seriously to the Congress which would authorize a Commission to 'study' clearcutting as a political rather than the biological and
ecological subject it is." If those opposed to clear-cutting wanted to "strike
a blow for the environment," Hagenstein urged them to appropriate
sufficient funds to reforest national-forest lands.[15]

Forest scientists working for public agencies approached the subject of
clear-cutting with greater caution. Jerry Franklin, a young scientist who

would be praised for his progressive understanding of forest ecology, co-authored a paper in 1972 with Crown Zellerbach forester Dean DeBell that addressed the complexities and appropriate harvest methods of various forest environments.[16] Two years later, in the midst of the clear-cutting debate, the *Wall Street Journal* published a front-page story referring to the Franklin-DeBell article as a major scientific challenge to clear-cutting. The *Journal* quoted one Forest Service researcher who remarked that the Franklin and DeBell paper was "the first major criticism of broad-based clear-cutting." The newspaper also cited well-known William Hagenstein, who called the Franklin-DeBell report "simplistic," the work of two "splendid young scientists without experience in forestry." He told the *Journal* interviewer, "If you want to grow Douglas fir in this area of Washington and Oregon, you've got to clear-cut."[17]

Franklin countered Hagenstein's charge with a carefully worded letter to the *Journal*. The study attempted "to clarify the relationship between the biological requirements for tree regeneration and cutting methods including clear-cutting." Those biological requirements, Franklin wrote, could be achieved through a variety of silvicultural approaches. On more productive lands, clear-cutting would normally meet those prerequisites for regeneration. The issue was far more complicated, however, in areas important for recreational or aesthetic values or where environmental conditions were sufficiently harsh to inhibit regeneration. Franklin reminded readers that clear-cutting was "a perfectly appropriate practice on lands where timber growing is the primary or sole objective." In selecting cutting methods, foresters were most interested in "economics, efficiency, management objectives, and overall environmental impacts."[18]

Although the clear-cutting debate was especially heated in Oregon, the state's elected officials generally supported clear-cutting. With timber production leading all economic indicators, state officeholders offered only minor opposition to clear-cutting. However, because clear-cuts were unsightly and provided environmentalists with powerful visual arguments, even conservative political figures occasionally questioned the practice. A 1972 incident involving a Woodard Logging Company clear-cut of old-growth timber along the Umpqua River corridor caused an uproar. Because a major highway paralleled the river most of the way to the coast, critics brought the clear-cut to the attention of the State Land Board (comprised of the governor, secretary of state, and state treasurer). When the board criticized state forester Schroeder for permitting the cut, Steve Woodard, (who owned the tract of timber) protested to Governor Tom McCall, defending the harvest because the trees were mature. "Old trees should be cut," he wrote the governor, "in spite of pressures from the do-nothing cult which seems to believe that trees can live forever." Oregon needed responsible managers for its public and private lands, Woodard concluded, especially individuals like Schroeder "who have the backbone to do what they know is right."[19]

Reedsport's Republican state senator Jason Boe, who was running for reelection in timber-dependent Douglas County, joined the Umpqua logging controversy as a mild critic. While he supported the state forester and thought the department was "doing a generally excellent job," he criticized the cut along the Umpqua corridor because its visibility made it more difficult for supporters "of clear cutting to defend against the onslaught of the know-nothing super environmentalists and other self-esteemed 'experts.'" Boe wrote Woodard that he fully supported clear-cutting as "the superior method of attaining a fast re-growth cycle," but he opposed cuts that would spur legislation to end the practice. In parallel remarks to the Roseburg *News-Review*, Boe called for a healthy balance between the state's economic livelihood and its environment: "We have to have a growing and viable economy to provide the necessary jobs, but we want a healthy, clean environment as well."[20]

Governors Tom McCall (1967–1975) and Robert Straub (1975–1979), both praised for their environmental records, never openly criticized clear-cut harvesting. McCall biographer Brent Walth claims that the governor's environmentalism had limits, especially when it might interfere with the state's economy: "No other area pitted McCall's principle against expediency as did timber." During McCall's two terms in office, the timber industry directly employed one of every eleven Oregon workers, a fact that prompted the governor to ask federal agencies to increase annual harvests to keep the state's economy afloat. "Timber was strictly an economic issue, not an environmental matter," Walth argues, "and he stuck with that view even as more environmentalists concentrated on protecting the nation's forests." In questions on preserving forests and cutting practices, McCall clearly sided with the timber industry. When several western governors asked for a moratorium on clear-cutting in 1971, McCall objected. He also opposed Senator Mark Hatfield's efforts to expand northeastern Oregon's Eagle Cap Wilderness, contending that it should be expanded only to prevent visitors from seeing the unsightly effects of logging. "His concern was with the scenery," Walth concludes, "not with preservation."[21]

Bob Straub's positions on lumber industry politics were similar to those of his predecessor. After graduating from Dartmouth College and serving in the military during the Second World War, Straub went to work for the Weyerhaeuser Company in Springfield. Over the years, according to Walth, "he became comfortably rich from timber and a contracting business." His wealth made possible his rise in politics from county commissioner to state treasurer and then his election as governor in 1974.[22] Given Straub's background in timber investments, it is no surprise that he would support clear-cutting as an appropriate silvicultural practice. Shortly after his only successful run for the governorship, Straub told the annual meeting of the Pacific Logging Congress that clear-cutting was prudent, intelligent forest management and the most effective way to convert old-growth forests to fresh stocks of fast-growing trees. A pending federal court decision to ban clear-cutting on the national forests, Straub contended, would be disastrous for the Pacific Northwest: "We must have this tool [clear-cutting] on most of our forest lands in order to manage them properly and reasonably and in the best public interest." The governor suggested, however, that clear-cuts be kept small to reduce their visual impact.[23]

The federal court decision that Straub referred to declared that clear-cutting violated terms of the federal management act of 1897, which permitted the sale of only "dead, matured or large growth of trees." In *Izaak Walton League* v. *Butz,* the West Virginia federal district court ruled that clear-cutting on the Monongahela National Forest breached the 1897 act. The Forest Service immediately appealed the decision to the Fourth Circuit Court of Appeals, which upheld the lower court in August 1975. The Monongahela finding put the nation's wood-products industry, the Forest Service and affiliated state agencies, and Northwest politicians in crisis mode. Straub was among those urging Congress to amend the 1897 law to allow foresters to continue the use of clear-cutting. What followed was a year of turmoil and legislative arm-wrestling between environmental groups, led by the Sierra Club, and timber-industry officials and the Forest Service. After acrimonious congressional debates and what Paul Hirt calls "the perceived need to legalize clear-cutting," Congress passed the National Forest Management Act in 1976. The measure repealed the 1897 law, preserved clear-cutting on national forests (with some limitations), and adopted a slightly more restrictive version of sustained yield.[24]

The timber industry and its supporters had other good reasons to feel besieged and under attack during the 1970s. For the first time since logging and lumbering surfaced as the most powerful element in Oregon's economy, the industry's silvicultural practices were seemingly being assaulted from every direction. While industrial interests and policymakers were debating clear-cutting, the use of chemical herbicides and insecticides on public and private forest lands emerged as another contentious point of conflict. For more than twenty-five years after the Second World War, timber landowners, Oregon State University scientists, chemical manufacturers, the Oregon Department of Forestry, and the Forest Service collaborated in using sprays to control insect infestations and fungi outbreaks, and to release shade-intolerant Douglas fir from competing vegetation. Although the application of DDT had come under broad attack after the publication of Rachel Carson's *Silent Spring,* in 1962, it remained the spray of choice for most foresters until the federal government banned its use in 1972.

The Forest Pest Control Act of 1947 was the first important congressional initiative to offer federal support for insect and disease protection on all classes of ownership. To take advantage of the new measure, public agencies and private organizations established the Northwest Forest Pest Action Council in 1948. The council spoke for commercial forestry interests and established a close working relationship with chemical companies and federal research agencies.[25] For the next twenty years, the council lobbied federal lawmakers and the Oregon legislature to increase appropriations for bug and disease work, but its most important function was to develop a common front and to coordinate strategies to promote insect- and disease-control activities. Like their agricultural counterparts, the Council and industrial foresters clung tenaciously to the use of DDT even after the Environmental Protection Agency banned its use.[26]

Because forest insects and diseases recognized no jurisdictional boundaries, the Forest Service and the Oregon Department of Forestry worked closely to coordinate spraying programs, with one of their first joint operations an effort to combat a large spruce budworm epidemic in northeastern Oregon in 1949 and 1950. To formalize the coordinated insect- and disease-control work, the two agencies signed cooperative agreements to share expenses of equipment, personnel, and materials. Future cooperative ventures between the two agencies produced studies indicating the safety of DDT spraying programs, and a 1964 report referred to the buildup of DDT in the fatty tissue of deer and elk as "a temporary condition." When spruce budworm infestations began occurring regularly in the 1970s, the Forest Service and the Oregon Department of Forestry used DDT until its ban in 1972. In the succeeding years, the sprays of choice—Malathion, Zectran, sevin 4 oil, and Orthene—were chemicals not specifically forbidden by the federal government.[27]

After the publication of *Silent Spring*, industrial foresters were increasingly assertive about the importance of pesticides in protecting the economic and aesthetic benefits of forests. But they also worried about the public's perception that modern pesticides were harmful to fish and wildlife. PNW Experiment Station ecologist R. L. Furniss acknowledged to the Portland Rotary Club in 1963 that outdoor people were worried about the effects of pesticides on wildlife. He assured the Rotarians, however, that the Forest Service strongly supported their continued use and always avoided harming fish and wildlife. He told the gathering that DDT was the most effective tool in the forester's kit, a progressive step in the control of bugs "similar to . . . antibiotics in medicine." Although pesticides were valuable, he agreed that they were potentially hazardous and should be used with care.[28]

Speaking to the Society of American Foresters in 1964, Weyerhaeuser's managing forester, Royce Cornelius, pointed out that timber owners depended on chemicals to limit forest destruction. But landowners also had to be wise and careful stewards to avoid "public resentment and restrictive legislation." Because forest environments differed from agricultural fields,

Cornelius thought some of the criticisms of pesticide use were unfair. The accumulation of chemicals in forests, for instance, was minimal because foresters used only a small amount of pesticides and sprayings were infrequent. "Pest control in the forest," he told his colleagues, "should not be confused with pest control elsewhere." Cornelius was concerned, however, about the continued reliance on DDT, the one chemical that offered unqualified success against a variety of forest insect pests. Because of the industry's "almost total dependence on DDT," he worried about those who were maneuvering to suspend the use of the chemical. Responsible insect control was "impossible without DDT."[29]

When the Northwest Forest Pest Action Council met in Seattle in the fall of 1965, several members were optimistic that they had turned the corner in the struggle to continue using pesticides. Minutes from the meeting reveal that committee chair Ernest Kolbe, forester for the Western Pine Association, thought the council was "on a more sound course regarding the use of pesticides." He noted that the emotional furor generated by Carson's book had "fallen into some disrepute," and government agencies had gone public in defending the wise use of chemicals. Kolbe commended Royce Cornelius's public efforts in defending the continued use of pesticides. Cornelius, who attended the meeting, thought the council "had rounded the corner on the public pesticide issue" and was in position to deal more constructively with public agencies. He told the council that the University of California was field testing new compounds and that foresters could expect "a steady flow of new chemicals."[30]

Efforts to control tussock-moth outbreaks have much to tell us about the biological sciences and our overweening hubris about the need to control the world around us. After the Second World War, chemical spraying operations became the standard response to tussock-moth infestations in widely differing environments. Tussock-moth populations were similar to other insects, increasing rapidly in one or two seasons, causing severe defoliation, and then naturally collapsing to inconsequential numbers in a second or third season. When insect numbers appear to be uncontrollable, a naturally reoccurring virus causes a sharp collapse in tussock moth numbers. Through the 1960s, most chemicals were applied during the declining stages of the outbreak, leaving scientists with a technical dilemma—to spray or not to spray. The justification for the spraying programs usually centered on the belief that an outbreak would spread far beyond a "hot spot" unless it was controlled—an assumption, according to a PNW Experiment Station report, for which there was "no sound evidence."[31]

The most celebrated agency effort to combat the tussock moth focused on an outbreak that spread through public and private lands in the Malheur and Ochoco National Forests in the mid-1960s. A Forest Service scientist estimated at the close of the 1964 season that defoliation had spread to 39,000 acres. The moth continued to wreak havoc through the area and the next year the Forest Service used helicopters to spray more than 65,000 acres because entomologists feared that the natural virus would not work quickly enough to prevent extensive tree mortality. When the Forest Service completed aerial spraying in July, the project was the largest helicopter spraying effort the agency had ever undertaken. A follow-up report called the operation "highly successful with virtually complete mortality of the target insect." Although aerial spraying with DDT brought an end to early-summer larval feeding, the survey also noted that the outbreak had been "simultaneously declining from disease." Because spraying reputedly stopped the outbreak from spreading into healthy timber, industrial foresters and federal-agency personnel praised the Burns project as a great success story.[32]

The ubiquitous moth would not go away, of course, nor would the ambition to control such outbreaks. The next large-scale event began in 1972 when a Forest Service aerial survey revealed that the tussock moth had "literally exploded" in the Blue Mountains, infesting 197,000 acres in Oregon and adjacent areas in Washington and Idaho. During the following winter the Northwest Forest Pest Action Council, the Oregon Department of Forestry, the Bureau of Indian Affairs, and the Forest Service began laying the groundwork for a carefully coordinated campaign to use the recently banned DDT to combat the outbreak.[33] Although the initial effort to gain EPA approval to use DDT failed, the state, federal, and forest-industry arguments favoring emergency use of the banned chemical are instructive. The Forest Pest Action Council encouraged its members to join a massive letter-writing campaign urging the Forest Service to recommend DDT as "the only chemical proven effective against this insect." William Hagenstein, representing the Industrial Forestry Association, thought the Forest Service was equivocating on the use of DDT. Anything less than using a proven chemical to control the moth, Hagenstein wrote the council, "would properly subject us to being branded as professionally derelict."[34]

Outbreaks continued during the spring and summer of 1973 and eventually covered thousands of acres in the Northwest—some of them well beyond the Blue Mountains. The Interagency Tussock Moth Steering Committee judged the infestation of more than 800,000 visibly defoliated acres "to be the most serious on record." The severity of the damage prompted the Forest Pest Action Council to pass a resolution holding the EPA accountable for permitting "the continuation of infestation damage."[35] At a series of EPA hearings in early 1974, the Interagency Tussock Moth Steering Committee, with the support of state and industrial foresters, newspaper editorials, and the Forest Service, presented wide-ranging testimony in favor of using DDT. The Forest Service argued that using DDT to prevent further losses outweighed "both the short-term and possible long-range adverse environmental effects." Using DDT would protect forest resources and what the agency referred to as "ancillary benefits": preserving watersheds, wildlife habitat, recreational landscapes, and aesthetic values; reducing fire hazards; and alleviating the suffering of people who were allergic to tussock-moth hairs. The EPA took the testimony under advisement, and then granted emergency permission in late February 1974 to use the chemical. Before the spraying got under way, project leaders estimated that insect populations threatened to defoliate 650,000 acres, and that the incidence of the native virus was unknown.[36]

Ultimately, the Forest Service sprayed DDT on approximately 421,000 acres in Oregon, Washington, and Idaho. David Graham, who headed up the project, told the Northwest Forest Action Council that the chemical proved very effective, killing 98 percent of the tussock-moth larvae. He also reported excellent cooperation between Forest Service officials and EPA observers who were assigned to each of the spray units.[37] But there were problems, and the euphoria over the success of the huge aerial spraying effort proved short-lived. Critics, including university entomologists and the Audubon Society, charged that the spraying was unnecessary because moth populations were declining from natural causes; the Forest Service, they charged, ignored research that indicated promising alternative control methods to DDT. There were also questions about high levels of DDT found in the fatty tissue of wildlife and in several thousand head of cattle on the Colville Indian reservation. Evergreen State College entomologist Steven Herman referred to the spraying effort as "an entomological My Lai," with the death of untold numbers of songbirds.[38]

When pressed by its critics, the Forest Service admitted there were shortcomings in the tussock-moth control effort. Many of its own biologists opposed the use of DDT, and it acknowledged that several EPA monitors accused the agency of doing inadequate research to meet the conditions of the emergency order. To David Graham, the project indicated that DDT could be used effectively in emergency forest situations. Ralph Peineke, resources director for Boise Cascade Corporation, told the *Wall Street Journal:* "We still have to keep DDT in our arsenal until a viable alternative is found."[39] But the Forest Service and its industrial supporters took a public beating over the spraying project. Private timber owners and their representatives objected to the spirit of public questioning, and they resented the unwillingness of state and federal foresters to speak out on the issue. In a letter to regional forester Theodore Schlapfer, William Larson, executive director of the Washington Forest Protection Association, urged supporters who favored the continued use of targeted sprays to "quickly regroup our forces" and defend the integrity of the tussock-moth spray program. If they failed to do so, the public would challenge their credibility "on all future controversial subjects."[40]

The effort to control tussock-moth and other insect outbreaks in the eastside forests ultimately proved a failure. The frequency and widespread nature of the infestations were linked to deteriorating forest conditions that were in turn directly related to the fir trees that had replaced the hardy ponderosa pines in many areas. The fire-nurtured ponderosa pine were capable of resisting most insect attacks, writes Nancy Langston in *Forest Dreams, Forest Nightmares,* a study of forest management in the Blue Mountains. Logging of old-growth ponderosa pine and a century of fire suppression had contributed to replacement by "thickets of fir trees" prone to insect attacks. In the Blue Mountains insect epidemics had spread through half the national forest stands by 1991; bugs infested 70 percent of the trees in some places. A Pacific Northwest Research Station report published in 1994 indicated that grand fir and Douglas fir had colonized much of the historic park-like ponderosa stands, a "conversion to insect- and pathogen-susceptible late-successional forests." Forest Service studies in the 1980s indicated that the worst of the Douglas-fir tussock moth outbreaks occurred in areas once dominated by ponderosa pine.[41]

Scientific evidence pointing to the stressed Blue Mountains forest ecosystems continued to mount. A Forest Service survey of the region's ecological health in the early 1990s revealed that damages from insects, diseases, and fire had increased across approximately 3.2 million acres of federal, state, and private forests during the previous twenty years. Ponderosa and lodgepole pine forests were densely stocked, lacking in vigor, and "susceptible to attack by bark beetles." The "above-ground forest biomass" of dead and down trees had spread over vast areas, creating incendiary conditions during the dry summer months.[42] In a 1997 syndicated column published in the *Oregonian,* Russell Sadler referred to the Blue Mountain forests as "a virtual buffet for armies of Tussock moths, pine beetles, spruce bud worms and other pests," its forest health problems centered in "100 years of orthodox industrial forest management." But the forest-health crisis was more political than ecological, as forest-industry and environmental groups offered alternative proposals to avoid catastrophic wildfire. While ecologists viewed fire in the Blue Mountains as a natural tool with the potential to restore a new forest of ponderosa pine and western larch, Sadler believed that restoring the region's ecological health would take a mix of approaches—removing some dead and dying trees and allowing fires to burn in some circumstances.[43]

Western Oregon's forests provided yet another setting where chemical spraying made national media headlines during the 1970s. The controversy centered on the use of a Dow Chemical Company product, 2,4,5–T, an herbicide used to kill fast-growing hardwoods and to release commercially valuable Douglas fir to sunlight. The U.S. Army had first tested 2,4,5–trichlorophenol for use as a defoliant during the Second World War, but the conflict ended before it could be put to actual use. Registered with the federal government and released for domestic use in 1948, 2,4,5–T quickly became a popular weed and brush killer.[44] Used as Agent Orange (a mixture of 2,4,5–T and 2,4–D) to defoliate forests in Vietnam, 2,4,5–T was suspect because scientific reports indicated that it contained a deadly contaminant, TCDD, a highly toxic dioxin that could potentially cause cancer, other sicknesses, birth defects, and miscarriages in humans. Scientists considered TCDD both carcinogenic and mutagenic. The Defense Department ended aerial spraying with Agent Orange in Vietnam following antiwar protests in the United States and contentious Congressional hearings over its continued use.

As early as 1966 the Bionetics Research Laboratory in Bethesda, Maryland, informed the National Cancer Institute that 2,4,5–T caused deformities and other birth defects in the unborn offspring of laboratory mice and rats. In the wake of several cautionary studies, the government proceeded with limited restrictions, suspending the sale of 2,4,5–T for use around the home and for spraying near ponds, lakes, and irrigation ditches. When a science advisory panel acknowledged the chemical's teratogenic (fetus-deforming) effects, the new EPA and its administrator, William Ruckleshaus, announced that the agency would continue to prohibit use of the chemical on food crops intended for human consumption. The Forest Service and industrial foresters, however, had no restrictions and concluded that the chemical was a valuable "tool" to facilitate the quick regeneration of Douglas fir. Although citizen protests to ban the use of 2,4,5–T occurred elsewhere in the country, western Oregon and especially the tremendously productive Siuslaw National Forest was at the epicenter of the dispute.[45] To add to the herbicide controversy being waged in the press and courtrooms, during the 1960s and 1970s the Siuslaw National Forest was producing more board feet of timber than any other national forest in the country.

Protests against the use of dioxin-contaminated phenoxy herbicides (including Silvex) began with a few objections to Forest Service spraying in the Siuslaw National Forest in the early 1970s and then escalated in 1978 when the first of two EPA studies revealed a close correlation between human miscarriages and aerial herbicide spraying in the vicinity of Alsea, a small logging community in the Coast Range. The debate escalated when the public became aware of a cluster of scientific studies indicating that the dioxin contaminant was the most deadly toxin known to humans. As early as 1972, Jean Anderson, a clinical psychologist who practiced in Eugene, began writing letters to Forest Service officials objecting to the agency's spraying program. Anderson and her husband, Ugo Pezzi, operated a cattle ranch with a twenty-mile border along the Siuslaw Forest near the small town of Swisshome. Anderson, who began collecting literature on 2,4,5–T and dioxin and contacting biochemists and scientists, complained that Siuslaw National Forest supervisor Spencer Moore had delayed sending her the agency's environmental impact statement (EIS) outlining its spraying plans for 1973. Moore reputedly told Anderson that his office did not distribute the document widely because it was costly to send copies out for public review.[46]

Anderson was persistent, protesting to the EPA's Seattle office and to the Oregon Department of Environmental Quality (DEQ). When the latter referred Anderson to the Oregon Department of Agriculture, she remarked that the agencies "all seem to have a wondrous way of stalling, avoiding and double speaking." Oregon Environmental Council (OEC) executive director Larry Williams added to the controversy when he told *Eugene Register-Guard* reporter Jerry Uhrhammer about the elusive EIS. When Uhrhammer requested a copy from the Siuslaw office, an employee told him they had only one available. An angry Uhrhammer then phoned Moore and the supervisor ordered a copy mailed to the *Register-Guard*.[47]

Acting on behalf of Anderson and Pezzi, state senator and attorney Ed Fadeley obtained a temporary restraining order prohibiting the Forest Service from spraying near their ranch during 1973. Anderson also wrote to Uhrhammer expressing her concern about a possible link between phenoxy herbicides and miscarriages. Citing the authority of several scientists and physicians, she warned that women should stay away from areas that had been sprayed with dioxin-contaminated sprays. Recent studies of the effects of 2,4,5–T, she told Uhrhammer, were "extremely alarming," producing birth defects and cancer in experimental animals. When Spencer Moore announced the resumption of the spraying program elsewhere in the Siuslaw forest that summer, Anderson worried about summer campers using Forest Service campgrounds, "particularly children and women."[48]

Aware that its silvicultural program was under public scrutiny, the Forest Service drafted an environmental impact assessment to explain its spraying programs for the Siskiyou, Siuslaw, and Umpqua National Forests in 1974–1975. Chastened by its failure to distribute EIS drafts the previous year, the agency placed copies in public libraries and distributed others to environmental organizations, including the OEC. The council took issue with both the premises and the details of the draft report; staffer MiMi Cutler criticized the Forest Service's criteria for choosing aerial spraying as its preferred alternative. The use of herbicides 2,4–D, 2,4,5–T, and Silvex was irresponsible, Cutler asserted, because numerous scientific studies indicated the dangers of dioxin. She concluded that little was known about the way chemicals affected humans: "We consider it backwards to apply chemicals to our Nation's forests, and then investigate whatever undesirable effects follow."[49]

With a citizens' organization, Citizens Against Toxic Sprays (CATS), also raising questions about herbicides, the controversy remained at the front of public attention for the next several years. CATS, the OEC, and Hoedads (an Oregon tree-planting cooperative) filed suit in May 1976 requesting a temporary restraining order to prohibit the Siuslaw National Forest from using certain sprays because the agency had failed to file adequate EIS findings for the years 1975 through 1977. The litigation in the case became more complicated when the Industrial Forestry Association intervened as a defendant on behalf of the Siuslaw forest. The legal battle lines were drawn: citizen environmental organizations aligned against a powerful federal agency and Oregon's most influential industrial trade group. Before Judge Otto Skopil delivered his decision, the plaintiffs filed a new motion specifying the herbicides to be banned: 2,4–D, 2,4,5–T, and Silvex. Judge Skopil issued a temporary injunction on March 7, 1977, prohibiting the Siuslaw National Forest from using 2,4,5–T and Silvex (but not 2,4–D) until the Forest Service effectively explained the effects of dioxin on human and animal health. The judge determined that the EIS needed to comply with the terms of the National Environmental Protection Act (NEPA).[50]

The Skopil decision was a limited victory for the plaintiffs. According to the court, the Siuslaw EIS did not violate the law protecting bald and golden eagles, and national forest officials had followed proper statutory procedures requiring public comment. Judge Skopil retained jurisdiction over the Siuslaw lawsuit pending submission of the revised EIS. In June 1977 an EPA official reported that the agency could take no action against 2,4,5–T unless there was conclusive proof that dioxin threatened human health by bioaccumulating in the food chain. With the temporary ban on the use of 2,4,5–T and Silvex in the Siuslaw National Forest, the Forest Service then voluntarily extended the order throughout Region 6. Siuslaw officials circulated the revised EIS draft early in 1978 and presented the document for Judge Skopil's review in March. In a parallel development the IFA held a "Pesticide Awareness Seminar" for industry representatives, warning that "anti-groups have made substantial gains with NEPA because of the procedural delays under it."[51]

Writing for the *New Yorker* in July 1977, Thomas Whiteside published a lengthy assessment of the scientific findings regarding dioxin-contaminated herbicides. In addition to reporting the research of Harvard scientists Matthew Meselson and Robert Baughman, Whiteside traced the EPA's long-standing circumlocutions on the safety of 2,4,5–T. He reviewed the Dow Chemical Company's many publications on the subject, including its special "Press Reference Manual," *The Phenoxy Herbicides.* Whiteside tracked the lingering controversy over the use of Agent Orange in Vietnam and follow-up studies indicating the accumulation of TCDD in animals. The *New Yorker* essay also recounted numerous accidents in West Germany (1953), Amsterdam (1963), Czechoslovakia (1965), England (1968), and Missouri (1971), and a big explosion in 1976 at a chemical plant near Seveso, Italy. Despite the accidents and the accumulated science on the toxicity of TCDD, Whiteside concluded that the EPA seemed powerless or unwilling to forbid the use of dioxin-contaminated herbicides. Because of "the almost unmatched toxicity of dioxin," Whiteside questioned why the government continued to accept the word of chemical manufacturers who argued that there was a safe, "acceptable" risk in the use of the chemicals.[52]

Oregon's public debate over herbicide safety intensified during the spring and summer of 1978. The Forest Service and the IFA gained a preliminary advantage in early April when Judge Skopil approved the revised EIS for the Siuslaw, Umpqua, and Rogue National Forests. Citizen groups immediately asked the Region 6 office to delay its spraying plans until the EPA had completed its Rebuttable Presumption Against Registration (RPAR), a mechanism to evaluate the health risks of spraying, in this case the toxicological effects of dioxin-contaminated herbicides. The Forest Service ignored the appeal and issued a news release declaring that it would proceed with its spraying plans. Region 6 forester R. E. Worthington announced that the agency would follow its preferred alternative filed with EPA, including the *"use of all herbicides available and registered for use"* (my emphasis). Worthington added that "neither 2,4,5–T nor Silvex are believed to be harmful to humans when used as registered by EPA." Of the 150,000 acres in Oregon and Washington to be sprayed, approximately 90,000 acres would be treated with 2,4,5–T.[53]

As the 1978 spraying season approached, newspapers from Seattle to Medford reported citizen accounts of miscarriages, birth defects, and drifting sprays. These were little more than "horror stories," according to the IFA's Michael Sullivan, "often with little or no medical facts." As the IFA's public face, Sullivan ignored a considerable body of scientific literature. Based on animal tests in 1970 indicating high rates of miscarriages in the early term of pregnancies, the EPA had prohibited the use of 2,4,5–T on food crops or where pregnant women were at risk of exposure. And then in early 1978 additional animal tests prompted the EPA to call for further review of the herbicide. Those tests again confirmed that laboratory animals exposed to TCDD showed common symptoms: dead and deformed fetuses, intestinal bleeding, and abnormal kidneys. Wilbur McNulty of the Oregon Regional Primate Research Center published some of the more important animal studies testing the relationship between TCDD and abortions and birth defects in rhesus monkeys. McNulty's research suggested that there was no "permissible" safe level of exposure for pregnant women.[54]

Because of continuing citizen appeals and questions raised by scientists, assistant agriculture secretary A. Rupert Cutler expressed his "great concern" to Forest Service chief John McGuire about the controversy. Because of the "conflicting reports and unanswered questions," Cutler wrote on

April 27, 1978, the Department of Agriculture was cooperating with EPA to review 2,4,5–T and Silvex. Until the EPA study was completed, Cutler wanted to oversee all Forest Service decisions to use such sprays.[55] In a policy memorandum released the same day, McGuire authorized the use of herbicides only after other alternatives had been considered; sprays containing TCDD should be used only when there were no other options. "Cost-effectiveness," the chief directed, "will not be the sole criterion" for using pesticides. Federal foresters were required to compare the "costs, safety, and effectiveness" of ground and aerial applications and "the possible consequences of drift to adjacent lands or water." A storm of protest greeted McGuire's directive.[56]

During the next several months the national media joined the ongoing debate. Respected nationally syndicated columnist Jack Anderson reported disturbing news from Oregon indicating that "eight young women exposed to the herbicide are having miscarriages at triple the national rate." Anderson, who thought these were legitimate concerns, pointed to a growing body of scientific evidence suggesting the high toxicity of TCDD. He referred to the recent laboratory studies indicating that the chemical contaminant caused miscarriages and birth defects in animals. The herbicide controversy also aroused concern in the Oregon governor's office when Bob Straub asked the Department of Forestry to tighten controls on the use of forest herbicides and to add new rules to the state's Forest Practices Act. The requested rules changes required the notification of adjacent landowners before spraying 2,4,5–T or Silvex; posting notices on the boundaries of spray areas; closing roads for 24 hours before and after spraying; and requiring a 200–foot buffer from waterways, public roads, and populated areas. When the Oregon Board of Forestry adopted the governor's recommendations during its regular August meeting, private timber companies greeted the decision with derision, charging that it was political and lacked scientific substance.[57]

Federal officials tightened the noose on the use of certain herbicides when Cutler issued a "rule of thumb" order prohibiting the use of 2,4,5–T within one-quarter mile of year-round streams and within one mile of any permanent home on the national forests. The new guidelines surpassed Oregon's requirement for 200–foot buffers on state and private lands. N. E. Bjorklund, vice-president of the IFA, thought the federal restrictions would eliminate the use of 2,4,5–T and Silvex on the national forests and "make a mockery of reforestation." It would be senseless, he argued, "to invest in tree planting if adequate brush control cannot assure its success." Because of protests from the governor's office and Oregon's congressional delegation, Cutler rescinded the restrictions in lieu of Oregon's recommended 200–foot buffer. Governor Straub issued a news release in September 1978 praising the decision, because the continued use of 2,4,5–T was "essential to increase Oregon's log production and provide jobs in the forest products industry."[58]

The *coup de grace* in the forest herbicide controversy involved nine women who lived in forested areas and whose miscarriages coincided with herbicide spraying near the small logging community of Alsea. The story begins with Bonnie Hill, an Alsea high school teacher, who read an article in the summer of 1977 summarizing the research of James Allen of the University of Wisconsin. Allen's inquiries with rhesus monkeys linked exposure to TCDD with adverse effects on reproductive systems and spontaneous abortions. Recalling what seemed to her a high incidence of miscarriages in the Alsea area, Hill gathered information from several women on the dates, location, and circumstances of their miscarriages. In collecting her data, she omitted all miscarriages that had not been handled by a doctor.[59] She then visited the offices of the Forest Service, BLM, the State Department of Forestry, and private timber companies and gathered information on the dates, locations, chemicals used, concentrations, and methods of application of all spraying operations. She drew up a chart listing the dates of miscarriages, the dates of the herbicide spraying, and maps that indicated an obvious correlation between spraying and incidences of miscarriage. Hill insisted in 1978 and in a personal interview in 2003 that her findings indicated a correlation only; she never claimed a cause-and-effect relationship. She drafted a letter asking federal agencies to look into the matter. The letter included the signatures of nine women who had a total of eleven medically documented miscarriages—eight of them occurring in 1976 and 1977, years of heavy spraying in the area.[60]

Hill mailed copies of the letter to public land managers, elected officials, and the EPA. Because of the gathering tempest over the controversial herbicides, the EPA sent an agency team with a detailed eighteen-page questionnaire to interview the Alsea women. The responses to the interview were reasonably consistent and convinced the EPA to turn over the questionnaire data to a team of obstetricians and gynecologists. Because several of the reviewers called for additional data, the EPA recommended an expanded study that would compare the incidence of miscarriages in the greater Alsea area with a comparable control group.[61] With little fanfare, the EPA carried out a much larger survey, covering a 1,600-square-mile area in Lincoln and Benton counties. The study, known as the Alsea II to distinguish it from the EPA's initial findings, examined hospital records in rural and urban areas and found significantly higher percentages of miscarriages for the study area, a statistic that was even higher yet following spray periods. The results for the rural and urban control group did not exhibit that pattern of miscarriages. With these data in hand, EPA issued a temporary order on February 28, 1979, suspending further use of 2,4,5–T and Silvex in forest environments. Because they were not informed of the ongoing Alsea II survey, the suspension took Hill and her cohort completely by surprise.[62]

The EPA's interest in Bonnie Hill's preliminary findings attracted intense media interest. Newspapers printed dozens of articles, radio and television stories aired accounts about the herbicide debate, and ABC's *20/20 News Magazine* broadcast two segments examining Agent Orange (July 25, 1978) and dioxin (August 1, 1978). Geraldo Rivera, then a respected investigative reporter for *20/20*, questioned scientists on both sides of the issue; reviewed the 1976 chemical plant explosion in Seveso, Italy, which released a cloud of dioxin-contaminated powder into the atmosphere near an urban area; and traveled to Alsea to interview Bonnie Hill. Because the highly rated documentary raised serious questions about the toxicity of TCDD, forest-industry personnel reacted negatively to the program. In a letter to ABC News president Roone Arledge, Industrial Forestry Association public-affairs director Michael Sullivan declared the show one-sided and made scurrilous remarks about Rivera. The Council for Agricultural Science and Technology (CAST), representing twenty-five food and agricultural organizations, criticized the *20/20* programs as lacking scientific evidence. On the question of spontaneous abortions associated with the use of 2,4,5–T, CAST accused the program underwriters of using "isolated persons . . . with no more evidence than the fact that spraying of the herbicide [2,4,5–T] was said to precede the events." The Council criticized EPA's Alsea II study of the flawed data collection and analysis.[63]

The EPA suspension brought a firestorm of criticism and court challenges. The forest products industry vigorously attacked the scientific underpinnings of the Alsea II study. Oregon State University faculty were especially critical of the EPA's analysis. Forest ecology professor Michael Newton, a vocal supporter of the continued use of 2,4,5–T and Silvex, pointed to "the near universal agreement among scientists *in the weed science and toxicology fields* [my emphasis] that 2,4,5–T is safe to use." Jim Witt, a professor of agricultural chemistry, thought the Alsea study was incompetent and dishonest. Following EPA's suspension order, the Dow Chemical Company flew Newton and his wife to its headquarters in Midland, Michigan, to join with the firm's officials in a strategy session. Newton later admitted that Dow paid for the trip and acknowledged that he had done consulting work for the company in the past. Critics, some of them on the OSU faculty, charged that the institution was sacrificing its credibility in assuming an advocacy role on the herbicide question. Writing for *Willamette Week*, editor Phil Keisling cited the university's long history of sponsoring studies that gave unstinting praise to herbicide use. Professor of forestry Bill Ferrell told Keisling that the forestry school was "too close to herbicide advocates" and was in danger of being "hooked on chemicals." As for Michael Newton, Ferrell observed: "Chemical companies haven't bought Mike Newton; his views are simply parallel to theirs."[64]

Within a week of the suspension order, Dow Chemical Corporation (manufacturer of 2,4,5–T) filed a lawsuit against EPA, charging the agency with violating the Federal Insecticide, Fungicide, and Rodenticide Act. The executive committees of the National Forest Products Association (NFPA) and the IFA engaged in a flurry of exchanges debating the efficacy of joining the Dow proceedings. The NFPA also solicited a friendly critique from the Battelle Research Center in Seattle, which found the Alsea research selective and inadequate. Scientific testimony in the Dow case, however, raised additional concerns about the toxicity of TCDD. University of Oregon molecular biologist George Streisinger testified that TCDD accumulated in the fatty tissue of elk, deer, and other animals exposed to 2,4,5–T in forested areas. He also thought it was difficult to determine the levels of TCDD that would prove non-toxic to humans: "Every dose so far examined has proven to be toxic in animal experiments." Respected biochemist Matthew Meselson affirmed the "extraordinary toxicity of TCDD" in environmental samples and suggested that it could accumulate in the food chain.[65]

Worried by successful ballot initiatives banning herbicides in two northern California counties, the powerful NFPA promised an educational campaign "to win at the polls" and stem the tide of anti-herbicide initiatives. As Crown Zellerbach, Weyerhaeuser, Georgia-Pacific, and Publishers Paper moved aggressively to oppose further bans on herbicide use, NFPA executive vice president Ralph Hodges proposed a strategy focused on "health risk information effectively delivered to voters." The effort would require a sizable commitment of financial resources; Hodges earmarked more than $500,000, with all of the funds allocated to defend the continued use of 2,4,5–T and 2,4–D. The NFPA initiative would support research into the controversial herbicides, make available health-risk information to the public, represent the industry's case to Congress, and assist local pesticide users in public-relations efforts.[66] But when the Ninth Circuit Court of Appeals upheld the EPA suspension in February 1980 (i.e., the Dow case), the decision rendered moot the NFPA "educational" campaign.

The EPA held formal cancellation hearings for 2,4,5–T and Silvex in the spring of 1980, taking voluminous testimony from numerous scientists and witnesses on both sides of the issue. The agency used two separate deposition venues: a "risk" section to take statements on the dangers the chemicals posed to human health, and a "benefits" section to determine the economic need for the herbicides. The hearings dragged on for several months, producing a huge volume of evidence. Ronald Reagan's coming to the presidency in early 1981 muddied the picture; his administration began secret negotiations with the Dow Chemical Company to settle the 2,4,5–T controversy.[67] When someone leaked hitherto concealed scientific data regarding the toxicity of TCDD in the Alsea study, both Dow and EPA announced the cancellation of any further registrations of 2,4,5–T.[68]

There is also a supply side to Oregon's modern forestry story, indications amidst the great production boom that the tide was beginning to ebb. There were expressions of concern, with some of the worries coming from districts of heaviest cutting. A Forest Service study of the Douglas-fir region published in 1972 cited rapid "inventory depletion" and inadequate reforestation and warned that "high rates of log production . . . have a limited future." By the 1960s, the huge harvests and increased mechanization in logging and manufacturing operations had already contributed to a reduction in the industry's workforce. There were also occasional mill closures, with company officials blaming diminished timber supplies and weak markets. An Oregon State University study in the mid-1970s predicted declining harvests in the state's western districts for the rest of the century. And then came the whirlwind—a rash of mill closures beginning in 1979 and continuing into the early 1980s, a social and economic crisis completely unrelated to rising environmental concerns about forestry issues.[69]

When the regional timber economy began to recover in the mid-1980s, Oregon's powerful senator Mark Hatfield and influential congressman Les AuCoin began amending appropriations bills to direct the Forest Service to increase federal timber sales in Oregon and Washington. Writing for the *New Yorker* in 1990, Catherine Caufield referred to Oregon's senior senator as "probably the strongest congressional supporter of continued high levels of cutting from public forests." A liberal Republican with good credentials on most environmental issues, Hatfield gained the ire of conservationists for his positions on federal timber policy. As the ranking member of the Senate Appropriations Committee beginning in 1981, Hatfield pressured the Forest Service to increase federal timber harvests far above the agency's planned cutting rates. Beginning in 1985, the Oregon senator also began attaching controversial riders to appropriations bills restricting citizen court appeals. While congressional restrictions on judicial review had traditionally been limited to national security matters, Hatfield's successful initiatives in the 1980s and early 1990s were used to override environmental laws.

Always mindful of voters at home, Hatfield told his constituents that the riders were necessary "to mitigate the unacceptable economic disruption that would result if entire areas of the state were denied access to national-forest timber."[70]

Forest Service officials who worked through the 1980s later acknowledged that they were pressured to maximize national-forest harvests during the decade. At the same time, federal foresters also struggled to draft comprehensive land-management plans mandated under the National Forest Management Act of 1976. The new forest plans, in the words of *Oregonian* journalist Kathie Durbin, would provide "pure drinking water and white-water rafting, campgrounds and wilderness solitude, deer and woodpeckers, wildflowers and clear-grain Douglas fir lumber." In the maelstrom of politics, the forest plans were subject to endless political bickering and heavy-handed pressure from Reagan administration officials and the Northwest congressional delegation. John Crowell, Reagan's assistant secretary of agriculture between 1981 and 1985, warned Forest Service planners not to place too much emphasis on protecting wildlife and scenery. Crowell's directives and those of Douglas McCleery, the agency's assistant director for timber sales, delayed the implementation of the Northwest land-management plans. James Torrence, who served as Region 6 forester until his retirement in 1989, believed the national forests would have been "better off" if lumber industry pressure had not slowed the planning process.[71]

Efforts to stall the forest plans through the 1980s and the politicking of Senator Hatfield and Congressman AuCoin boosted national-forest timber production to unsustainable levels. The rapid harvest rates during the decade worsened slide conditions on the rain-saturated slopes of the Coast Range and accelerated the fragmentation of remaining old-growth forests. An increasing number of federal and state biologists, geologists, and soil scientists criticized the effects of road-building and logging practices on fisheries and wildlife populations. Fishery biologists argued that removing timber along small streams silted salmon-spawning areas and jeopardized the survival of all fishes. But the environmental issue that loomed the largest as the decade drew to a close centered on *Strix occidentalis caurina*, the northern spotted owl, a species heavily dependent on old-growth forests. The controversy over the owl evoked heated and passionate debate across the

Northwest; it raised questions involving the reauthorization of the Endangered Species Act, the economic viability of timber-dependent communities, and the larger meaning of indicator species.[72]

The northern spotted owl muddied the debate waters and became a surrogate for the lumber industry and environmental organizations, both jockeying for influence in shaping the region's forest policy. Biologist Eric Forsman was the first scientist to bring attention to the dwindling number of spotted owls through the publication of his research findings in the mid-1970s.[73] Using radio-tracking devices, Forsman's careful study indicated that the owls lived almost exclusively in old-growth forests and that each breeding pair needed between 1,200 and 3,000 acres of such forests to successfully feed, nest, and propagate. The National Forest Management Act also figured in the debate, because the measure required the Forest Service "to identify and protect 'indicator species' (such as the spotted owl)" to maintain the health of forest ecosystems and to protect minimum populations of vertebrates. It quickly became apparent to many in the environmental community that diminishing stands of old-growth timber on the national forests threatened the spotted owl. Although old-growth forests were not synonymous with spotted-owl habitat, the birds usually lived in such areas.[74]

Because Oregon's forest-products economy had become increasingly dependent on federal timber supplies during the 1980s, the furtive owl became the subject of heated public debate. Invoking the Endangered Species Act (ESA) and recent scientific findings regarding the owl, the Sierra Club Legal Defense Fund filed the first of three spotted-owl lawsuits in 1987 on behalf of several environmental organizations. In *Northern Spotted Owl v. Hodel*, the environmental organizations challenged the U.S. Fish and Wildlife Service for its unwillingness to find the owl threatened or endangered under the ESA. Finding that "expert opinion is to the contrary," the federal district court in Seattle directed the Fish and Wildlife Service to reconsider its decision. Environmentalists appeared to be moving toward success in the spring of 1989 when the Fish and Wildlife Service announced that it was considering the owl a threatened species. Reflecting on the course of events, University of Washington ecologist Jerry Franklin praised the advocacy groups for "picking up on the science and using it in their lawsuits" and bringing "us out of the closet."[75] The threatened court actions, in effect, created a framework for bringing scientific evidence to bear on national-forest management decisions.

Industry representatives complained as early as 1985 that protecting the spotted owl would jeopardize the Northwest's timber supply. Because the federal government enjoyed a "virtual monopoly" on the region's harvestable trees, economist Dan Goldy accused the federal government of threatening Oregon's leading industry. In apocalyptic language, Oregon's former director of economic development charged that federal set-asides for wilderness areas and to protect spotted-owl habitat would remove hundreds of thousands of forested acres from the commercial timber base and force private landowners to liquidate their own second-growth stands to meet market demand. Goldy urged the Forest Service to return to the drawing board and revise the forest plans "to avert the economic and forestry disasters that threaten Oregonians." He accused the agency of grossly undercounting spotted-owl populations and ignoring evidence that the birds were "nesting happily in young second growth timber." According to Goldy, the owls were concentrated along forest roads, "in cut-over areas, and just about everywhere but in the old-growth wilderness areas which they are supposed to prefer."[76] With their own hired biologists, forest-industry groups continued to insist that Forest Service scientists were misrepresenting the owl's actual numbers. As the spotted-owl debate became more bitter, economist Randy O'Toole, an industry critic, noticed a curious reversal in roles: "In the past it was the environmentalists who resorted to emotionalism to make their case and industry which relied on technical arguments. Now the timber industry is running a grassroots campaign based on emotional, not technical appeals, and that's because the technical data no longer support their position."

While the spotted-owl debate simmered and the national-forest planning process continued to move forward, Forest Service officials were striving to reduce the huge harvests of the 1980s to more reasonable levels. Industry spokespersons countered with proposals to increase harvests to meet market demand, pointing out that Oregon's national forests totaled more than 15 million acres, provided more than half the state's annual harvests, and were vital to its economic future. "The problem today is timber supply," Dan Goldy wrote in the spring of 1989 as the spotted owl was moving toward an ESA listing. The forest-products industry would, he feared, be "put to death because radical environmentalists" were using legal statutes to prohibit federal timber sales. The real issue for environmentalists was not the spotted owl, according to Goldy, but their desire to place all old-growth forests off limits to timber harvests.[77]

Matters moved quickly in the late 1980s. Jack Ward Thomas, a Forest Service wildlife biologist who chaired the Interagency Scientific Committee in 1988, was mandated to draft a scientific strategy to protect the owl. Published in 1990, the committee's report recommended establishing Habitat Conservation Areas (HCAs) at varying intervals across the owl's range, connecting habitats to reduce the extinction of isolated owl populations. Just as the committee's report was released to the public, the Fish and Wildlife Service listed the owl as a threatened species. And then in May 1991, District Judge William Dwyer issued an injunction blocking all national-forest timber sales involving potential owl habitat until the Forest Service satisfied the requirements of the National Forest Management Act. Dwyer sensed that something more was involved than the agency's failure to follow the law:

> The most recent violation of the National Forest Management Act exemplifies a deliberate and systematic refusal by the Forest Service and the Fish and Wildlife Service to comply with the laws protecting wildlife. This is not the doing of the scientists, foresters, rangers and others at the working levels of these agencies. It reflects decisions made by higher authorities in the executive branch of the government.[78]

Dwyer directed his sharply worded decree at high-level officials in the Reagan and George H. W. Bush administrations and to prominent Oregon politicians whose heavy-handed directives pushed annual harvests to levels that were widely believed to be unsustainable.

The spotted-owl and related endangered-species issues continued to move through the courts during the 1990s. Congressional proposals were floated to provide additional protections for old-growth forests, and other initiatives were launched to override the Endangered Species Act and the National Forest Management Act. At the request of the House of Representatives, a team of four scientists (the "Gang of Four") convened a meeting of professionals to map and examine species dependent on old-growth timber on national forest and Bureau of Land Management lands. Delivered to Congress in July 1991, the report offered a variety of alternatives, all of them predicting some level of job losses. All the recommendations were sharply criticized by industry officials, and none made it out of committee. The election of President Bill Clinton in 1992 led to the highly publicized April 1993 "Forest Summit" in Portland and the unveiling of the Northwest Forest Plan that summer. The Clinton plan promised to follow the law, end courtroom gridlock, and provide a modicum of logs to timber-dependent communities. It would also reduce federal timber harvests by 75 percent from the record levels of the 1980s.[79]

Despite sharply diminished national-forest harvests, the Pacific Northwest continued to be the leading lumber producing region in the nation. Oregon ranked first in the mid-1990s with 5.7 billion board feet and Washington second with 4.2 billion board feet. While trade-association officials continued to praise the impressive production figures, they worried that long-term log supplies would remain a problem. Western mill production had declined from 62 to 51 percent of the nation's total during the previous decade, and Southeast and Canadian mills were taking larger shares of the market. Reduced federal timber sales contributed to pockets of unemployment, the Oregonian reported in the spring of 1995, but job losses were less than one-third of industry predictions. Between 1989 and 1995, the three Northwest states lost 16,695 direct wood-production jobs, far below the Northwest Forest Research Council's prediction of more than 65,000 job losses. But there were significant human hardships; the new environmental restrictions had their greatest impact away from the prosperous Interstate-5 corridor, especially in small eastern Oregon communities— La Grande, Lakeview, Chiloquin, Burns, Klamath Falls, and Prineville.[80]

There is still more to this story of environmental restrictions and lumber production. During the booming national-forest harvests of the 1980s (and before the spotted-owl rulings), nearly 200 sawmills closed in Oregon and Washington, and forest-products employment dropped by 25,000 workers. Technological changes in the woods and mills made it possible for fewer workers to turn out an ever increasing volume of forest products. New logging equipment, especially yarding machines, sharply reduced labor as a factor in logging operations. But the most significant techno-logical changes during the decade took place in mill operations, where laser-controlled cutting machines and highly automated equipment drastically reduced employment in lumber manufacturing.[81]

As Oregon moved into the twenty-first century, forest-related issues con-tinued to divide citizens. Locally heavy precipitation in November 1996 caused several landslides along the fragile slopes of the Umpqua Basin, killing four people on Hubbard Creek thirty miles northeast of Roseburg; another person died when a landslide swept her car into the swollen Ump-qua River east of Reedsport. Thirty other people along Highway 38 were trapped in their cars, some of them for more than eighteen hours. Two years later human tragedy was narrowly averted when a huge mudslide along Highway 34 near Alsea slammed into a small log-cabin house, trapping Andrea Burnett and her sixteen-year-old son, Josh, in mud and darkness. Crawling through broken glass and aided by the headlights of a westbound vehicle, Burnett and her son crawled through an opening formed by the roof and escaped to safety. The Umpqua and Alsea incidents reopened pub-' lic discussions about the relationship between forest practices and land-slides in western Oregon.[82]

Although landslides are not new to western Oregon, it is clear that roads and harvest practices have greatly increased the potential for slides, espe-cially during the region's heavy seasonal rains. The tragic landslides trig-gered legislative hearings, investigations, and appeals to the governor to declare a moratorium on steep-slope logging above dwellings and major roadways. The Oregon Board of Forestry hosted a public hearing on the landslide question in March 1997, with industry representatives, environ-mentalists, and rural residents offering testimony. At the close of the hear-ing, the forestry board asked logging companies to *voluntarily* agree to avoid logging steep slopes. The landslide issue, however, remained before the public in the form of additional hearings, task-force studies, and Board of Forestry oversight of logging on sensitive slopes.[83]

In retrospect, the landslide controversy is merely an additional element in the stew of twenty-first-century forestry politics. The great promise of the Clinton Northwest Forest Plan has been immersed in endless court challenges, and the question of federal timber harvests in Oregon has persistently reverberated through state and national politics. Timber-related issues continue to divide rural and urban Oregon—with a heavily weighted advantage to the Portland metropolitan area and the Willamette Valley, where citizens increasingly care about quality-of-life issues and the amenities that old-growth forests offer. Conversely, rural communities suffer from many decisions that work to the advantage of urban desires.

IV
OF RIVERS
AND LAND

7

RICHARD NEUBERGER'S

CONSERVATION POLITICS

It seems evident that one of the prolonged controversies during the
next 50 years ... will be how much of the public lands in the American
outdoors to leave as primeval wilderness and how much to adapt to
supervised use. Both sides already have mounted strong and convincing
arguments regarding a Wilderness Preservation Bill, now before Congress.
One pervasive theme threads through my own mind. Once wilderness
is mined or grazed or logged, it never can be true wilderness again. This
should induce Americans to proceed slowly when they alter the character
of their few remaining primitive realms, because such a process inevitably
becomes irreversible.

Nature has done well by our United States. It is man's part which needs
constant attention and improvement.—RICHARD L. NEUBERGER (1959)[1]

Two major figures in both journalism and politics, Richard Neuberger
and Tom McCall, played influential roles in Oregon's conservation
and environmental communities in the third quarter of the twentieth century. Differing in journalistic talent and political temperament,
Neuberger and McCall gained national reputations—the former as an
ardent conservationist, the latter for his environmental accomplishments.
Before his election to the U.S. Senate in 1954, Neuberger had already earned
prominence as a crusading journalist, publishing articles in the nation's leading magazines on issues ranging from celebrations of Hells Canyon to the

engineering successes of Bonneville Dam's fish-passage facilities. McCall
began in print journalism and migrated to television broadcasting, where
he gained renown as a news commentator with special emphasis on environmental problems. Both Neuberger and McCall eventually used their public personas to leverage themselves into state politics, as U.S. senator and
governor respectively.

Backwoods and metropolis mingle along the shores of the Columbia
River, Richard Neuberger wrote in 1949. Nowhere in the United States did
the forces of nature and civilization mix so indiscriminately as they did
in the Pacific Northwest, "one of the last frontiers, still in the dawn of its
impact on civilization."[2] By the mid-1950s, the talented Neuberger had
occupied center stage in Oregon's journalistic discourse for nearly two
decades. When he defeated incumbent U.S. senator Guy Cordon in 1954,
Neuberger believed that an activist government would best serve his constituents. An old-style New Dealer who believed in planning when it was
becoming less fashionable, Oregon's junior senator was an ambivalent mix,
believing in developing and protecting nature, supporting large public-
works projects, and setting aside public lands in national parks and
wilderness areas. Neuberger wanted to win *all* those battles: to vanquish
private-power interests, to build huge dams in scenic canyons, and in the
same breath to offer his support for wilderness preservation. The balance
of his writings and the positions he took in the Senate on important natural-resource issues place him in the mainstream of liberal democracy. Neuberger supported clean-water initiatives, voted to clean up the nation's rivers,
promoted good forestry practices, and urged the full development of the
Columbia River to meet "the needs and the responsibilities which aggressive Soviet competition imposes upon us."[3]

Although Neuberger was deeply appreciative of the Columbia River
country's beauty, he was equally passionate about building dams and about
the benefits of cheap public power. Inexpensive public power, he argued
on one occasion, was Oregon's best hope for new payrolls and provided a
way to avoid economic downturns. When the Eisenhower administration
proposed public-private partnerships for future multipurpose dam projects,
candidate and then senator Neuberger took the offensive. His differences

with the administration began early in 1954 when Interior secretary Douglas McKay, a fellow Oregonian, suggested that the proposed John Day Dam be considered as a partnership project. When Senator Cordon and eastern Oregon congressman Sam Coon introduced companion bills to build John Day under a partnership proposal, senate candidate Neuberger denounced the plan. After narrowly defeating Cordon, he challenged Congressman Coon to a series of debates on the merits of the partnership scheme. "It seems to me the public is entitled to facts and not to epithets and . . . name-calling," he wrote the congressman.[4]

Neuberger prepared a thirty-five point statement supporting public power development. The Columbia's "falling water," he declared, was the region's greatest asset: "It provides our only natural resource of energy to expand industry—to work for us in our homes, farms, and factories." Partnership power, on the other hand, would be high-cost power. Neuberger then turned his attack on the private-power lobby and Secretary McKay, equating their support for partnerships with mediocre power sites and utility monopolies over exceptional locations such as Hells Canyon.[5] In the debates that followed, Neuberger charged that the John Day partnership proposal did not "play fair with future generations" and would hurt rural consumers. Toward the end of their nightly exchanges, when Congressman Coon admitted that private power would be more expensive than public power, Neuberger responded: "The debate has ended here and now."[6] Political victory eventually went to Neuberger: Congress defeated the partnership bill and John Day Dam was completed as a federal project.[7] The Army Corps of Engineers began construction in June 1958 and completed work a decade later as the waters pooling behind the big dam extended seventy-five miles upstream to McNary Dam.[8]

Richard Neuberger firmly believed that Northwesterners could enjoy both cheap hydropower and plentiful salmon runs. Early in his term in the Senate, he joined his senior colleague, Wayne Morse, in recommending that power revenue from federal dams be used to fund irrigation development. Reclaiming arid lands, Oregon's senators argued, would serve as "sources of food for present and future generations and . . . provide new farming opportunities." More specifically, Neuberger and Morse urged more funding for federal projects in the Columbia Basin. A growing population, they asserted, threatened to move the nation "from an era of food surplus to food deficit." Neuberger, who firmly believed in federal action, regularly took political jabs at Interior Secretary McKay who favored private development of Hells Canyon.[9]

Senator Neuberger supported most federal water-development proposals and opposed only those ticketed as private ventures. Although public and private power-development issues influenced his support for particular projects, scenic vistas and spectacular waterfalls did not always influence his thinking. Hells Canyon, a spectacular geographic defile that Neuberger celebrated in a 1939 *Harper's* magazine essay, is a case in point. More than any other public figure of his time, Neuberger was linked to Hells Canyon, even to the point of convincing the Oregon Board of Geographic Names to begin the process of making official the commonly used name. "Neuberger's prolific writings about Hells Canyon," writer Steve Neal argues, "made the name stick." In his 1939 article, Neuberger referred to the canyon as the narrowest and deepest of its kind in North America; more than any other American stream, the Snake was "primarily a wilderness river" draining "our last frontier." Although Neuberger could write about the river flowing through a mostly pristine hinterland, he also acknowledged that the Snake watered "one of the principal irrigated sections of America." It was the canyon, however, that captured his attention, and he suggested that at some future date it might be suited for national-park status. "On windless days," he wrote, "the only sound from Hells Canyon is the roar, faint and far off, of the mountain river, gnawing at the rocks the earth spewed up long ago."[10]

"Few stretches of American rivers," historian Keith Petersen wrote, "have endured as much controversy as the middle Snake." Army Corps of Engineers surveys in 1947 proposed two huge dams, one immediately downstream from the Salmon River and another upstream at Hells Canyon Creek. The combined dams, according to the Corps, would produce enormous amounts of electricity and provide flood-control storage. To move the federal program along, Morse and Idaho congresswoman Gracie Pfost introduced companion bills in 1952 to build a gigantic 720–foot dam at the Hells Canyon site, a structure that would rank second in height to Hoover Dam on the Colorado River. The election of Republican Dwight D. Eisenhower to the presidency in 1952, however, shifted momentum away from public power and placed advantage with those who supported private development. The ensuing political battles turned the middle Snake River into the nation's most publicized skirmish between public- and private-power advocates.[11]

Shortly after taking office in the spring of 1953, McKay suggested that the Hells Canyon project recommended in the original Corps of Engineers 308 Report be turned over to the Idaho Power Company. As part of a broader move toward privatizing water-development projects, the U.S. Solicitor's office ordered the Federal Power Commission (FPC) in May 1953 to withdraw the Interior Department's "1952 petition of intervention" to the Idaho Power Company's application to build five "run-of-the-river" dams on the Snake River. The private projects would be in lieu of the government's huge Hells Canyon dam and would have no value for flood control. With positive feedback from the president's State of the Union message and with Secretary McKay's support, Idaho Power filed permit applications with the FPC to build three dams on the middle Snake.[12] The debates that took place from the 1940s through the 1960s, therefore, centered not on whether dams would be built, but who should build them.

The move away from New Deal public-works projects offended Neuberger, who became increasingly critical of the Idaho Power proposals. Even as he positioned himself to run against Cordon, he used every opportunity to promote public-power development, especially a big Hells Canyon dam. In that sense, he was at one with other public officials with conservation leanings who supported building a high dam in Hells Canyon. Idaho's youthful Frank Church, elected to the U.S. Senate in 1956 in the midst of the Snake controversy, argued that the Hells Canyon dam would boost the entire Northwest and bring to Idaho the benefits that Bonneville and Grand Coulee delivered to Oregon and Washington.[13]

Neuberger's opposition to Idaho Power Company's application for a license to build dams in Hells Canyon had everything to do with public-power development and nothing to do with protecting the Snake River gorge. In an address to the Albany Kiwanis Club in December 1953, he upbraided Secretary McKay for withdrawing "the federal government's claim to the magnificent power site in the heart of that great gorge." The interior secretary was "giving away to the Idaho Power Company . . . a site that should belong to all the people." Low-cost public power had spurred the Northwest's industrial development, Neuberger told the audience, and the Columbia River had the potential to do even more "under a program of long-range federal financing." McKay and congressional supporters like Cordon were wrecking the potential for a federal project in Hells Canyon on behalf of private utility corporations. Neuberger was especially critical of Idaho Power's scaled-back plan to build three run-of-the-river dams because they failed to "fully develop" the site. Finally, he asked his Albany audience, "if Hells Canyon is not safe, is another site safe?"[14]

Building a big federal dam in Hells Canyon became Neuberger's noble cause—his heroic fight against private monopoly, his struggle for public power. As a journalist and as a senator, Neuberger was convinced that federal hydropower development could take place without environmental consequences, that electricity generated from the region's abundant "power" streams would produce clean energy. If Republican opposition could be overcome, Hells Canyon promised to be a wonderful public enterprise. The problem, Neuberger wrote in *The Progressive* shortly before his run for the Senate, was the opposition of "the gold-dust twins of the anti-public power forces," Cordon and McKay. Once elected to the Senate, his position on public power remained steadfast. When he attended dedication ceremonies at Hills Creek dam in the Willamette Valley in September 1955, Neuberger enthusiastically backed river-development projects: "Without destroying any basic resources and without using up any quantity of raw materials, Hills Creek Dam will be providing flood-control benefits, low-cost water power, and recreation for countless numbers of Oregon's people yet unborn."[15]

Shortly after taking his seat in the Senate, Neuberger joined Morse in introducing a bill to authorize the construction of a federal dam in the great canyon of the Snake. "Hells Canyon is the greatest natural hydroelectric power site left on our continent," Neuberger told his Senate colleagues. Turning the canyon over to "private monopoly" would mean "far-less-than-full development." To properly serve the public's interest, it was necessary that the canyon be developed to its fullest capacity. Neuberger urged senators to uphold the progressive Republican principles of Theodore Roosevelt and Gifford Pinchot to "retain Hells Canyon for all the American people," and he appealed to Congress to adopt a water policy for the Columbia Basin that would "realize the full possibilities" of a magnificent resource.[16]

Oregon's journalist-turned-politician pulled out all the rhetorical stops in a news release just before the Hells Canyon hearings. Repeating many of the same remarks he had made on the Senate floor, Neuberger praised the canyon's natural beauty and its utility to human welfare:

> Hells Canyon is not an academic issue. It means a vital natural resource. I have walked through much of the marvelous canyon. I have ridden horseback down its side gorges. My wife, Maurine, and I have camped alone on its towering rimrock. I have voyaged by boat on the foaming river. I would hate to see this magnificent resource taken from the people of America and turned over to private monopoly.

Neuberger saw a parallel between those who opposed Grand Coulee Dam because it was a federal project and politicians who favored private development in Hells Canyon. Because the public project at Grand Coulee turned into "the greatest power project ever erected," surrendering the Snake River to private development would be unfair to future generations. Hells Canyon's remote, rock-walled chasm, Neuberger argued, offered "an unparalleled location for a water storage reservoir."[17]

In a "PERSONAL & *CONFIDENTIAL*" letter to Charles Sprague, publisher of the *Oregon Statesman,* the Oregon senator told the newspaperman that three Rocky Mountain senators had promised to vote for the Hells Canyon bill if Neuberger would drop his opposition to the controversial Echo Park Dam on the Colorado River. Critics who charged him with being a zealot for public power might raise their eyebrows, Neuberger wrote, "if they knew of this situation."[18] In the end, Neuberger refused to compromise, cosponsoring an amendment to the Colorado River Storage Project that eliminated Echo Park Dam (which would have inundated Dinosaur National Monument). When Congress approved funding for the Colorado storage project in 1956, the authorization did not include Echo

Park. In a bittersweet victory for David Brower and the Sierra Club, however, Congress authorized dams at Flaming Gorge and Glen Canyon. While
lawmakers deliberated over those authorizations, the FPC was moving to
approve the Idaho Power Company's proposal to build three dams on the
middle Snake.[19]

Although he could make claims to preserving spectacular Dinosaur
National Monument, Neuberger pursued much more ambiguous policies
for the Pacific Northwest—supporting a big federal dam in magnificent
Hells Canyon and opposing with a preservationist's fervor dams on the
Clearwater and Salmon Rivers. The latter two dams, he argued, would
"mean the virtual end of the commercial and sports fishery of the Columbia River." No project, in his view, would more adversely affect salmon
migrations than damming Idaho's Salmon River. But when it came to his
pet project for Hells Canyon, Neuberger was adamant. The deep gorge simply provided "superior advantage" for a high dam.[20]

The Hells Canyon controversy proved to be one of the most heated political struggles of its time. Despite Neuberger's tireless efforts to fund a federal project, shifting political winds kept the proposal off balance. Hostility
from most of the region's governors, Idaho's divided congressional delegation, the Eisenhower administration's opposition, and ultimately a U.S.
Supreme Court decision and the nation's emerging wilderness movement
conspired to kill the federal project. Although the Federal Power Commission awarded the long-sought license to Idaho Power in July 1955, Neuberger continued to criticize the "larceny at Hells Canyon" and the
unwillingness of the Eisenhower administration to authorize another
"Grand Coulee" on the Snake River. The FPC decision, he charged, would
rob the Pacific Northwest "of its major asset in attracting new payrolls."
Although he was able to get the Senate to pass an authorization bill for a
federal dam in Hells Canyon in 1957 (with the cosponsorship of Senators
Morse and Church), the measure failed in the House. A frustrated Neuberger wrote to James Marr of the Oregon Labor Council that the Hells
Canyon issue "occupied overwhelmingly the largest single fraction of my
time and energies through three sessions of Congress." When the Corps
of Engineers dropped the Hells Canyon project from its revised 308 Report

on the Columbia Basin (released in 1957), Neuberger accused the agency of retreating from its earlier flood-control plan.[21]

Although Northwest senators lost the fight to build a public-power facility in Hells Canyon, private-power interests achieved only a partial victory. Idaho Power Company subsequently constructed three smaller dams on the middle Snake: Brownlee (1958), Oxbow (1961), and Hells Canyon (1967). Within a year of the completion of Brownlee—the uppermost of Idaho Power's three dams—Albert Day, director of the Oregon Fish Commission, reported problems with upstream and downstream fish passage. The failure of a fish trap downstream from the Oxbow construction site at the peak of the 1958 run killed about half of the potential spawners. "The offspring of that ill-fated run" were in even greater trouble, Day reported, when surface temperatures in the Brownlee pool forced the downstream smolts to seek the cooler water in the bottom of the reservoir. And then the migration urge sent them "under or through the Brownlee net and through the turbines in large numbers." Research barges captured hundreds of chinook fingerlings in August 1959, most of them dead. To add to the debacle, seagulls hovered over the tailrace diving after the small chinooks. "This situation," Day remarked, "has the fishery agencies seriously worried." Photos of thousands of dead fish on the middle Snake attracted national attention and also pointed to the shortcomings of dam technology.[22]

In a February 1958 letter to the Forest Service Region 6 chief, Lyle Watts, Neuberger blasted the Republican-dominated Federal Power Commission for failing to protect migratory fish runs. Producing more power without endangering salmon and steelhead, he wrote, "is one of the most critical issues facing our region," and yet the FPC's recent decisions had failed to protect fish and wildlife. Idaho Power Company's careless construction activities at Brownlee Dam best illustrated the Commission's disregard of fishery issues, he charged. Although the dam was nearly completed, Neuberger pointed out that fish-passage facilities still had not been approved: "If such laxity on the part of the Commission is permitted to continue, what assurance do we have that future projects will not imperil the existence of valuable Columbia fish runs?"[23] Neuberger's political opportunism is striking; the big federal dam proposed for Hells Canyon would have dwarfed any of the Idaho Power dams and permanently blocked all migratory fish runs.

Within a year after Oxbow Dam was completed, salmon had disappeared from the upstream sections of the river. Writer Tim Palmer, who traveled the river from April to November 1988, estimates that the dams closed off approximately 3,000 stream miles to anadromous fishes. Moreover, the runs were extinct even before Idaho Power completed Hells Canyon Dam in the southern part of the canyon. That structure, 213 feet high, was far short of the proposed federal project's 582 feet.[24] For the remainder of his time in the Senate, Neuberger continued to oppose dams for the Clearwater and Salmon, referring to them as salmon sanctuaries, streams necessary for the survival of the fish. Putting up dams on those streams, he contended, "would mean irreparable damage to migratory fish." Except for Dworshak Dam on the North Fork of the Clearwater River, built by the Corps of Engineers, no dams were licensed for the Clearwater and Salmon systems. But Neuberger's speeches and writings about protecting salmon were considerably less passionate than his arguments for public power, a line of reasoning that he aggressively pursued on a broad scale during his years in the Senate.[25]

The long-running Hells Canyon debate continued into the era of modern environmentalism and the emergence of "quality-of-life issues." Both public and private interests jockeyed for position to build a dam near the confluence of the Snake and Salmon Rivers, downstream from the Idaho Power dams. When the Federal Power Commission granted a license in 1962 to the Pacific Northwest Power Company to build a dam at the High Mountain Sheep site upstream from the mouth of the Salmon River, both the Washington Public Power Supply System and the Department of Interior filed appeals. The case eventually made its way to the U.S. Supreme Court and a landmark decision in 1967, in which the justices remanded the issue back to the FPC. In a decision written by Justice William O. Douglas, the Court ordered the FPC to consider other alternatives—and "whether any dams should be built at all, not just which one." The Supreme Court decision, Sara Ewart observes, "dramatically refocused the Hells Canyon debate" and recognized the amenity values of the Snake River. Finally, Sen-

ator Church and his colleague, James McClure, joined with Oregon senators Mark Hatfield and Bob Packwood to create the 662,000-acre Hells
Canyon National Recreation Area in 1975.[26]

Cold War and national-defense requirements shaped much of Richard Neuberger's thinking about Columbia Basin water-development projects. In
that sense, he shared views similar to his senatorial colleague, Wayne Morse,
and Washington's influential senators Warren Magnuson and Henry
"Scoop" Jackson. Although he could write passionately about protecting
salmon, most of the development measures he supported—high dams,
building revetments, and dredging and channelizing streams—were popular with the public and detrimental in the long run to anadromous fish.
When The Dalles Port Commission asked Neuberger to look into the possibility of deepening the ship channel between Vancouver and the Bonneville Pool, the senator worked with the Army Corps of Engineers and
congressional committees to gain funding for the work. Because dredging
and maintaining a twenty-seven-foot navigation channel was vital to the
upstream shipment of alumina to the big plants above Bonneville, the matter also involved national defense. Neuberger wrote the Port Commission
that he would "do everything possible to help toward making The Dalles
an ocean port on the upper Columbia River."[27]

The Oregon senator reminded Samuel D. Sturgis, chief of the Corps,
that it had completed the authorized channel, 27 feet deep and 300 feet wide,
between Vancouver and Bonneville in 1949. By 1955, however, the channel
had filled with silt and the Corps was maintaining a depth of only fifteen
feet. Because the area above Bonneville was "entering a new phase of economic development," Neuberger wrote, it was necessary "to restore the
channel to project depth." The prospect of large ocean-going vessels moving upstream to The Dalles justified the work, because alumina production facilities were "essential to national defense." When Emerson Itschner
succeeded Sturgis, Neuberger told the new chief that he was seeking appropriations for the next budget cycle to deepen the channel. In the interim,
he asked if the Corps had the funds to begin dredging operations. "The
Harvey Aluminum Company and other concerns," Neuberger pointed out,
"anticipate a heavy volume of ocean traffic upstream to The Dalles," where

another $80 million aluminum smelter was under construction. Beginning the work immediately, he contended, would benefit "the commerce and prosperity of our region." Although the Corps lacked the funds until the next budget cycle, the agency dredged the channel to the required depth within a year of Neuberger's request.[28]

Richard Neuberger's term in the U.S. Senate coincided with a revolution in commercial shipping, a trend toward building ever larger ships requiring deeper and wider channels. When the Corps of Engineers conducted hearings in Astoria in 1958 to consider channel modifications for the Columbia River, Neuberger offered his full support. Because of changes taking place in inland waterborne commerce across the nation, the Oregon senator supported "immediate action" to dredge a deep-water channel upriver to the Portland-Vancouver area. The completion of the St. Lawrence Seaway Project and improvements proposed for the Gulf, East Coast, and Mississippi River shipping lanes were suggestive of the plans for bigger ships and increased traffic. Of the Columbia River's many "natural assets," Neuberger testified at the Astoria hearing, "the potential for navigation development is only equaled by the value of the energy which it can provide." Improving the river for navigation would provide a "great impetus" for industrial expansion in the Pacific Northwest. What the Columbia needed, Neuberger told the panel, was a deeper, wider channel to enable shippers to reduce their unit costs of transportation. He urged the Corps of Engineers to strive for such a channel so that the Columbia could "achieve its full potential as one of America's greatest rivers."[29]

Developing the region's hydropower capacity ranked first and foremost among Neuberger's political ambitions when he joined the Senate. When President Dwight Eisenhower announced the need for an intensified defense program in November 1957 to counter similar moves in the Soviet Union, Neuberger warned that national-defense needs would require cuts in federal social programs. Because he feared that the administration would target Columbia River projects, especially the construction of John Day Dam, Neuberger urged the president to include full funding for the John Day facility. The "newly intensified competition with the Soviet Union," he argued, called for accelerated development of the Columbia River's hydroelectric potential. Pointing to the Second World War and Bonneville and Grand Coulee's contributions to airplane production, he told the president, "energy is a fundamental measure of national strength."[30]

Well-planned Columbia Basin power development, Neuberger reasoned, "would add a margin of safety to the nation's energy reserves" and enable the United States to keep pace with the Soviet Union. Framing his argument in classic Cold War rhetoric, the Oregon senator warned that communism's greatest challenge was its ability to promote rapid, forced industrialization in backward countries. The United States was well equipped to meet this challenge if it continued its "public programs for developing basic economic strength and for underwriting minimum social standards." Neuberger believed the nation could build both missiles and dams and that water projects made a significant contribution to national strength. He agreed with Eisenhower that the American people were prepared for the "heavy burdens of national security and . . . international leadership." For its part, the River of the West had the potential to bolster the nation's defense and strengthen the region's prosperity. By ordering "full progress on construction of the Federal John Day project," the president could promote defense and security and sustain the economic health of the Northwest.[31]

Richard Neuberger's term in the Senate coincided with one of the most contentious and controversial issues in modern Oregon history, the termination of the Klamath Tribe and the conversion of the reservation's rich pine forests into the Winema National Forest. The culprit in this human tragedy was a misguided Congress whose ambition—to "get out of the Indian business"—was motivated by the seemingly lofty desire to end the federal trust relationship and to fully integrate Native Americans into American life. That goal was expressed in House Concurrent Resolution 108, passed in 1953, and Congress moved to sever the federal trust relationship with tribes deemed ready for termination. The targeted groups included several Oregon tribes with modest land holdings—and the Klamath, who held 880,000 acres of valuable timberland. Wisconsin's Menominee, another large tribe scheduled for termination, also held significant forest lands. Neuberger's part in the Klamath termination debacle was limited to the brief period when the Klamath Termination Act, Public Law 587, was working its way through Congress.[32]

Klamath termination took place in the midst of the postwar construction boom, a period when sawmills without significant forest holdings of their own were continually on the prowl for fresh timber. Along with many of his constituents, Neuberger feared that cheap timber would flood the market and bring economic chaos to the Klamath Basin and beyond. He told his friend Malcolm Bauer of the *Oregonian* that the state or the federal government should purchase the reservation to avoid having the timber "liquidated under extremely unfavorable conditions." Neuberger feared that the public would neither understand nor appreciate state or federal purchase, a move that would be "far less popular politically than a tax rebate or some other kind of melon-cutting." To forestall the rush to place reservations lands on the open market, he offered an amendment to the Klamath Termination Act to provide more time to look into the issue.[33]

Neuberger then prepared a bill to amend the Klamath Termination Act; his measure required federal purchase and management of the reservation lands by the Forest Service and the Fish and Wildlife Service. The Oregon senator held a three-day hearing in October 1957 to allow various interest groups to express their preferences. While the Western Forest Industries Association and the Weyerhaeuser Company promised higher prices for reservation timber, Klamath Falls civic groups and conservation organizations supported Neuberger's proposal. Interior Secretary Fred Seaton offered an alternative amendment in early 1958 that was similar to Neuberger's. When a sustained-yield clause was added to the Interior bill, the Oregon senator dropped his bill and the Senate approved the Klamath purchase measure. Because of differences in the House of Representatives version, however, the matter remained in limbo for the next few months.[34]

Two prominent Oregon newspaper editors, Robert Sawyer of the *Bend Bulletin* and Bill Jenkins of the *Klamath Falls Herald and News,* joined Neuberger in the fight for federal purchase of the reservation forests. Writing companion pieces for the March 1958 issue of *American Forests,* the two editors made extended arguments for federal purchase. According to Sawyer, the unresolved Klamath situation threatened "calamity to a region and the virtual destruction of supremely important natural resources." If the timber were auctioned on the open market, "an orgy of timber cutting" would take place followed by economic desolation. Federal purchase would protect sustained-yield harvest practices, maintain high log and lumber prices, provide family stability, and preserve the watershed. "The answer, and the only sensible answer," Sawyer wrote, "is the outright and undisguised purchase of the reservation, payment to the withdrawing Indians of their share of the price, [and] addition of the timber lands to the national forest reserves." The reservation marshlands, he added, should be turned over to the Fish and Wildlife Service.[35]

Klamath Falls' Bill Jenkins was even more emphatic in arguing for federal purchase of reservation timberlands. And he went even further, condemning the Klamath Termination Act as "that ill-advised document" hastily pushed through Congress that would wreak irreparable harm to the Klamath watershed. If the forest lands fell into private hands, Jenkins feared that rapid timber sales would contribute to a "boom-and-bust era of logging activity" and ultimately economic and social disaster. Equally important, rapid timber cutting would threaten the upper basin watershed, because half of the water flowing into Upper Klamath Lake originated in reservation forests. And then there was the question of Klamath Marsh, "one of the major nesting and resting areas for migratory waterfowl" that the government had an obligation to protect. The Klamath journalist offered a comprehensive and utilitarian assessment of the basin, emphasizing the need to protect the upper watershed because of its importance to irrigation agriculture. Plans to add another 200,000 acres of irrigated land in the upper basin by means of pumping and drainage "cannot be brought to fruition without the water to work with." Although he did not propose "the total preservation of the timber" to protect the watershed, Jenkins believed that federal purchase and sustained-yield practices would build the "industrial, agricultural and recreational potential of southern Oregon and northern California."[36]

Oregonian staff writer Ken David thought the unresolved Klamath issue had "assumed the shapeless form of a monstrous nightmare." Because the Klamath Termination Act had fomented controversy both inside and outside the tribe, it was necessary "to correct its errors." The values of the reserve, especially its giant marsh, David observed, could not be measured in hard cash. Klamath Marsh was a strategic feeding and nesting grounds for millions of migratory waterfowl. Moreover, the marsh and the reservation forest served as the source for much of the abundant water vital to power production, irrigation, and waterfowl and wildlife habitat in the lower basin. Because no one was satisfied with the termination act, David thought it critical for Congress to follow Senator Neuberger's lead and seek a solution that would avoid dumping reservation lands on the open market. With the current political climate in the nation's capital, he wrote, few were optimistic that Neuberger would succeed with Interior's federal-purchase bill.[37]

To rally House support for federal purchase, Neuberger proposed that an unofficial bipartisan group of Oregon legislative leaders and former Governor Charles Sprague visit the nation's capital. In a June 1958 letter to Malcolm Bauer, he suggested that an aggressive campaign was needed in the House to push the measure along. In the meantime, he told Bauer, his Senate Interior and Insular Affairs subcommittee would sit on the House-passed legislation so that he would have some bargaining power with members opposed to the Klamath purchase bill. "Klamath is so urgent to our state," he wrote, "that it would be a tragedy to have it bog down after our successful effort in the Senate." In a subsequent letter to Bauer, Neuberger complained of the "shameful" conduct of the National Lumber Manufacturers Association, the powerful trade group opposed to the measure.[38]

After a series of negotiations with Eisenhower administration officials, the House finally passed the purchase bill and the president signed the Interior bill—Public Law 731—in September 1958. As a sign of its gratitude, the Klamath County Chamber of Commerce held a community banquet in Neuberger's honor. The Oregon senator gave thanks to his senate colleagues James Murray of Montana and Clinton Anderson of New Mexico; to Oregon's Second District congressman Al Ullman; and to Governor Robert Holmes. Passage of the original termination act was "a grievous mistake," Neuberger told the Chamber, and "could have led to ruinous results." In addition to the promise of sustained-yield management that would protect "the moisture-retaining characteristics of the Indian forests," he told the Klamath Falls business group, the measure would bring other conservation benefits—adding 15,000 acres of reservation marshland to Klamath Basin migratory-waterfowl refuges.[39]

The Klamath purchase bill and the creation of Winema National Forest had everything to do with the rich stands of ponderosa pine, a point that historian Donald Fixico emphasizes: "In the Klamath Basin, the lumber companies became the real beneficiaries of the termination." Federal officials and humanitarians may have expressed concern for the welfare of the Klamath Tribe, but ultimately the real winners were the lumber firms, local banks, and other businesses dependent on Klamath timber. Although conservationists such as Richard Neuberger opposed placing the reservation forest on the market, Fixico correctly argues that they were more concerned with the land "than with the people who inhabited it."[40] After the passage of the Klamath legislation, Neuberger moved immediately to assure that local mills had access to a steady supply of logs. Writing to Herbert Stone, head of Forest Service Region 6, he asked that federal timber sales in the Klamath Falls area be accelerated while the details of Klamath termination were being resolved. Neuberger reported to Klamath Falls mayor Lawrence Slater in June 1959 that the Forest Service was indeed accelerating its sales to meet log requirements in the Klamath Basin. Because the log-supply situation "was especially acute," the senator thought the increased sales would help avoid mill closures.[41]

In an article published in *Harper's Magazine,* Neuberger claimed victory for both the Klamath Tribe and the local economy. Like most American Indian tribes, he wrote, the Klamaths had "an old history of sorrow," a legacy that promised to worsen in the early years of the Eisenhower administration when the tribe "seemed fated for new disaster." What promised to be a tragedy, however, "turned into one of the few almost-bright chapters" in the government's Indian policy. Amidst the mounting anxiety in the basin over the disposal of reservation timberlands, Neuberger wrote, Interior Secretary Douglas McKay appointed a panel of personal friends, "all staid and reliable Republicans," to look into the issue. The panel shocked McKay, Neuberger declared, when it recommended that "the government itself had to buy the Klamath Indian reservation, and fast." The senator applauded his colleagues for standing firm in the face of the National Lumber Manufacturers Association's "bitter and abusive attack" against purchase. He concluded the *Harper's* essay with praise for Oregon's citizens, especially the press, labor unions, and conservation groups.[42] Neuberger was convinced that he stood at the center of a public consensus in support of Public Law 731.

Laurence Shaw was one of the local lumbermen who benefited from the creation of Winema National Forest. The founder of the Modoc Lumber Company—soon to become one of the largest lumber firms in the Northwest—Shaw was instrumental in organizing local businesses in Klamath Falls to support federal purchase with its sustained-yield implications. In his thoughtful study of the basin's water difficulties, writer William Kittredge points out that Shaw "did good work on behalf of the Klamath Basin timber industry, and on behalf of his own business." The Modoc Lumber Company brought lots of jobs to Klamath Falls and Shaw himself became known for his local philanthropy. But, as Kittredge observes, "his was also a ruling-class career" where a small economically privileged group made all the major decisions. Times have changed in the Klamath Basin, and the hegemony of logging and milling has passed; today water-allocation fights threaten the area's once booming agricultural economy.[43] Richard Neuberger's vision of the conservation and prudent development of the basin's resources is a far cry from the experiences of Klamath residents in the twenty-first century who quarrel over scarce water supplies, endangered species, and surrounding forests that no longer promise economic prosperity.

During Neuberger's term in office, one of the more controversial proposals to emerge was that by the Eugene Water and Electric Board (EWEB) to build a $7 million hydropower project at Beaver Marsh on the upper McKenzie River. The elaborate scheme called for building a diversion dam at Clear Lake, the headwaters of the McKenzie, and an 8,400-foot tunnel that would carry the water past spectacular Koosah and Sahalie Falls to power two generators. The project included a second dam at Fish Lake and a "reregulating" dam farther downstream to maintain a constant flow in the river. At the June 1955 hearings on the proposal, Joseph Hayden, the Federal Power Commission's counsel, described the project as a trade-off between power and recreation benefits. Eugene Chamber of Commerce manager Fred Brenne told the hearing panel that the proposal was essential to community growth. In his opening statement, EWEB attorney Windsor Calkins assured everyone that the Beaver Marsh undertaking would not mar the area's natural beauty, nor would it affect fish life. Led by attorney Rollin Bowles of Portland, opponents of the scheme argued that the Beaver Marsh project would destroy "one of the world's greatest beauty spots" and do great damage to fish life.[44]

The Beaver Marsh proposal moved forward when the state water engineer approved the EWEB application in September 1955. Although the permit allowed a small dam at Clear Lake, it limited surface fluctuation of the lake to two feet during the recreation season (May to September) and eight feet during the remainder of the year. EWEB was also required to maintain a minimum flow through Koosah and Sahalie falls during the recreation season's daylight hours. Although there might be some turbidity during the construction phase, the state engineer believed the project would not detract from the area's scenic values. The *Oregonian* urged the State Water Resources Board to review the engineer's ruling, calling it "wholly unjustified" and a threat to beautiful Clear Lake and the scenic falls of the upper McKenzie. Responding to a letter from a Beaver Marsh opponent, Neuberger agreed that the development would irreparably harm Clear Lake and the upper McKenzie. "Some people seem to think," he wrote to University of Oregon biology professor James Kezer, "that a waterfall which can be turned on and off like the spigot of a beer barrel is the same as a natural cataract." He commended Kezer for exploding that argument in his testimony to the FPC.[45]

When FPC examiner Francis Hall issued a favorable report on the Beaver Marsh project in late October 1955, he remarked that "no timely exceptions" had been filed with the agency. The commission ruled that objections filed by the Save the McKenzie River Association had arrived too late. In a message to the FPC, Neuberger again voiced his disapproval, asserting that the project would do "irreparable damage to the recreational and scenic assets of one of America's extraordinary fishing, boating and hiking areas." Because Beaver Marsh would hurt the tourist trade and the citizens of Eugene, Neuberger vowed to block what "will prove to be a tragic mistake." It was unbelievable that EWEB "would imperil the magnificent McKenzie river wilderness" to gain a mere 30,000 kilowatts of power. With its huge

stake in the tourist trade, Neuberger wrote, Lane County would be "making a major blunder" if it permitted the upper McKenzie River waters to
pour into penstocks.[46]

Famed outdoor photographer Ray Atkeson also voiced his opposition
to Beaver Marsh, arguing in a letter to the *Oregonian* that the scenic beauty
of the upper McKenzie would be "spoiled forever by construction of a power
dam, diversion tunnel and transmission line." He urged citizens to write
the FPC to protest the licensing of the Beaver Marsh project and to protect spectacular Clear Lake and Sahalie and Koosah Falls. Atkeson regretted that comparatively few Oregonians were aware of "the scenic grandeur
of the upper McKenzie." He ridiculed the EWEB promise to protect the
area's scenic qualities by permitting water to flow over the falls on weekends to entertain sightseers. It reminded him of the old joke "of turning
the water on and off at Multnomah Falls" on the Columbia River. In his
work as a photographer of scenic places in the West, Atkeson ranked the
upper McKenzie one of the most beautiful he had ever seen.[47]

At a meeting in Neuberger's Portland home in late December 1955, out-
door and conservation leaders pledged to support the senator's efforts to
cancel the FPC license for Beaver Marsh. Representatives of the Oregon
Wildlife Federation, Isaak Walton League, Nature Conservancy, Mazamas,
Sierra Club, and Wilderness Society agreed to work through the Save the
McKenzie Association to achieve their objectives. Rollin Bowles of the Isaak
Walton League told Neuberger that the Beaver Marsh proposal was only
the first step in transforming the McKenzie from a scenic and recreational
stream "into a series of pipes and penstocks." In a letter to Luvilia Richards,
who lived in Blue River on the McKenzie, Neuberger said that he would
"do all possible legislatively" to stop the Beaver Marsh development. The
success or failure of his bill would depend on mobilizing public opinion
in its behalf. "This is where you come in," he told Richards, asking her to
spread the news among her friends.[48]

The bill to revoke the Beaver Marsh license brought Neuberger into
another confrontation with Interior secretary McKay,[49] who was in the
process of resigning his position to run against Wayne Morse. Testifying
in favor of Neuberger's bill before the Senate Public Works subcommittee
in March 1956, McKay accused the Oregon senator of attempting to embar-

rass him when Neuberger asked the Interior secretary about the timing of his sudden opposition to Beaver Marsh. Denying that he was trying to embarrass the secretary, Neuberger replied: "You are the one who used the word. I didn't." McKay shot back: "You've been very handy at trying to embarrass me, but you haven't succeeded." Despite the sharp exchange, McKay testified that the upper McKenzie was one of the places in Oregon "that aren't worth spoiling for a commercial project"—a remarkable admission for an Interior secretary who had gained a reputation for wanting to sell federal lands and resources to the highest bidder.[50]

To finance the Beaver Marsh hydroelectric project, EWEB needed public approval of a bond measure. The Eugene City Council called a special election for March 1956 to seek approval of an $8.5 million revenue plan, a ballot item described as "one of the most controversial issues to face Eugene voters in a number of years." To those debating the Beaver Marsh question, the differences were clear: EWEB argued that the project would help Eugene meet its future power requirements, while the Save the McKenzie Association cited the area's scenic and recreational values. When the votes were tallied on election day, March 27, citizens had rejected the bond by a narrow margin. The *Oregonian* praised Eugene voters for meeting "their civic responsibility in refusing to sanction a self-serving power raid on the incomparable beauties of Clear lake and the wild upper McKenzie river." The newspaper sensed that something else was afoot as well, a shift in public sentiment that favored "conservation and wise . . . use of our natural advantages."[51] Simply put, Beaver Marsh was not essential to Eugene's power requirements. The Eugene *Register-Guard*, which belatedly opposed the project, declared that the McKenzie in its natural state "provides enjoyment and satisfaction that is increasingly hard to find in our complex way of life." The river would not have those "inspirational qualities if it is marred with hydroelectric projects."[52]

But the proponents of Beaver Marsh would not go quietly. Although Eugene citizens had turned down the bond measure, the license remained active. One can imagine Neuberger's alarm when he learned that the utility had requested a two-year extension to begin construction of its McKenzie facilities. He immediately fired off a letter to the Federal Power Commissioners opposing such an extension, because it would give legitimacy to a project with only "dim prospects of becoming a reality." Beaver Marsh had been highly controversial from the beginning, and there was nothing to indicate that Eugene voters had changed their minds. He told the commission that he had already urged Forest Service chief Richard McArdle to classify the upper McKenzie as a recreation area. In light of those developments, he advised the commission to reject the application to amend the license. In a telegram to Robert Frazier, publisher of the *Register-Guard*, Neuberger repeated his opposition, noting that voters had already passed judgment on the proposal.[53]

Although the Federal Power Commission granted an eighteen-month extension, the grandiose project was dead. Fourth District congressman Charles Porter, who called the FPC move "an affront to the citizens of Eugene," urged EWEB to assure the public that no attempt would be made to exercise the permit. The FPC's decision prompted the *Oregonian* to urge Oregon citizens who fought to protect the upper McKenzie to renew their activism. When the Eugene utility submitted a drastically scaled-back hydropower proposal for the upper river in late 1957, there was only a ripple of opposition. Conservationists testifying at a Eugene hearing were divided. Dan Allen, director of the Oregon chapter of the Izaak Walton League, called the new Carmen-Smith plan "a sincere effort to give consideration to the scenic and recreational as well as the economic values of development." Eugene's Karl Onthank, president of the Western Federation of Outdoor Clubs, was opposed to any structures on the river between the town of McKenzie Bridge and Clear Lake.[54]

The new plan, which "wisely has moved downstream," drew none of the opposition stirred by the former proposal, primarily because it did not interfere with Clear Lake or Koosah and Sahalie Falls. The *Oregonian* called the Carmen-Smith works "a sensible alternative to the Beaver Marsh project." The plan also called for a reregulating dam further downstream at Trail Bridge to protect water flows in the lower river. EWEB expected to build recreation facilities at the lower reservoir to accommodate the boating, fishing, and camping public. Although the *Oregonian* strongly opposed the Beaver Marsh scheme, its editorial staff was pleased to congratulate local citizens for recognizing the aesthetic values of the spectacular upper McKenzie. As for Neuberger, he thought Carmen-Smith was a reasonable compromise; it protected Clear Lake and the beautiful upper falls and added to Eugene's power supply.[55]

An even more controversial water project reached a turning point during Richard Neuberger's years in the Senate: the Portland General Electric Corporation's (PGE) decision to build a hydropower dam in the Deschutes Gorge. The most scenic and spectacular waterway entirely within the state of Oregon, the Deschutes River has its beginnings in snowmelt and glacial streams in the central Cascades. As it winds northerly through open grasslands, it is joined by the Little Deschutes, and then descends through lava-created cataracts and falls. Below the city of Bend, the river enters a deep gorge where the Crooked River, which drains the Ochoco Mountains and the semi-arid fringes of Oregon's high desert, augments its flow. A short distance further downstream, the clear, cold Metolius, which emerges from the ground near the base of Black Butte, adds to the Deschutes' volume. Famous for its native trout and salmon and steelhead runs, the Deschutes was a renowned destination for sports fishers, especially with the explosion of automobile travel after the Second World War. The so-called Pelton Dam, proposed for the deep canyon, became an object of public debate when PGE first floated the project in the late 1940s.

PGE rested its decision on Department of Interior "withdrawals" of hydropower sites in the early twentieth century. The Pelton site and another one just upstream at Round Butte were among the reservations in the state of Oregon. The Pelton case, which eventually wound up in the U.S. Supreme Court, involved private-power development interests, national defense needs, a reputed Northwest energy shortage, federal-versus-state jurisdiction, and the maneuvering of PGE consultant Thomas Robins, a retired Corps of Engineers officer. PGE first floated its proposal in early 1949 when a consortium of regional power companies (PGE, Pacific Power and Light, and Washington Water Power) formed the Northwest Power Supply Company and advanced plans to move ahead with the Pelton proposal. Located in the 800–foot-deep Deschutes canyon and bordered by federal lands (including the Warm Springs Indian Reservation), the proposed dam would flood approximately eight miles of the gorge. Writing to the *Oregonian* in defense of the project, Colonel Robins praised the

Pelton location because it was suited to quick construction and would help meet the region's power shortage.[56]

The debate that followed was heated and lengthy. Ralph Cowgill, a retired Bureau of Reclamation engineer, pointed out that state law prohibited dam construction unless affiliated state agencies such as the Oregon Fish Commission and the Oregon Game Commission granted their approval. Neither agency had sanctioned the Pelton project. Cowgill singled out the grave dangers such a dam posed to the celebrated Metolius fishery; once the dam blocked the Deschutes, "the Metolius will be just another tributary to another man-made lake." Building Pelton, Cowgill contended, would be another example of "the progressive destruction of one stream after another." Robins responded that Oregon needed new industries which would mean "some changes in the character of the state—in the way its citizens live and enjoy themselves." Although the Pelton project meant changes for the Deschutes River, scientific advances suggested that such dams could be built without hampering the commercial and recreational fishery. Robins pointed to PGE's offer to include a modern fish hatchery downstream from the dam site, a development that he believed would enhance fishing in the Deschutes. The region's increasing population, Robins argued, meant more consumers of electricity: "Power dams on our streams are an inevitable consequence."[57]

By the time the Pelton proposal was moving toward the licensing phase, the United States was embroiled in the Korean War and national-defense needs overrode other priorities in natural-resource decision making. PGE had assumed sole sponsorship of the Deschutes enterprise, with Robins directing strategy. In its application to the Federal Power Commission, PGE asked for a "certificate of necessity," a move that would speed the licensing process in the interests of national defense. Colonel Robins informed a former colleague in the Corps that PGE would proceed with the request for a "necessity certificate" and FPC license "regardless of opposition by the fish people." Before the FPC or Oregon officials approved the Pelton project, PGE also requested federal permission to build a dam at the Round Butte site—a dam that would back water eight to ten miles up the Deschutes, Metolius, and Crooked Rivers. At the FPC license hearing for Pelton, conservation and fishery groups cited an earlier federal agreement to build no more dams below McNary. W. J. Smith of the Oregon Wildlife Federation opposed the Pelton project because there were so few Columbia tributaries accessible to salmon. He told the commission that no hatchery had ever succeeded in maintaining salmon runs.[58]

The *Oregonian* also opposed the Deschutes proposals, as well as Tacoma City Light's companion application to build two dams on Washington's upper Cowlitz River. In its effort to turn the Deschutes River over to power production, the newspaper warned, PGE intended to "override state authority, unless legally restrained." It accused PGE of disregarding state law and relying solely on the Federal Power Commission's licensing authority. The *Oregonian* reminded readers that the departments of Commerce, Agriculture, and Interior, the Corps of Engineers, and the governors of seven Pacific Northwest states had agreed in 1947 to prohibit hydroelectric projects on both the Cowlitz and the Deschutes. The legal issue, the Portland newspaper asserted, was whether a federal agency could "override the water rights reserved to the states." It urged the seventeen western land-grant states to join with Oregon and Washington to defend their long-established water rights. Nonetheless, PGE expected little difficulty with its second license, for the much larger Round Butte project. The utility reasoned that the Pelton decision would "open the door to progress on the Round Butte project without any serious objection." PGE also noted that the Confederated Tribes of the Warm Springs Indian Reservation supported the projects.[59]

At this point, the Oregon Hydroelectric Commission joined the fray, denying the PGE permit because Pelton Dam would eliminate anadromous fish runs above the project. The commission based its decision on the findings of fisheries biologist Willis Rich, who reported that the dam would block the Deschutes to migrating fish, thereby violating Oregon law: no practical or viable plan could be devised to "preserve and maintain the run of salmon and steelhead presently breeding in the Deschutes river and tributaries above the site of the proposed Pelton dam, should it be constructed." The Pelton matter moved directly to the courts in July 1952, when the state of Oregon challenged the Federal Power Commission's constitutional authority to license projects on navigable waters wholly within the boundaries of a state. The U.S. Supreme Court agreed to review the case in late

1954 after the Ninth Circuit Court of Appeals in San Francisco upheld Oregon's objections to the license on the grounds that it would destroy the Deschutes fishery. The Pelton issue was basic, the *Oregonian* declared, addressing the question of "whether the state or the federal government has controlling authority to determine the best use of non-navigable waters wholly within a state."[60]

In its June 1955 decision the Supreme Court upheld the Federal Power Commission's license for Pelton Dam. In a seven-to-one vote, with only Justice William O. Douglas dissenting, the high court reversed the Ninth Circuit Court, arguing that such licenses were federal constitutional matters and that state authority did not extend to federal lands set aside as hydropower sites:

> [There is] no question as to the constitutional and statutory authority of the FPC to grant a valid license for a power project on reserved lands of the United States, provided that as required by the act, the use of the water does not conflict with vested rights of others. To allow Oregon to veto such use, by requiring the state's additional permission, would result in the . . . duplication of control.

In blustery rhetoric, the *Oregonian* thundered that the Pelton case was "a staggering blow to sovereignty of the western states over their internal, non-navigable waters." The Pelton decision meant that Oregon and other western states had no jurisdiction in those cases where lands "have been 'reserved' for specific uses by an administrative order of a federal official." There were no legal options at this point, but the *Oregonian* thought the state could regain control over its internal waterways by seeking redress in Congress. With the assistance of populous California and other western states, "Oregon's congressional delegation should go to work on this immediately."[61]

Shortly after the Supreme Court handed down its Pelton decision, Richard Neuberger was hard at work seeking legislative action to reverse the verdict. He received a flood of letters from constituents urging him to back any proposal that would give Oregon control of its waterways. The big question, one writer observed, was "whether or not the states are to be allowed to regulate entirely domestic matters." The Izaak Walton League's Rollin Bowles thought the Supreme Court decision placed Oregon "completely at the mercy of the Federal Government" in determining its water policies. "Several pieces of corrective legislation" would be necessary, he told Neuberger. Oregon attorney general Robert Thornton, who argued the case before the high court, wrote Neuberger that the decision would have "tremendous and far-reaching effects on all of the western states." While he considered filing a petition for a rehearing, Thornton thought the best course would be a legislative solution. To move the process forward, Neuberger asked a Senate subcommittee to determine how the Pelton decision would affect state administration of intrastate and non-navigable waters.[62]

Neuberger acted on several fronts to thwart the Deschutes projects, joining with Senator Morse in introducing a bill to revoke PGE's license to build Pelton Dam. He received both praise and criticism for his efforts. Portland attorney Jerome Bischoff commended the junior senator for introducing a bill "to put an end to the P. G. E. Pelton Dam Project" and disputed the Portland Chamber of Commerce position that there was a relation between industrialization and people's happiness. "There is very little left of Oregon in its natural state," Bischoff wrote, "and what little remains, ought to be well taken care of." J. L. Waud of Madras, a small town only a few miles from the proposed dam, asked Neuberger and Morse to consider "the little fellow" in central Oregon, an area where people were not wealthy and needed an economic boost. Readily accessible sections of the Deschutes, he wrote the senator, were already "owned by Portland people who have locks on gates and no trespassing signs on them." Building Pelton and Round Butte Dams would create more public recreational space and help develop central Oregon. Waud accused Neuberger of attempting to "stop development east of the mountains." He asked the senator to "think of us . . . as being part of Oregon and not to be used by Portland."[63]

William Stollmach, affiliated with the St. Charles Memorial Hospital in Bend, wanted to register a "gripe" about the senator's position on Pelton Dam. Although he owned no stock in PGE, Stollmach thought the "Commercial Fisheries and Sportsmen's lobby" were giving the company "a rough deal." He belittled the notion that there was an appreciable salmon run in the Deschutes and pointed out that the proposed Pelton site was "almost inaccessible to fishermen." Because it was not a multipurpose dam, Stoll-mach could not understand Neuberger's opposition. "There is a time when even the P. G. E. is entitled to a break." In response, Neuberger noted that the FPC had licensed the project despite the objections of Oregon agen-cies concerned with preserving fish runs in the state's rivers. His congres-sional bill, the senator wrote, "would reverse the legal basis on which this disregard of state law was upheld." Neuberger responded more forcefully to another Stollmach letter, telling the Bend resident that he had long believed "that we should not blockade our small rivers with dams" until the federal government had "fully tapped the power potential" of the Columbia and Snake Rivers.[64]

Oregon's senators were fighting time in their effort to pass a bill deau-thorizing the Pelton project. Through an agreement with Washington sen-ator Warren Magnuson, Neuberger gained a limited hearing before the Senate Interstate Commerce Committee to withdraw the Pelton license. Magnuson agreed to Neuberger's request because "several interested wit-nesses from Oregon" would be in the nation's capital to testify on fishery bills before his committee. Despite the Commerce Committee's crowded schedule, Neuberger thought it "worthwhile to have a beginning on hear-ings." Even though testimony would be limited, Neuberger believed it was important "to initiate the collection of a committee record" on the bill. He urged his friend Rollin Bowles to submit a written statement and to pro-vide a brief oral summary if he planned to be in Washington. Among the exhibits submitted to the Magnuson committee was a letter from Erskine Wood—son of Oregon's famous raconteur, poet, lawyer, and military personage C. E. S. Wood—questioning the FPC's wisdom in licensing the Beaver Marsh and the Pelton projects. Wood thought the commission should review the Deschutes proposal and give adequate consideration to fish and wildlife.[65]

In opposing the Pelton project in the spring of 1956, the *Oregonian* cited "a new and powerful segment of public opinion" favoring the prudent use of the state's natural resources. PGE did not need the kilowatt capacity of the Pelton project, and public opposition to the undertaking was growing. More important, if there were "a statewide plebiscite on PGE's program to turn the Deschutes into a series of fluctuating pools," such a measure would suffer an overwhelming defeat. There would be no balloting on the Pelton issue, and Neuberger's effort to void the PGE license was proving futile. He learned in May that the company was proceeding with construction, although he insisted that the action was "in defiance of state law." The *Oregonian* reported in June that there was little hope Congress would act on Neuberger's efforts to revoke the PGE license. Most damaging to his initiative, Neuberger thought, was the lack of support from the Oregon governor's office. Republican governor Elmo Smith, who succeeded to the office on February 1, 1956, when Governor Paul Patterson died in a plane crash, supported private power-development and refused to testify against the Pelton project.[66]

The Neuberger bill to suspend construction until state agencies approved the Pelton project was the target of praise and condemnation before the Senate Interior Committee. Portland Chamber of Commerce spokesman Harold Say accused the senator of violating the spirit of long-established legislative principles in his efforts to rescind a measure already approved. "If the various states were to adopt a practice of outlawing licenses after they granted them," Say told the committee, "there would be chaos." The City of Tacoma's director of utilities, with an interest in licensing projects of his own, opposed Neuberger's idea because it would divest Tacoma of its rights to build two dams on the Cowlitz River. Approving Neuberger's bill, he testified, was not in the best interests of the states or the nation.[67]

In a June letter to his former assistant, Herb Lundy (who had joined the *Oregonian* staff), Neuberger said he thought his bill would have succeeded except for "the total absence of the state whose sovereignty presumably had been violated in the Pelton case." He criticized the testimony of two PGE officials for minimizing Pelton's adverse affects on fish runs—"as did their stooges from the various Chambers of Commerce." He told Lundy that he still held out hope that there "would be a very close vote on the Neuberger amendment." With the bill still lodged in committee, letters protesting his opposition to Pelton continued to arrive in his Washington office. A Lakeview woman reminded the senator that cheap electricity would help farmers "far more than turning the Deschutes into a sportsman's fishing park." She accused Neuberger of wanting "to turn Oregon in to a sports man and fisherman's paradise." The Madras Chamber of Commerce cited a lengthy list of business enterprises that had invested heavily in central Oregon with the expectation that Pelton Dam would be completed. The Chamber pointed to "the large amount of local employment" that "should merit some consideration by you." The Confederated Tribes of Warm Springs tele-graphed Neuberger, opposing his efforts to scuttle the project. The tribes had an agreement with PGE guaranteeing an annual minimum payment of $90,000 from hydropower sales.[68]

With construction in the Deschutes canyon proceeding apace and his efforts to rescind the FPC license going nowhere, Neuberger issued an ambiguous and petulant news release in September 1956. Damming the Deschutes, he wrote, would now "stand as a symbol for the insecurity of Oregon farmers' water rights." The real issue was no longer the dam but Oregon's ability to control the streams within its boundaries. The PGE claim that it had "a paramount right" to build Pelton Dam (upheld by the Supreme Court) meant that it did not need to comply with state water law. Neu-berger claimed he had taken "a consistent attitude from the beginning": the construction project would bring only a temporary and local boom, which he thought "an exorbitant price to pay for jeopardizing eternally the water rights of farmers and imposing another barrier to the passage of fish." His effort to restrict FPC's licensing power, he now argued, was to prevent Pelton from becoming "a precedent for further imperilment of irri-gation water rights." Finally, he again blamed Governor Smith, who had taken "a stand against the irrigators on Oregon farms when he opposed my amendment."[69]

The Pelton controversy slowly faded from public attention as con-struction continued deep in the Deschutes Canyon during the late sum-mer and fall of 1956. In December, the *Oregonian* reported that the dam was assuming "recognizable shape" as workers drilled and placed metal structures in the bedrock, poured concrete around the giant penstocks, and laid the foundations for the powerhouse. The completed project would be "a cinch to lure tourists and vacationers," especially to see the world's longest fish ladder winding along the eastern wall of the canyon for about three miles. At the upstream end of the fishway, engineers had designed a "skim-mer" that would create a current "to attract the surface-swimming fry into the ladder." The recreation area, complete picnic facilities, camping sites, and a modern boat ramp, would be accessible to thousands of Oregonians and, especially to people living in Portland, a mere two-and-a-half-hour drive away.[70]

Portland General Electric was proud of Pelton's fish-passage design and praised its engineering features in its monthly circular, "Electric Light and Power." PGE chairman Robert Short invited Neuberger to visit the facilities when they became operational in May 1958 and he boasted that the success of the fish-passage facilities would be a major step toward "ending the fish versus dams controversy." Before the end of the year, Neuberger expressed his support for Round Butte Dam, as long as state wildlife authorities approved. Although he had opposed the downsteam Pelton project, he acknowledged that he was "overruled, . . . and damage already has been done to the Deschutes as a source of fish life. It stands to reason that our state now should obtain the jobs and kilowatts from Round Butte Dam." Because he considered Round Butte "proper for construction by a private utility company," he announced that he would urge the FPC to license the project.[71]

When Portland General Electric filed its Round Butte permit with the Oregon Hydroelectric Commission, its application asked permission to build a 440-foot-high rock-filled dam just downstream from the Metolius River. If it were approved, it would be the largest hydroelectric project within the state's borders. Because of the height of the dam, Neuberger wrote to PGE official Thomas Delzell expressing hope that the fish-passage design would not compound the problems that caused him "to oppose your original Pelton project." Although he would still prefer the Deschutes and Metolius preserved for sports fishing and recreational purposes, the Pelton decision had "placed this question behind us." And then he added: "I hope that my attitude toward the Round Butte dam will serve at least as evidence that my attitude toward proposed projects is not determined by any doctrinaire public vs. private power test."[72] Neuberger neglected to mention that he originally opposed Pelton Dam in part because its electricity would be more expensive than the big federal projects on the Columbia River.

Fish passage through the Pelton facility experienced difficulty from the beginning. Albert Day of the Oregon Fish Commission reported that warm surface waters flowing through the three-mile-long passageway delayed fish movement in the summer of 1959. To correct the problem, PGE officials began pumping cooler water from the depths of the reservoir into the fish ladder to encourage upstream movement. PGE's aquatic biologist, George Eicher, declared the experiment a success in September when he announced near-record numbers of anadromous fish passing above Pelton Dam. Although there was "a certain amount of ladder rejection" at first, Eicher declared that "water pumped from the bottom of the reservoir, free of algae growth," had solved the problem. However, while some fish mastered the ladder, still others were trapped below the dam and had to be hauled upstream. Testifying before the Oregon Water Resources Board in 1960, Eicher was even more confident about the success of the fish passageway at Pelton Dam. He told the board that the construction of the second dam should pose no additional fish problems; indeed, there was every likelihood that Round Butte would have a beneficial effect on fish life. Charles Campbell, a biologist with the Oregon Game Commission, replied that he knew of no Columbia Basin hydroelectric project that had increased migratory fish populations.[73]

Despite his public support for the Round Butte project and his assurances to PGE officials that his concern was for the fishery, Neuberger's private correspondence continued to criticize building dams on the Deschutes River. Still recovering from cancer surgery in late 1959, he wrote Gervais science teacher Bill Aldridge that Round Butte would destroy "the spectacular Cove Palisades State Park, . . . a most regrettable and unfortunate development." Neuberger told Aldridge that in his testimony before the Federal Power Commission's examiner, he had voiced "strenuous opposition" to the project. In another letter, to Cordelia Murphy of Portland, he thanked her for sharing her opposition to dam building in the Columbia Basin. But the sands were running out, both for the opponents of Round Butte and for Richard Neuberger's frail health. The senator died on March 9, 1960, and the Federal Power Commission's examiner recommended approval of Round Butte Dam in June.[74]

The giant construction project, employing nearly 1,500 workers during the peak of construction activity, was completed in 1964. The combined production from the two dams—300,000 kilowatts at Round Butte and 124,000 at Pelton—was equivalent to four-fifths the output of Bonneville Dam in the mid-1960s. Equipped with a tram to lift upstream-migrating fish from a trap at the base of the dam and a "skimmer" to trap ocean-bound smolts, Round Butte Dam was a marvel of technical efficiency. Or so it seemed. Within a season, it became apparent that smolts were not finding their way through Lake Billy Chinook, the huge reservoir behind the dam. The fish-passage facilities were an abysmal failure, and PGE shut them down in 1966. The ancient salmon and steelhead runs in the upper Deschutes Basin were extinct.[75] There was no attempt to resurrect the runs until PGE applied to relicense the dams in 1999. When the Confederated Tribes of Warm Springs submitted a competing license application, PGE quickly generated a new program to reestablish salmon and steelhead runs above the dams. Still in the exploratory stage, the new approach involved experiments with tracking devices to help biologists understand fish movement through Lake Billy Chinook. A PGE biologist observed that there were no assurances that the new experiment would prove successful.[76]

If the Deschutes River gained regional and national prominence during the 1950s and 1960s, western Oregon's meandering and troubled Willamette River, flowing through the state's most productive agricultural land and its major population centers, captured even more attention. Eventually, the charismatic and ambitious Tom McCall used the Willamette's story to thrust himself into state politics.

8

TOM McCALL AND THE
STRUGGLE FOR THE WILLAMETTE

Pollution in the Willamette River system is a state shame. . . . This magnificent river is at present in part an open sewer in which tremendous quantities of untreated human sewage and industrial wastes are disposed.—*The Fishes of the Willamette River System in Relation to Pollution* (June 1945)[1]

Tests conducted by *The Oregonian* reveal dangerous chemicals in Portland-area fish.—*The Oregonian*, December 17, 2000

For much of the twentieth century, nearly three-fourths of Oregon's population has lived in the Willamette Valley. Its 11,250 square miles have harbored the state's three largest cities, its major industries— sawmills, food-processing facilities, pulp and paper mills, flax-retting plants, packing houses, creameries, dehydrators, and textile mills—and, beginning with the Second World War, a growing number of technical and chemical manufacturing plants. The Willamette River, which had long served as a trafficway for commercial shipping and moving logs to mills, was increasingly treated as an industrial "commons" as the twentieth century advanced. The combined effects of dumping human and industrial wastes into the basin's waterways had turned the river's main stem into an open sewer by the 1920s. In many ways, the Willamette was a proper fit for Lewis Mumford's classic remark in *The City in History* that the industrial waterway was "less a river than a flood of liquid manure."[2] Towns, cities, and industries from Cottage Grove and Eugene downstream to Portland

248

poured untreated effluent into the Willamette system, treating valley water-ways as one of the externalities of doing business. Until the 1930s, there was little concern for the tertiary effects on the environment and public health. This story is one of the darker sides to the valley's history in the twentieth century.

Not until 1926 did the Oregon Board of Health and the Fish and Game Commission call the first official meeting on stream pollution, a gather-ing that included public engineers and officials from most of the larger Willamette Valley communities. The *Oregonian* compared the Illinois River in the Midwest with the Willamette and urged citizens to never let their river become "so shamed and shameful a stream, . . . a reproach to our civilization." Although the Willamette had carried away "the waste materials assigned to it" thus far, the newspaper warned that population numbers would soon bring the river to "the point of saturation." Evidence presented at the stream-pollution conference suggested that the river was nearing that point. In another study of the river's condition, the Portland City Club reported in 1927 that the river was "filthy and ugly" and cited Portland as the worst offender.[3]

George Gleeson's significant 1972 study of pollution in the Willamette River concluded that by "1929 the Willamette River could be considered to be polluted and grossly so in the lower reaches." The state's epidemiol-ogist reported in the early 1930s that "the odor, stench, and filthy appear-ance of the river violate our aesthetic sense." The increasing number of investigations and reports indicated that the river was most toxic during the low-flow months of late summer and early fall, when the volume of water in the stream was insufficient to dilute the urban and industrial effluents discharged directly into the waterway. To compound those difficul-ties, some industrial polluters insisted that wastes were simply part of doing business. Providing jobs, after all, was an important contribution to the community's welfare.[4]

When a particularly egregious fish kill took place in the Willamette Slough below Portland in October 1935, outraged public officials accused the St. Helens Pulp and Paper Company of dumping untreated effluent directly into the slough. State game warden Walter Ryckman told the *Ore-gonian* that boiling-hot, obnoxious-smelling effluent from the mill was being poured into the slough. Staff chemists at the mill, however, denied that the company was responsible and accused fisheries people of taking samples inside the company's holding dike. The game commission's Fred Halde-man countered that his samples were taken from the slough proper. As for the dead fish, the chemists reported that they could find none in the vicin-ity of the plant. State officials argued that mill personnel were in denial, because there were numerous reports that thousands of dead silver salmon and cutthroat trout were seen floating in the slough. After reading staff reports about the fish kill, state game supervisor Frank Wire ordered the St. Helens company to immediately stop the "dangerous flow" of pulp wastes into the slough.[5]

The Willamette Slough fish kill, similar incidents elsewhere, and the river's growing notoriety finally convinced the state legislature to pass a pollution-control law. When the Oregon Anti-Stream Pollution League brought the measure to the 1937 legislative session, opponents charged that the bill would put paper-manufacturing plants in the state out of busi-ness. One legislator argued that it would force the two big plants at Wil-lamette Falls to close operations. Despite those admonitions, the legislature passed the modest bill—only to have conservative Democratic governor Charles Martin veto the measure. The Izaak Walton League and the Ore-gon Wildlife Federation countered the governor's veto and backed an initiative—the "Water Purification and Prevention of Pollution Bill"—that voters approved in November 1938 by more than three to one (247,685 to 75,295). The measure established the Oregon State Sanitary Authority and marked the painfully slow beginnings of a statewide pollution-abatement program.[6] Because of Depression-era constraints, the legislature appro-priated little money for the agency during the 1930s, and then the Second World War put on hold the state's efforts to clean up its waterways.

Writing for the *Oregonian* in January 1945, Virgil Smith observed that the big water-pollution battle crossed a major hurdle when Portland vot-ers approved a $12 million bond issue to build new sewer lines and a sewage-treatment plant. Reviewing projects planned elsewhere on the Willamette system, Smith predicted: "The rivers of Oregon are going to flow sweet, clear and clean again." With the city of Portland promising to mend its ways, upriver communities would be forced to stop dumping sewage and industrial by-products into the river. Although it was difficult to interest the public in water pollution, the *Oregonian* writer was certain that people were interested in jobs, especially with the end of the war looming and returning servicemen on the hunt for work. Installing sewer systems and building treatment plants would serve a civic purpose and—according to the *Oregonian*—as postwar make-work projects. Everyone understood that manufacturing sewer pipe and sewage-treatment machinery, laying sewers, and building plants would mean "all kinds of jobs, from the professional engineer down through all the ranks of wage-earning effort."[7]

As the transoceanic conflicts were slogging toward their gruesome ends, Governor Earl Snell ordered the Sanitary Authority to survey the condition of the Willamette River and its tributaries. The governor directed the Authority to compare its findings with a 1929 study, to assess the magnitude of sewage and industrial wastes still being dumped into the river, and to determine the effects of pollutants on fish life. The survey results were not encouraging: Oregon's major waterway was "seriously contaminated" during the summer and fall months in several sections below Cottage Grove and "badly polluted" downstream from Salem. Even more significant, the pollution load had dramatically increased since 1929, especially in upstream sections. In addition to its findings for the main stem, the Authority found that several tributaries were "severely contaminated." Its survey found that discharge pipes dumped effluent along the riverbanks of several cities during low-flow months. To compound the difficulties of treating urban sewage, cities mixed domestic sewage with industrial wastes from canneries and textile plants.[8]

The most striking feature of the Sanitary Authority survey was its recognition that the dilution method of treating sewage was no longer an option. It was unfortunate, the agency reported, that "the period of minimum stream flow is coincident with the period during which the heaviest seasonal waste loads are discharged." Building upstream reservoirs to store winter precipitation would provide some relief during the summer and fall, but the only effective solution would be the installation of sewage-treatment facilities. The present condition of the river, according to the report, "seriously affected" fish life, especially species intolerant of water with a low oxygen content. The Willamette River, the survey concluded,

was in a "septic condition" during the late summer when high temperatures and heavy seasonal wastes severely depleted oxygen in the stream.[9]
A parallel fisheries report was even more blunt: "Pollution in the Willamette
River is a State shame." Although Oregonians had long prided themselves
on living in one of the nation's premier recreational locations, the state's
largest internal waterway was "an open sewer in which tremendous quantities of untreated human sewage and industrial wastes are disposed." Pollution was so severe that it had seriously damaged a world-famous
recreational and commercial fishery and had ruined chinook salmon
spawning grounds. The condition of the Willamette River, the fisheries survey concluded, was a civic matter, because "the people of Oregon have
sacrificed a heritage in the aesthetic value of clean water."[10]

With the return to peacetime, *Oregonian* editorials took a more critical
position on water pollution issues. In a May 1946 editorial, the newspaper
called pollution "our crime against the Willamette and its fish life," viola-
tions against nature that provides "its own penalities for culpability and
innocence." The river's condition was "a cancerous thing to which we contribute and in which we persist." The condition of the lower Willamette,
the editorial charged, warranted "our social indictment." Restoring the river
was a matter of civic, even biblical redemption:

> The return of the silver salmon in their former numbers to the Willamette,
> a restoration which can be attained only through our own reformation, would
> mean more than the revival of an economic asset, far more. It would mean
> the restoration of our self respect, with all that this connotes. For we shall
> be judged by our river as we are judged.[11]

If swimming in Portland Harbor was inadvisable, the same could be said
for points far upstream, including the University of Oregon's recently refur-
bished millrace. After students, alumni, and Eugene residents had successfully raised thousands of dollars, workers repaired the old millrace and
water once again passed beneath the overhanging trees to the east of campus. "Looks beautiful," the *Oregonian* editorial declared, a setting for
canoes, swimming, and romantic moonlight evenings. But there was an illu-
sory reality to that bucolic setting, the newspaper reported: "The mill race

is an open sewer. And wham! There goes romance." How many students would get starry-eyed "floating down the open sewer?" Because Springfield was the only sizable settlement above Eugene and the river was unfit for "a casual dip" at the University of Oregon campus, the *Oregonian* asked "what must it be" by the time it reached Portland?[12]

Following the river surveys requested by Governor Snell, the Sanitary Authority developed a series of strategies to control pollution. As a first priority, the agency required municipalities to construct primary sewage-treatment facilities and to chlorinate effluent. In lieu of more intensive levels of treatment, the authority believed that Willamette Valley Project reservoirs would provide increased stream flow during the critical low-water months, thereby flushing contaminants downriver and out of sight (and smell). In the years following the war, towns and cities along the Willamette system began installing primary sewage-treatment plants: Newberg and Junction City (1949), Portland (1951), Salem (1952), and Eugene (1954). After Harrisburg completed its sewage plant in 1957, every sizable Willamette River community had primary treatment facilities. When the Corps of Engineers completed the two largest Willamette Project dams in 1953 and 1954 (Detroit and Lookout Point), the reservoirs provided storage to significantly increase the river's summer flow, a technical improvement that officials hoped would serve as a flushing mechanism to dilute contaminants.[13]

The most visible symbols of the troubled Willamette River were the two large pulp and paper mills across the river from each other at Willamette Falls. Although there were three other pulp and paper mills at other locations on the river, the Willamette Falls facilities had been operating since the 1890s. The Crown Zellerbach Corporation and Publishers Pulp and Paper Company plants were physical representations of industrialism; the *Oregonian* suggested in January 1948 that the buildings and smokestacks reminded one "of a manufacturing city of New England." When Crown Zellerbach announced plans to add four new buildings on the western side of the river, the *Oregonian* commented that the expansion would "add strength to the industrial sinews of Oregon City." Although sawmills, grist mills, and woolen manufacturing had contributed to Oregon City's welfare, pulp and paper manufacturing had been the lifeblood of the local economy since the turn of the century.[14] The two big plants, both pouring untreated sulfite liquors directly into the water, would continue to serve as public reminders of the river's difficulties for another two decades.

To the Oregon division of the Izaak Walton League, the pulp and paper industry was Public Enemy Number One. The most relentless group pressing for immediate action to clean up the Willamette River, the League's stream purification committee accused the industry of "stalling" to put off the renovations needed to treat sulfite liquors. The League repeatedly asked the Sanitary Authority to take more aggressive action against the industry, even to the point of closing the mills until they installed treatment equipment. The driving force behind the League was David Charlton, a longtime conservationist who operated Charlton Laboratories, testing facilities often involved in water-quality assessments. In a 1947 letter to Harold Wendell of the Sanitary Authority, Charlton admitted that municipalities were making some progress in building sewage-treatment plants, but he accused the pulp mills of doing nothing. Crown Zellerbach officials, he charged, "were saying about the same thing today that they told the State Planning Board . . . in 1935."[15]

In a December 1948 report from the League, Charlton argued that little had been accomplished in the ten years since the passage of the state's antipollution act. The *Oregonian* paraphrased Charlton's report and declared the Willamette River "more grossly polluted than ever chiefly from pulp and paper mills and food processing plants." Despite the Sanitary Authority's encouragement, Charlton charged, those industries had accomplished little in pollution abatement. In their defense, industry public-relations people responded that research had not yet identified an economic and practical method for disposing of sulfite liquors. Harold Wendell told the *Oregonian* that the Sanitary Authority was trying "to work out a reasonable and practical program without throwing thousands of workers out of employment." He disputed Charlton's comment that there were profitable outlets for the by-products of the paper-manufacturing process.[16]

When he sent a copy of the League report to the Oregon State Game Commission, Charlton warned the agency that the Sanitary Authority was not enforcing the state's laws against pollution. The head of the Authority, he charged, had taken a complacent attitude toward the pulp and paper industry, leaving the impression that mills would not be required to take action until the city of Portland completed its sewage-treatment project. "Nothing in the law justifies this position," Charlton wrote. "We have all been more or less lulled to sleep by certain types of propaganda and frequently recurring reports of progress." The Sanitary Authority and other state agencies should not accept at face value the industry's "shadow" research program, because it had little merit.[17] If David Charlton was a burr in the Sanitary Authority's hide, he had lots of support among Oregon's growing conservation community and commercial and sports fishers who wanted the state's waterways freed of pollutants.

Commodity values permeated discussions about pollution in the Willamette system. For sports and commercial fishers, clearing toxic substances from the waterway meant returning salmon and steelhead, steady incomes for people who serviced sports fishing, and more fish for the big canneries on the lower Columbia River. The League used similar arguments: clear, clean waters in the Willamette system would protect against disease, improve fishing, and attract tourists to the valley. J. W. Smith, a League member, told Portland's commissioner of public works that eliminating pollution in the Willamette and lower Columbia would benefit "our entire business community, from the standpoint of increased tourist and recreation facilities of all sorts." Harold Say of the Portland Chamber of Commerce was another who believed that restoring streams would bring material benefits. Pollution was a "shameful thing," because it meant "a loss in a fundamental recreation revenue that could be built up on the Willamette and other waters if they were clean."[18]

As one might expect, the cannery and pulp and paper industries put a very different spin on the pollution issue. Treating effluent before it entered waterways was a huge expense that hitherto had been borne by treating the river as a vast commons; constructing holding ponds and treatment facilities threatened profits. Pulp and paper company representatives repeatedly told the Sanitary Authority that they lacked the revenue to comply with the agency's pollution-control directives. Charlton and the League urged the companies to seek ways to turn the sulfite waste liquor into salable by-products. Charlton sent a copy of the League's 1948 report to Walter Murphy, editor of *Chemical and Engineering News,* drawing attention to the canneries and pulp mills' egregious dumping practices. In a return letter, Murphy agreed that turning sulfite wastes into alcohol or some other by-product held promise, but there was no legitimate argument for continuing to dump wastes into streams "simply because there is a slight difference in manufacturing costs."[19]

The sulfite-waste question became more heated when the Sanitary Authority discussed the possibility of ordering the industry to stop dumping wastes into the Willamette system. An attorney for Publishers Paper told the agency that it would be impossible for his company to comply with the order because there was no economically feasible method for disposing of wastes. If the order were issued, Publishers Paper would have to go to court to protect its interests. Harold Wendell responded that litigation might be a good thing. While Oregon's pollution law had been in effect for ten years, the state had been hesitant to enforce the regulation because it had been sensitive to the companies' problems. Other powerful individuals also pleaded the pulp and paper industry's case. Charles Sprague, publisher of the Oregon Statesman and a former governor, told Charlton that the mills faced a heavy capital outlay; he doubted that by-products would be profitable. The Sanitary Authority, he wrote, had to be lenient: "We must maintain our economy; and none of us wants to restore the virgin purity of our streams by reverting to a state of nature."[20]

As the state's most widely circulated newspaper, the Oregonian exercised a powerful influence in shaping public attitudes toward environmental issues. Following the Second World War, the Portland paper became increasingly critical of municipalities and industrial firms who continued to use the state's rivers and streams as dumping grounds for wastes. Taking some pride in what it saw as Portland's accomplishments in passing bond levies to clean up its share of pollution, the Oregonian praised the Sanitary Authority's stiffening position toward municipal and industrial polluters. "The situation has been soft-pedaled too long," the paper charged in a December 1949 editorial. When the Sanitary Authority ordered pulp and paper companies to cease polluting Willamette waterways by December 31, 1951, the move indicated that the agency was "through with temporizing." Acknowledging that the pulp and paper companies faced a serious problem, the newspaper noted that the industry failed to recognize that dealing with pollution should be part of the cost of production. Even worse, the companies had "unconscionably delayed their efforts to solve it." The editorial praised David Charlton and the Izaak Walton League for their reasonable suggestions about waste-treatment processes.[21]

The industrial groups were sensitive to the public-relations consequences of their activities. When Crown Zellerbach's lawyer told the Sanitary Authority, "if you're fixing a deadline for us, we're planning to shut down," company officials accused newspapers of misinterpreting his testimony. Company vice president F. N. Youngman issued a statement to let the public know that no threat was involved. Crown Zellerbach was working hard to solve the sulfite waste-disposal problem and would continue to do so:

> But if lawful authorities insist that pollution caused by discharge in the Willamette river of such industrial wastes shall be abated in accordance with minimum requirements of the state sanitary authority pertaining to stream cleanliness not later than December 31, 1951, and a workable method is not available by that time, Crown would have no alternative but to obey the order of the lawful authority and shut down its operations.

The company was not threatening to close its mills and did not want to leave that impression, Youngman concluded. If the mill closed, it "would result only from Crown's inability to comply with orders made by lawful authority."[22] Nice public wordsmithing, but the threat remained.

The pulp and paper industry continued to play a waiting game, alternately asking for more time, because of the complexities and costs of treating wastes, and threatening to close operations if the state insisted on an arbitrary date for ending pollution. When the industry's nationwide trade affiliate, the National Council for Stream Improvement, became concerned with sulfite wastes in the Willamette system, it contracted with the University of Michigan's School of Public Health to investigate conditions in the waterway. The results were not encouraging; pollution was increasing in the vicinity of Salem and conditions were worsening in the pool above Willamette Falls. "Under the 1950 pollution loads," the study indicated, "oxygen depletion approaching complete exhaustion can be expected in the critical reach between Newberg and Willamette Falls at a summer drought flow." In its summary findings, the University of Michigan survey found the river's condition to be in worse shape than in any previous year.[23]

Following another set of hearings in February 1950, the Oregon Sanitary Authority set a new deadline, May 1, 1952, for the five paper companies on the Willamette system to end their practice of dumping wastes into the Willamette system. The agency extended its original deadline from December 1951 to May 1, 1952, to allow the companies more time to determine appropriate disposal processes. In a finding of fact, the Sanitary Authority reported that the mills had made "little or no actual progress" toward reducing pollution. The agency also revealed that pulp and paper mills at Newberg, Lebanon, Salem, West Linn, and Oregon City were responsible for 84 percent of the total pollution load in the main river (exclusive of that from tributary streams and the City of Portland). The Lebanon mill alone contributed more than 90 percent of pollutants in the South Santiam River. When the ban went into effect, mills would be prohibited from dumping oxygen-depleting sulfite liquors into waterways from July through October and whenever the rivers fell below a specified minimum flow. The Sanitary Authority also directed pulp mills to submit treatment and disposal plans for its approval. Agency officials reminded industry representatives that pulp mills elsewhere in the United States were using a variety of strategies to mitigate their pollution problems: storing wastes in lagoons during low-water periods, storing for purposes of evaporation, treating the effluent with lime to remove solids, burning the by-products, and a variety of other techniques.[24]

Despite the Sanitary Authority's efforts, the pollution-abatement dance continued as the agency periodically granted one-year extensions to the pulp and paper mills to allow the companies to install treatment facilities. The Sanitary Authority reported in early 1952 that the state had achieved only minor gains in its industrial waste-abatement program. But the delays and extensions continued as the Authority struggled to bring pressures against recalcitrant mills and lagging municipal authorities. In March 1952, the Sanitary Authority again pushed ahead the deadlines for the mills; the agency agreed to monitor progress on a monthly basis and granted the companies more time to work through a variety of waste disposal experiments. The *Oregonian* reminded readers in the spring of 1952 that the state's increasing population meant greater recreational use of its waterways, circumstances that ran counter to the discouraging fight against pollution. "State authorities," the newspaper confessed, "seem to have fallen behind public opinion in the matter of enforcement of the antipollution law."[25]

The state Sanitary Authority was active on another front, threatening legal action against ten municipalities to force them to complete sewage-treatment facilities. The agency required Newport, Lebanon, Mount Angel, Vale, Nyssa, Coquille, Myrtle Point, Toledo, Wheeler, and Nehalem to complete construction by specific dates or face legal action. State sanitary engineer Curtiss Everts reported that some major metropolitan centers—Portland, Eugene, and Salem—had voluntarily installed facilities to keep contaminants and wastes from streams. Thirty-seven sewage-disposal and treatment systems had been built since the Second World War, and another eighteen would be completed by the end of 1952. But municipalities facing legal action had delayed construction despite repeated warnings. Cities argued with the state over technicalities. Lebanon, which dumped its untreated sewage and industrial wastes into the Santiam River, objected to installing a secondary filtering system. And the Sanitary Authority cited the small farming community of Mount Angel for dumping effluent from its thirty-seven-year-old septic tank directly into the Pudding River.[26]

The ever cautious *Oregonian* praised the Sanitary Authority for moving "slowly and wisely to achieve its ends through persuasion and diplomacy." As the agency neared the end of its long and difficult task, it deserved "the gratitude and congratulations" of all citizens for attempting to enforce laws against stream pollution. The Portland paper touted the state's progress in cleaning up its waterways: the biggest polluter of them all, the City of Portland, had completed a $16 million sewage-treatment plant, and the pulp and paper plants along the Willamette were showing an increasingly cooperative attitude toward the Sanitary Authority. Under the agency's patient pressure, Publishers Paper Company had discovered that sulfite liquor could be used as a binder to settle dust for unpaved roads. On the west side of the river, Crown Zellerbach Corporation had built a big lagoon to hold sulfite wastes during the low-flow months of the year. The pulp wastes, the *Oregonian* speculated, could then be dumped into the river during the winter months. Further up the valley on the Santiam River, Crown Zellerbach's Lebanon operation was experimenting with evaporating and burning wastes. Although those combined developments were promising, the newspaper warned that the Sanitary Authority's patience was not endless: "No one has a right to pollute the public waterways. . . . The law is clear, and the people will insist it be observed."[27]

Oregon's laws prohibiting pollution may have been clear, but the municipalities and industries' lack of compliance continued to bedevil public officials. The state's political leadership during the 1950s lacked the will to force compliance with anti-pollution legislation. The Sanitary Authority would threaten local governments and targeted industries, but without the governor's support, progress seemed glacial. The agency's director repeatedly complained that municipal authorities sometimes delayed putting bond issues before the public; when they did, the voters often rejected such measures. The pulp and paper companies were more direct, threatening to shut down their plants if they were forced to comply with the Authority's directives. At the same time, citizen protests against individual polluters in a variety of industries continued to mount. Residents along Salem's Pringle Creek, for example, petitioned the Sanitary Authority and a Marion County court in 1957 to prohibit the United Growers' cannery from dumping wastes in the stream.[28]

Despite the *Oregonian's* claims, the City of Portland still had not resolved its problems with the Willamette River. State and federal officials met in Portland at a wide-ranging U.S. Public Health Service conference, in September 1958, to determine the sources of pollution in the Columbia River between Bonneville Dam and Cathlamet, Washington, some sixty miles downstream from Portland. The City of Portland and pulp and paper mills at Camas and Longview, Washington, drew the brunt of the criticism at the gathering. According to testimony, Portland contributed about 90 percent of the bacterial contaminants flowing into the big river, while the mills were indicted for dumping industrial wastes. Other witnesses indicated that some outlying sections of Portland continued to dump untreated sewage into the Willamette and Columbia Rivers. Water immediately below Bonneville Dam, according to one federal official, was clear and of good quality, but in the Portland-Vancouver area the concentration of coliform organisms was ninety-seven times the level considered safe for swimming. Portland's public-utilities manager told the conferees that the city had already spent more than $19 million to build interceptors and sewage treatment works and that voters would be asked to approve an additional $5 million bond measure later in the year to complete the interceptor system and two small treatment plants.[29]

Harold Wendell of the Sanitary Authority reprimanded Portland for the design of its bond measure, pointing out that the city included the sewage issue with a package of $34 million in other requests. But he assured the Public Health Service officials that his office would use "its power to see that Portland does what it is required by law to do." It was important that the city be brought into compliance, Wendell testified: "We don't propose to let this violation continue." Murray Stein, conference chair and the federal official charged with enforcement of interstate water standards, summarized the discussions by calling for collective action to eliminate pollution in the lower Columbia. Bacterial contamination from untreated sewage, he concluded, was a hazard to recreation and fishing in the river and jeopardized its use as a potential water supply.[30]

The *Oregonian*, which had been assuring readers that Portland was doing its part to clean up the Willamette River, defended the city's effort, citing the expenditure of millions of dollars and the construction of interceptors, pumping stations, and the primary treatment plant. The newspaper acknowledged, however, that more needed to be done, especially providing sewage facilities for small nearby settlements not linked to the intercepting system and treatment plant. The Sanitary Authority would, it warned, soon act to enforce compliance. To no one's surprise, the agency ordered the city to stop dumping its sewage wastes into waterways in mid-

October 1958. Because of the pending bond election, the Authority offered the city a grace period of several months to submit an action plan. The order followed another Sanitary Authority report indicating that domestic and industrial wastes were being discharged directly into the Willamette River from twenty "outfall sewers" along the Portland waterfront. Because of bacterial contamination, the Portland Harbor area and downstream "cannot be considered safe for recreational use."[31]

In his defense of the city, Portland mayor Terry Schrunk told hearing officials that the upcoming ballot measure indicated the city's "good faith." Other city officials were less cooperative, charging that the Sanitary Authority had never established water-quality standards for the lower Willamette and Columbia. City engineer Gordon Burt questioned whether the Authority had considered "the dollar value of recreation on the lower Columbia compared to the dollar cost to keep it completely free of pollution." One Sanitary Authority commissioner shot back: "What is the value of human life?" Burt responded that the city should be praised for providing parks and swimming pools so that citizens could avoid the "polluted, treacherous Columbia River." Portland's city attorney also rejected the Sanitary Authority's charge that the city was responsible for the polluted Columbia Slough.[32] Before proceeding with corrective measures, city officials had to wait on public approval of the capital-improvement levy.

As might be expected, voters turned down the substantial catch-all levy and left Portland's elected officials embarrassed, facing the possibility that the Sanitary Authority would haul the city into court. While other Willamette Valley municipalities had made major advances in pollution abatement, Portland stood alone, an outcast in the state's efforts to clean up its waterways. The *Oregonian* published a cryptic editorial about the city's dilemma:

> Whatever the city's legal position might be, from the moral and emotional standpoint it is certain to look bad if it becomes a defendant in such a case. The State Sanitary Authority is fighting for clean rivers, for motives of public health and welfare. Thus it automatically is on the side of the angels. One who opposes it plainly is aligned with the forces of evil, whether he likes it or not. The only smart way out of such a situation is to take the necessary steps to avoid getting into it.

Although the newspaper defended Portland's record in fighting pollution ("until recent years"), it urged the City Council to hold a special election with the sewage-disposal measure as a single ballot item. Portlanders appreciated the need for clean rivers, the *Oregonian* reasoned, and would "not want to be on the wrong end of an argument with the State Sanitary Authority."[33]

While Portland's city council sorted through its options, state and federal agencies continued to coordinate their efforts to crack down on industries and municipalities responsible for polluting the lower Columbia. The Sanitary Authority, Washington's Pollution Control Commission, and the U.S. Public Health Service adopted a program in August 1959 to end industrial pollution of the waterway by March 1961: no pollution would be tolerated that interfered with water quality suitable for domestic or commercial purposes. Industries would be required to submit regular reports indicating their pollution-control plans, and all municipalities were ordered to install sewage-treatment facilities sufficient for "effective disinfection" of wastes. The state and federal agency report once again cited the City of Portland for its "unsatisfactory" pollution-abatement efforts.[34]

Despite threats of lawsuits and the anguished pleading of city officials, Portland voters did not approve a bond measure to fund the long-delayed sewage treatment facilities until 1960. By then, the Sanitary Authority was requiring cities and towns along the Willamette River to build secondary treatment plants to further purify wastes before they entered the waterway. Portland's new facility came on line in 1973 at the cost of more than $16 million.[35] But pollution problems with Oregon's largest interstate waterway were far from being resolved. Economic considerations, population increases, industrial expansion, and other environmental disturbances slowed the pace of improving water quality in the Willamette. An Oregon Department of Commerce investigation of Willamette Basin water pollution in the early 1960s turned up the usual findings—the most serious problems occurred during low-flow periods, especially in the lower reaches of the river. While water quality "remained essentially static" above Newberg, it was "definitely worsening" from Newberg downstream to Oregon City. Below Willamette Falls through the Portland Harbor area, conditions were "far below acceptable water quality standards." Of the major sources of water pollution, industrial wastes were responsible for most of the problems. The Commerce Department study warned that "the market process cannot be relied upon" to achieve water purity.[36]

Before the Army Corps of Engineers completed the full complement of Willamette Valley Project dams, seasonal low flows and years of lower-than-average precipitation provided little flushing action to move pollutants downstream into the Columbia. During a period of higher-than-average precipitation patterns between 1953 and 1963, water quality above Salem was "quite satisfactory" for the critical months of July, August, and September and "satisfactory" during the same months for the stretch between Salem and Newberg. Although the Willamette's condition remained problematic between Newberg and Portland Harbor, the Sanitary Authority proudly boasted of improvements in many sections of the river. But the agency's pollution-abatement claims were severely tested in the late summer months of 1965, when the Willamette River ran unusually low and slow through the flat valley landscape. The low-water conditions reduced the river's ability to carry its still considerable pollution load, created crisis conditions on the lower sections, and gained the attention of Interior secretary Steward Udall—who warned that Willamette Basin dams alone were not the final solution to pollution control. Other politicians, including state treasurer Robert Straub (who had eyes on the governor's office), criticized the Sanitary Authority and Governor Mark Hatfield for their inability to curb industrial pollution. The Army Corps of Engineers responded to the crisis by releasing additional flows from its upstream reservoirs to help dilute pollutants in the lower river. The *Oregonian,* ever willing to moralize on issues involving the Willamette River, observed that diffusion did not effectively move contaminants through the waterway: "Pollution must be stopped at the source, for health reasons, for recreation, for fish and wildlife, and for industrial use of water."[37]

Although it still had enforcement problems, the Sanitary Authority persisted in its efforts to tighten water-quality standards for the Willamette River and its tributaries. It ordered pulp and paper mills to implement year-round procedures to remove solid wastes and to provide secondary treatment during the low-flow summer months. Despite the increasing number of storage dams on the Willamette system and the augmented flow of the river during the summer and early fall, however, water quality showed no appreciable improvement.[38] Because primary sewage treatment did not prove adequate, the Sanitary Authority directed Eugene, Salem, and Newberg to build secondary treatment facilities, and in 1967 the agency implemented a policy of universal secondary treatment. The Authority included a proviso that additional processing might be necessary for certain industries. By the mid-1970s such practices were universal throughout the basin.[39]

George Gleeson's 1972 study of the Willamette found that water quality was below acceptable standards during the low-flow months of 1967.[40] Although municipalities and industries were reducing the volume of wastes dumped into the waterway, the sheer volume of human numbers and industrial activity in the basin compromised those gains. The river had also become the center of more intense state politicking by the 1960s with the emergence of two progressive political figures, Robert Straub, a Democrat, and his Republican counterpart, Tom McCall. Straub, who became wealthy through a timber and contracting business, began his political career as a Lane County commissioner and then won a seat in the Oregon Senate before his successful run for state treasurer in 1964. A working journalist with Portland's KGW-TV, McCall had served as Governor Douglas McKay's top aide between 1949 and 1952 and then lost a closely contested election in 1954 to Edith Green for Oregon's Third Congressional District. Ten years later, at the age of fifty-two, McCall was elected secretary of state. For both Tom McCall and Bob Straub, the health of the Willamette River was a political priority.[41]

During his years with KGW-TV, McCall's evening commentaries gave increasing attention to Oregon's pollution problems, especially the persisting difficulties with the Willamette River. With the support of his boss, Tom Dargan, McCall began working on a documentary to reveal both the volume of pollutants still being dumped into the river and the failure of the Sanitary Authority to enforce its own regulations. According to his biographer, Brent Walth, McCall "was stunned by what he found." His findings proved that Oregon was developing the same symptoms of enviromental destruction that had been seen in other states. McCall spent nearly a year on the project, and KGW-TV aired the expensively produced one-hour documentary, *Pollution in Paradise,* on November 21, 1962. Although he never appeared as a talking head, McCall narrated the entire script in his distinctive and familiar voice, saving his sharpest criticisms for the pulp and paper industry whose effluent "churns at river's bottom, forming into rafts that rise to the surface in sluggish, foul-smelling masses of filth." Using film footage showing the Crown Zellerbach mill at Oregon City dumping thousands of gallons of bubbling pulp wastes directly into the Willamette, McCall posed the question: "Do we have a right to ask why more hasn't been done by more people?" Brent Walth provides a pithy assessment of the documentary: "McCall had produced a smashing report void of the timidity and equivocation rampant in local television reports of that era." He might have added that McCall had staked out a moral position in the pollution debate that placed Oregon's livability at the forefront of public discussion.[42]

Pollution in Paradise was a tour de force, pressing home the powerful argument that there was no contradiction between jobs and the state's livability. A healthy economy was consistent with quality of life and a clean environment, and it was time for the Oregon Sanitary Authority and state politicians to recognize the connection. The *Oregon Journal* called the film "something of a shocker," saying that it had bite and would step "on some toes." Copies of the documentary were sent to educators and community groups, and a second showing was scheduled in January 1963, shortly before the state legislature was to convene. Because of the momentum generated by *Pollution in Paradise,* state senator Ted Hallock introduced a successful bill giving the state the authority to shut down polluting companies. Brent Walth argues that the measure signaled a new political spirit in Oregon, ultimately placing environmental concerns at the forefront of Oregon politics for the next two decades.[43]

His presence as a television commentator vaulted Tom McCall to the pinnacle of statewide politics, first as secretary of state and then to the governor's office in 1966. In his successful campaign against Bob Straub, his erstwhile political friend, the Willamette River provided background for election commentary and served as a focal point for political debate. Straub attacked outgoing Republican governor Mark Hatfield for being "grossly derelict" about the Willlamette's problems; McCall, attempting to avoid an intra-party fight, focused on the future, arguing that Oregon should avoid the example set by southern California. When Straub tried to use the Willamette River's persisting pollution problems to gain an edge, the charismatic McCall simply agreed with him. In the end, the governor's race was an uneven although polite contest, with the flamboyant television journalist easily besting his opponent. Because the two agreed on so many issues, McCall later remarked, "he and I had the 'Tom and Bob show.'"[44]

Time has not diminished the significance of Tom McCall's coming to Oregon's governorship. While he shares credit for several wide-ranging environmental achievements, it was the Willamette River that brought the governor to national attention. In truth, McCall's two terms in the governor's office were serendipitous, paralleling significant improvements in water quality in Oregon's major river and the completion of the last of the Willamette Valley Project dams. Only three months after his inauguration, McCall appointed himself to the Sanitary Authority, signaling his determination to enforce the agency's regulatory powers. With broad public support, the Oregon legislature strengthened the state's anti-pollution laws, creating a mandatory waste-permit program that allowed officials to monitor and limit the volume of wastes being discharged into the state's waterways and to establish compliance schedules. But Oregon did not act alone in its efforts to clean up its rivers. Federal initiatives aided the state's effort. Congress passed legislation in 1965 requiring states to meet certain standards for interstate waters; in Oregon's case, this included the Willamette River upstream to Oregon City. Another congressional initiative in 1965 established the Federal Water Pollution Control Administration, the immediate predecessor to the Environmental Protection Agency; the new federal agency's reports provided strong support to spur industries to control pollution. Coupled with those new initiatives, the federal government also provided municipal grants to fund the construction of waste-treatment plants.[45]

The Oregon legislature further strengthened the state's pollution regulatory authority when it incorporated the Sanitary Authority into the newly created Department of Environmental Quality (DEQ) in 1969. DEQ assumed the Sanitary Authority's jurisdiction over water quality and enjoyed expanded statutory powers:

to provide that no waste be discharged into any waters of this state without receiving the necessary treatment or other protective action to protect the legitimate beneficial uses of such waters; to provide for the prevention, abatement, and control of new or existing pollution; and to cooperate with other agencies of the state, agencies of other states and the Federal Government in carrying out these objectives.

The agency's administrative rules and regulations included provisos that the state should strive for the "highest and best practicable treatment and control" of its waters and that it restrict sewage and industrial discharges affecting water quality; the new agency had the authority to set scientific standards for water purity.[46]

Two years after the legislature established the new department, McCall appointed irascible and outspoken L. B. Day to head the agency. "L. B.," as he was known to friends and detractors alike, became McCall's point man in pressing the governor's environmental program. Day's most notable public moment took place in the spring of 1972 when the DEQ director ordered Boise Cascade Corporation's Salem plant to close until it met the state's regulatory standards for waste treatment. Employing 650 workers, the plant was an important contributor to Salem's economy; the downtown pulp mill also spewed noxious odors into the atmosphere and dumped untreated wastes directly into the Willamette River. When DEQ investigators found Boise Cascade in gross violation of the state's waste-discharge regulations in July 1972, Day (with McCall's support) had agency lawyers draw up an injunction to close the plant. Before a court could issue the injunction, Boise Cascade closed the plant on July 25, strongly hinting that responsibility rested with McCall and an over-zealous DEQ. When angry union workers marched on the Capitol the following day, McCall told the infuriated crowd that management was using the employees as scapegoats for its failures to install effective pollution controls. And then, within hours after DEQ filed its request for an injunction, Boise Cascade agreed to meet the state's requirements for air and water discharges. The confrontation, Brent Walth argues, "made for great political theater" but was probably unnecessary.[47]

In retrospect, it is obvious that Boise Cascade officials had been putting off pollution abatement improvements for its aging mill as long as possible. Built in 1920 and producing business, writing, and printing paper, the plant was a visible reminder of industrial blight in the Willamette heartland. Two years before its celebrated confrontation with the state, Boise Cascade proudly announced a $6.4 million water-pollution abatement renovation that would be completed in June 1972. The water-quality improvements, which involved burning sulfite liquor (hence, worsening air pollution), promised to remove 80 percent of the spent wastes discharged into the Willamette River. At the news conference announcing the effort, resident manager C. J. Fahlstrom admitted that Salem's air quality would suffer for a short period of time until the company could experiment with equipment to reduce the new airborne pollutants. The company also promised to store the waste water in a large holding pond for an additional eight months to further treat it with oxygen before discharging the water into the river.[48] For two years following the 1970 news conference, the company cited an array of excuses for its failure to meet the state's pollution-abatement requirements: equipment failures, faulty pollution-control technology, and the costs of installing new and untried apparatus.

Even with the advantage of hindsight, events seem to have moved at a dizzying pace during the McCall years. In an interview with Brent Walth, Barney McPhillips, who had served six previous governors on the Sanitary Authority Board, reflected upon the spirit of the early McCall administration: "No governor before Tom ever cared much what we did. Tom was the only one to take an interest in what we were doing and push us. The only thing Tom didn't understand was that these things took time. He wanted action immediately." But progress was evident on every front in cleaning up "point-source" pollution during McCall's years in office. All municipalities along the Willamette system provided both primary and secondary sewage treatment by 1972; with one exception, pulp and paper mills and canneries were meeting the state's requirements for waste treatment. An equally important variable contributing to the river's improved condition was the greater volume of water released from upstream reservoirs during the low-flow months—a technological fix that provided a valuable cleansing action, diluting pollutants and flushing them downstream. The cubic-feet-per-second (cfs) figures for September on the lower river were striking: 5,400 cfs (1966); 7,000 cfs (1968); 10,000 cfs (1970); and 12,000 cfs (1972). Without the augmented flow, a contemporary study cautioned, "investments in treatment plants on a river such as the Willamette would not produce the same improvement in water quality." The report also warned that the river's healthy condition was no guarantee that it would remain so into the future.[49]

Even as the McCall administration was waging war against pulp and paper mills, two downriver chemical plants that produced tons of pesticides were regularly dumping waste material into sloughs and shallow ponds adjacent to the Portland Harbor area. James "Dick" Forney, a truck driver for one of the plants, remembers dumping drums of waste into Doane Lake, a small wetlands area on Portland's industrial west side, during the 1960s. Some time later, company workers pumped water from the lake into the Willamette River and then used bulldozers to fill the lake with dirt. The Environmental Protection Agency would eventually declare the fill in the old lake bed among the most polluted in the entire harbor area. At another nearby chemical-production site, similar activities during the 1960s contributed to a dioxin buildup that seeped into groundwater and began migrating toward the river at levels that would be lethal to fish.[50] As time would prove, the chemical materials entering the soil and water during those years were a time bomb in the making, creating conditions much more insidious than the biodegradable sewage that state officials were battling during the McCall years.

Despite its long history of fits and starts, Oregon's fight to cleanse its waterways of pollution began to attract national attention by the early 1970s. James Agee, the regional director of the newly created Environmental Protection Agency, told a Portland audience in the fall of 1971 that although the state's pollution-control program was still incomplete, Oregon had been the most active of the three Northwest states in cleaning up its air and water. Because the state was proactive, he thought it would take a rare situation for his agency to intervene. Agee added, "I don't know of a water quality program in the nation that is better" and pointed out that Washington now trailed Oregon in setting deadlines for industrial waste treatment along the Columbia. While Oregon limited aluminum-mill fluoride discharges,

Washington had no limits. But all three Northwestern states, Agee asserted, faced problems establishing and enforcing standards for agricultural and solid waste disposal.[51]

Ethel Starbird's article in the June 1972 *National Geographic* provided the most celebrated praise for Oregon's efforts to clean up the Willamette River. Ignoring the decades-long anti-pollution fight, the author extolled the McCall administration for bringing the waterway "from shame to show-case." One of the most polluted rivers in the Northwest in the 1960s, the Willamette was now "free of 90 percent of the wastes that once drained into it." Kessler Cannon, McCall's adviser on natural-resource issues, told Starbird that the river was a depressing scene in the 1950s, "the filthiest waterway in the Northwest and one of the most polluted in the Nation." Because oxygen levels dropped to near zero in places, "fish died," and the threat of disease brought an end to swimming and discouraged water ski-ing. But with fishermen and conservationists leading the way, Cannon declared, the cleanup effort began to gain momentum. The turning point came when Tom McCall made his "real shocker of a documentary film." The ebullient Cannon told Starbird: "It's mighty hard to refute evidence everyone can see."[52]

David Charlton, who had devoted his professional and private life to the good fight to clean up the Willamette River, tells a much more com-plex story. A pioneer conservationist closely linked to the Izaak Walton League, Charlton took issue with "the McCall version [that] appeared in the *National Geographic* and . . . other versions of an historical review type such as was prepared by E.P.A." The real stream-pollution struggle had been a long one and some of the early activists and organizations, he charged, were being neglected: the State Board of Health; the Game and Fish Com-missions; organized sportsmen, especially William L. Finley and Ed Aver-ill; and the Izaak Walton League. Governor McCall's claim that a "unique attitude toward land, air, and water . . . surfaced in the 1960s when the state turned its efforts towards restoring and preserving its waterways" was an empty boast, according to Charlton. Those concerns originated in the 1920s, gained momentum in the 1950s when municipalities built primary sewage-treatment plants, and reached fruition when pulp mills kept sulfite liquors out of the river and reservoir releases in the summer and early fall doubled the flow of water in the Willamette River. "The stories of the Willamette River's restoration by the Governor or by others (e.g. National Geographic article)," Charlton argued, ignored recent history.[53]

While the *Oregonian* applauded the state's celebrity status in its anti-pollution fight, it warned that Oregon's livability faced continued threats from haphazard urbanization and sprawl into farm and forest lands. Intelligence and constant vigilance would be required to protect Oregon as a healthy place to live. The state had received a great deal of favorable national acclaim for its achievements in cleaning up the Willamette River, but it would be wise "that we do not let this go to our heads."[54] Words of caution were not, however, the center of public attention during those years. Oregonians were justly proud of the state's environmental accomplishments, especially the improved water quality in the slow-moving Willamette. Swimmers and boaters returned to the river in droves from Eugene downstream to Portland as the waterway once again became a pleasant recreational playground for valley residents.

What Oregon had accomplished in its "cleanup" of the Willamette River was the elimination of "point-source pollution," the most egregious and readily identifiable types flowing from pipes and sewers. Primary and secondary treatment of wastewater (and in some instances tertiary procedures) and augmented stream flows dramatically improved water quality in the Willamette system. In its summary assessment of Oregon's pollution-control achievements in the 1970s, Governor John Kitzhaber's Willamette River Basin Task Force reported in 1997 that neither the pollution-abatement effort through primary and secondary treatment nor augmented stream flow "would have been sufficient alone." The combination of releasing stored water during the summer and early fall and the state's requirement for treating municipal and industrial wastes, however, placed Oregon in the national limelight in the fight against pollution. When Congress passed the federal Clean Water Act in 1972 requiring secondary treatment of sewage, the state of Oregon stood as a model for pollution-control guidelines that were more stringent than federal requirements. The Kitzhaber task force concluded: "By 1972, Oregon had already achieved in the Willamette River Basin what Congress required the rest of the nation to achieve by 1977." Although the

Willamette River was in relatively good health at the close of the twentieth century, the report warned that serious problems persisted, putting both the watershed's condition and human health at risk.[55]

Other than occasional sewage treatment plant malfunctions in every Willamette Valley metropolis and problems with overtaxed systems during periods of heavy rainfall, nearly two decades had passed since the river's condition made headline news. And then, at the close of the twentieth century, accounts about the troubled river came with a rush. The Kitzhaber task-force report explained why. The valley's increase in population from 1.49 million people in 1970 to 1.94 million in 1990—and an estimated 2.2 million in 1995—placed additional burdens on the water quality of the Willamette River. Increasingly sophisticated scientific instruments also began to detect petroleum residues, toxic chemicals, metal byproducts in urban areas, and pesticide and nutrient contaminants in rural sections of the stream. The task force emphasized that non-point-source pollution entering the Willamette system during periods of high precipitation far exceeded point-source contamination. The report also pointed out that urban areas contributed a higher pollutant load per acre than rural lands, although the latter produced a larger overall quantity because of the infinitely greater acreage involved.[56]

Following release of the task force report, an *Oregonian* editorial asked: "How could this happen?" After all, only yesterday Oregon was drawing national attention for cleaning up the Willamette. That now seemed "the stuff of history." The task force reported, continued the editorial, that dioxins, PCB's, industrial and municipal effluent, riverbank development, and farming were poisoning the Willamette, causing deformities in fish, contributing to dwindling salmon and steelhead runs, and threatening domestic water supplies. "Oregon's dirty little secret," outlined in the Kitzhaber report, should serve as a "wakeup call," the newspaper declared, urging industry and citizens to take action to implement its recommendations. Urban runoff from lawns, driveways, streets, and parking lots; crowding of the Willamette's banks with crops and houses, and over-use of pesticides for domestic and agricultural purposes were factors contributing to the river's deterioration. Citizens, the *Oregonian* reflected, likely engaged in too much breast-beating and bragged too loudly about the cleanup story.

Primary and secondary treatment of industrial and municipal effluent restored healthy oxygen levels to the river but "achieved little else." The task force report would remain "just talk," the editorial concluded, "unless the governor and the Legislature, with citizen support, turn it into a battle plan for another war on Willamette River pollution."[57]

The sixty-page task-force report recommended more than 100 wide-ranging "actions" to restore the river's health. The more prominent included restoring wetlands and managing flood plains along the river to "mimic, to the extent possible, natural runoff conditions"; compiling inventories of pesticide use and reducing toxic runoff; establishing water quality for fish and wildlife as "priority uses"; continuing to manage Willamette Valley Project dams to provide minimum flows; and developing a better understanding of the relationship between population growth and the implications for the basin's ecosystems. To reduce non-point pollutants, the report urged the state to protect streams from the detrimental effects of pesticides and fertilizers, to control soil erosion from construction sites and agricultural and forest lands, and to establish greater vegetative buffer zones along streams and wetlands to filter contaminants. Finally, the task force recommended that financial resources be invested to correct "the most significant sources of the contaminants."[58]

As the twenty-first century neared, the Willamette River was seemingly beset with all sorts of problems. While deformed fish in the long, meandering Newberg Pool attracted media interest, state and federal officials began to devote increased attention to contaminants in the Portland Harbor area. Writing for the *Oregonian* in November 1998, Brent Walth observed that decades of industrial enterprise along the lower Willamette had "sullied the land and polluted the river bottom" to the point that the Environmental Protection Agency thought a significant section of the Portland Harbor area would qualify as a Superfund site. Oregon's own Department of Environmental Quality had already identified several contaminated industrial sites, all of them subject to an ongoing EPA investigation. Polluting landowners, including the City of Portland and the Port of Portland, joined in an effort to block the threatened Superfund listing. Emphasizing voluntary effort to avoid the costly, time-consuming, and cumbersome Superfund program, the group wanted DEQ to direct the cleanup work. Walth quoted Nina Bell of Northwest Environmental Advocates, who argued that DEQ was part of the problem: "DEQ's overriding desire to get along with the industries it regulates too often interferes with the job regulators ought to do. The companies don't want the EPA because it will be harder to avoid paying for the cleanup the harbor deserves."[59]

The National Marine Fisheries Service put the state of Oregon (and the entire Columbia Basin) on notice when it invoked the Endangered Species Act in the spring of 1999 and listed as "threatened" nine Northwest fish populations. The listing included steelhead and chinook salmon runs in the Willamette River. The federal hammer fell again the next year when the Environmental Protection Agency designated a large section of Portland Harbor as a Superfund cleanup site. The endangered-species and Superfund listings took place in the midst of a booming economy and sharp population increases throughout the Willamette Valley. In the wake of those developments, the *Oregonian* issued a series of recommendations to help restore the river's health: prohibit building along streams and tributaries in the Portland metropolitan area, restrict farmers from plowing land draining directly into waterways, and urge homeowners from Eugene to Portland to "cut chemical and fertilizer use on lawns and [to] stop washing cars in areas that drain directly into streams." The newspaper urged citizens to remember the region's special heritage, to honor the natural world, to stress livability, and to protect the environment.[60]

In a replay of the acrimony that took place between the city of Portland and the state in the late 1950s, Portland and the DEQ differed over the proper approach to solving the city's problem with sewer overflows during the rainy season. Portland's agreement with DEQ required it to reduce overflows by 96 percent by 2011, an expensive process that would involve installing larger sewers and treatment lagoons. When city commissioners Erik Sten and Dan Saltzman asked for a delay in the sewer-overflow project to study more cost-effective ways to deal with storm-water contaminants, Governor Kitzhaber questioned the city's commitment to resolving its sewage problems in remarks to the Oregon Logging Conference. While it might be good politics to bash city folks in rural parts of the state, the *Oregonian* thought the governor's message should not be ignored. The city council, the newspaper reasoned, was right to question the cost-effectiveness of the sewer works as the sole solution to the pollution dilemma. Despite pressures from DEQ, the *Oregonian* contended, Portland had good reason to reassess its strategy to cleaning up the river: "You can't clean up part of the Willamette. You have to do the whole thing. City dwellers will, indeed, have to do more. But so will everybody else."[61]

In an op-ed piece a few days later, Portland city commissioner Dan Saltzman argued that fixing only the sewer overflow problem would not dramatically improve water quality in the Willamette River. Instead, the city should broaden its strategy and work toward more comprehensive improvements for the entire watershed. Saltzman pointed to the huge cost of the sewer upgrade, expenses that would be borne entirely by Portland ratepayers. The city was not retreating on the issue, he said, but was broadening its focus in an effort to achieve "multiple improvements in water quality and salmon habitat." In a companion column, Nina Bell accused the city of "pointing fingers and making excuses in the way of small children." Portland was failing to take responsibility for the huge volume of raw sewage it dumped into the river every year. Bell charged the city with raising the old argument that it would be fruitless to resolve its sewer-overflow difficulties because the water was already polluted when it reached Portland.[62]

Repeating its position of the late 1950s, the *Oregonian* defended Portland's sewage-abatement efforts. It accused DEQ commissioners of sticking it to the city "with what can only be imagined as the governmental equivalent of glee." The state agency had made it clear that only more and larger pipes could prevent sewage and storm-water overflows from emptying into the Willamette. The dispute, according to the newspaper, centered on different visions of cleaning up the river: that of state environmental commissioners, who seemed narrowly focused on cleansing the river of sewage bacteria by 2011, and that of the city which wanted to pursue a more comprehensive approach to deliver "a truly improved Willamette River nine years after the state's deadline—by 2020." The *Oregonian* cited the support of Mike Houck, an urban naturalist with the Audubon Society, who favored a broader approach that included purchasing land along streams, planting trees, and offering incentives to divert storm water from roofs and parking lots. The answer seemed clear to the newspaper's editorial writers: it would require courage for DEQ to push Portland's deadline back "and admit the unpopular truth: Even the largest city in the state can sometimes be right."[63]

In one important respect the Portland city council was correct in directing attention to other pollution sources. Dean Marriott, director of Portland's Bureau of Environmental Services, pointed out that activities carried out by residents in their gardens, driveways, and streets ends up somewhere in the Willamette system: "The challenge is there are hundreds of thousands of yards and driveways. Collectively, it can make a big difference." Although toxic material from urban runoff had not been widely studied, public agencies were increasingly recognizing urban storm water as a pollution threat. Thirty streams in the Tualatin River Basin were in violation of federal clean-water standards in 1999. Fanno Creek, one of the basin's most urban streams, carried a heavy load of bacteria, heavy metals, an abundance of algae, low dissolved oxygen, and high summer water temperatures. A local activist who regularly hiked along the stream told a reporter that Fanno Creek was severely degraded and devoid of aquatic insects.[64]

Evidence about the Willamette system's unhealthy condition continued to mount as scientists used increasingly sophisticated techniques to measure water chemistry. Scientists planted trout eggs in the Newberg Pool in 2000 and found that one-third of the baby fish had crooked spines and some had deformed heads. Commissioned by DEQ, the Newberg Pool study suggested that the water's chemistry, with trace elements of pesticides and metals, was responsible for the deformities. Of the Newberg Pool sample, 34 percent of the fry had deformities compared with 5 percent or less elsewhere in the Willamette system. According to the *Oregonian*'s Brent Hunsberger, the samples provided "the most compelling evidence that the chemistry of the pool—the 30-mile segment of the Willamette from Newberg to Oregon City—is to blame."[65]

While those findings were being made public, the *Oregonian*, in cooperation with Oregon State University scientists, was in the midst of its own investigation of the Willamette River. In a wide-ranging four-part series, "River of Risk," reporters looked at the issue of contaminants in the waterway, possible connections between fish consumption and public health, known sources of toxicants in the river, and the larger meaning of scientific findings. The "dirtiest fish," those with high levels of PCBs (polychlorinated biphenyls), were taken from the Portland Harbor area. Widely used in electrical transformers before Congress prohibited their use in 1976, PCBs are potent and persistent chemicals similar to DDT. Fish in the lower river also had varying levels of DDT and other associated compounds. The findings of both the DEQ and the *Oregonian* studies caused state health officials to reissue and strengthen warnings against consuming fish from the Willamette. Because of the high level of contaminants, Ken Kauffman of the Oregon Health Division thought the lower river "would need a more protective advisory." Despite such alarming findings, the *Oregonian* noted that Willamette system pollutants were comparable to or below levels found in other Northwest urban streams. Somewhat smugly, the Portland paper reported that contaminant traces found in fish were "hundreds of times lower than levels found in fish in the polluted Great Lakes and New York's Hudson River." Toxic substances in the lower Willamette were also far below levels found in other high-profile Superfund sites in the eastern United States.[66]

Bill Monroe, a contributor to the *Oregonian* series, pointed out that endangered fish such as salmon and steelhead had captured all the headlines, but year-round resident species would serve as "the better bellwethers of the Willamette's health." With the assistance of OSU scientists, the *Oregonian* team selected fish that occupied different habitats in the lower river: carp, omnivorous bottom feeders that ingest plants and insects in bottom sediment; black crappie, a species that swims in schools and feeds on insects and the small fry of other fish under the cover of piers and pilings; and smallmouth bass, top-river feeders that eat everything from insects to frogs and other fish. Monroe indicated that the fish in the study clearly showed the effects of bio-accumulation, the way chemicals such as PCBs and DDT work their way up the food chain through a series of digestive systems:

> Bio-accumulation is magnified by nature's food chain. As small creatures fall prey to larger or more aggressive animals, their chemicals migrate into the predators' fatty tissues. As an animal eats more tainted food, its chemical load grows. The highest concentrations are found in top predators such as eagles and humans.[67]

As a group, the Willamette River fish studies produced even more controversy, with much of the discussion addressing the lack of definitive information: the standards that a cleanup should strive for; scientific data on creatures in bottom sediments; the degree to which contaminated fish were harmful to humans; and the effects of pollution in Portland Harbor on downstream migrating salmon smolts. The research findings for the lower river indicated that PCB levels in the fatty tissue of fish exceeded both state and federal standards and were four times greater than the levels found in upstream fish. The contaminated fish strongly suggested an ecosystem under severe stress, with the degree of chemical pollution subject to heated debate. The *Oregonian*/OSU study also found, much to the surprise of some DEQ scientists, that fish with trace elements of chemicals from banned pesticides had quantities below state and federal standards in every section of the Willamette River.[68]

The absence of convincing data promised continued controversy and heated exchanges among regulatory agencies, companies and landowners liable for pollution damage, and environmentalists. William Fish, a professor of environmental sciences at Portland State University, pointed out that chemically produced river pollutants "are nightmares to clean up. In almost every case you can't return it [the river] to a pristine case, and who knows what a pristine case is anyway in a harbor that's been used industrially for 100 years." As harbor landowners challenge both the cleanup costs and the science behind the regulatory directives, the future suggests a repeat of the debates waged during the 1950s. When the 2001 session of the state legislature balked at taking action, Bob Jenson, chair of the House Stream Restoration Committee, suggested to a reporter that the problem was too large and too expensive for the legislature to deal with. "Government simply can't do everything," he remarked, "and this is a huge thing." The solution, he thought, should be vested in collective action involving the state, industry, and environmentalists. Matt Blevins, legislative director for the Oregon Environmental Council, charged that the real problem was procrastination: "We always lament, the Willamette is dying a death of a thousand studies."[69]

For the last 150 years, the relative condition of the Willamette River has always reflected agricultural, forestry, and urban activities throughout the greater Willamette Basin. Land-use practices—how people work the land, where they build their homes and industries, their provisions for open spaces and recreational needs, how they manage wetlands and protect against natural disaster, and their stewardship of the valley's vital resources—have affected all citizens. The larger public interest in land-use issues became increasingly apparent with Oregon's skyrocketing population following the Second World War. By the close of the twentieth century, those concerns had spread to central Oregon, the state's fastest growing region. Like the requirements for cleaning up the Willamette River, imposing restrictions on how people use their land inflamed long-standing tensions between property-rights claims and concerns for the greater public good.

9

ECOLOGIES OF SPRAWL

THE LAND-USE NEXUS

Around metropolitan centers, real estate developers are pushing suburbia farther and farther into the countryside. Out in the deserts and wood-
lands, people who want vacation homes are scrambling to pick up the
pieces of the good earth: They are being joined by speculators, who have
rediscovered in real estate the fast-buck thrills that a droopy stock market
rarely provides.—*Time,* Oct. 1. 1973

We are in dire need of a state land use policy, new subdivision laws,
and new standards for planning and zoning by cities and counties. The
interest of Oregon for today and in the future must be protected from
the grasping wastrels of the land.—GOVERNOR TOM MCCALL[1]

In a reflective commentary on the western American appetite for uninhibited boosterism, historian and essayist Patricia Nelson Limerick
offered this witty statement: "If Hollywood wanted to capture the emotional center of Western history, its movies would be about real estate."
While the celebratory sprawl that characterized the Los Angeles Basin, the
Silicon Valley, and Phoenix and Tucson dwarfed similar developments elsewhere in the region, every western state, including Oregon, experienced
rapid, unplanned growth during the postwar years. Oregon's slowly diversifying economy, labor-saving technological changes in the agricultural and
wood-products industries, and an increasing number of skilled professionals
in light manufacturing contributed to a growing middle class and the pro-

liferation of bedroom communities, mini-farms, and ranchettes across the countryside. Those developments reflected the changing character of Oregon's population and a rapidly expanding transportation infrastructure that shifted the utility of agricultural and forest land to commercial, residential, and recreational development.[2]

Land issues in postwar Oregon continued to resonate with timeworn historical tensions associated with property rights and considerations for the greater common good.[3] A classic expression of those conflicting interests took place in the late 1990s, approximately twenty miles up the meandering Willamette River from Portland, amid the nurseries and commercial berry farms in the Peace Grove neighborhood. The area made headlines when the *Oregonian* announced that Hollywood Video millionaire Mark Wattles had requested an exemption from Oregon's land-use laws to build a 50,000-square-foot home on high-quality agricultural land along the Willamette. When the video executive purchased the thirty-acre farm in 1995, the property carried an agricultural exemption that limited property taxes to $545 a year. The title to the land, including a farm-management plan, restricted residential buildings and landscaping to one acre. In mid-1996 Clackamas County planners gave Wattles permission to build and landscape up to eight acres of rich riverfront farmland.[4]

The plans for the monster home, which included a full-sized basketball court and a 4,000-square-foot swimming pool, placed Oregon's heralded land-use planning system in jeopardy. Governor Tom McCall, who signed Oregon's land-use law in May 1973, saw the legislation as a way to end "the unfettered despoiling" of the state. Sagebrush subdivisions, runaway coastal development, and sprawling suburbs in the Willamette Valley, the governor warned, threatened to ruin the state's reputation for environmental stewardship. Oregon's pioneering legislation required every city and county to develop comprehensive land-use-planning measures consistent with state-mandated guidelines. The Wattles project directly challenged the spirit of the state's planning principles. After further conflict with the Army Corps of Engineers over his efforts to cordon off a section of the Willamette River, the video executive abandoned the construction project and moved his family to Las Vegas.[5] Although the Wattles case is a crude example of an effort to evade Oregon's land-use-planning legislation, it raises questions about the public's welfare, unrestricted use of land, and the extent to which the free market is a viable mechanism for serving the common good.

The debates over property rights, takings initiatives, and the public good date from the inception of the American republic. With more than 50 percent of its land in federal ownership, Oregon, along with other western states, has been involved in some of the more contentious disputes involving property rights and the larger interests of the public. Those issues boiled to the surface with the population explosion following the Second World War, a rapidly expanding state highway system, and seemingly runaway development beyond Oregon's urban centers. Issues of unregulated growth, quality of life, and the greater common good underscore the persisting tensions between the right to the exclusive use of property and the public interest. During the second half of the twentieth century, urban encroachment on farm and forest land, grazing rights on state and federal lands, deteriorating habitat for spawning salmon, and the proper management of public and private forests became increasingly divisive public issues. Although conditions in postwar Oregon differed from earlier property-rights debates, those disputes still resonated with references to the free market, individual liberty, and the general welfare.

The decennial census reports, the spatial expansion of cities, and the booming growth of urban counties document much about Oregon's changing landscape in the decades following the Second World War. The state's population more than doubled between 1950 and 2000, increasing from 1,521,341 to 3,421,399. But increases across the state were disproportional: some eastern Oregon counties showed net decreases in numbers, while others remained static. By far the biggest gains took place in Willamette Valley counties, home to approximately 70 percent of the state's population. Of Oregon's ten largest cities in 2000, only two (Medford and Bend) were located outside the Willamette corridor. The 2000 census listed 1.3 million people in the three counties and twenty-four cities in the Portland metropolitan area.[6] The exploding population numbers since the Second World War have persistently posed problems with air and water quality, traffic patterns, urban sprawl, and the disappearance of agricultural and forest land. As a consequence, land-use questions have been among the most contentious issues before the Oregon public in the last half century.

As a flood of veterans returned home after 1945 and a booming home-building industry followed, many of Oregon's smaller communities began to expand into the surrounding countryside, often beyond the reach of water and sewage facilities. Although planning in the state dates to legislation passed in 1919 and 1923 authorizing cities to draft zoning ordinances, law-makers did not respond to the chaotic postwar growth until 1947. The leg-islature granted counties the authority to appoint planning commissions with powers to recommend and zone for future development. In an essay published in the *Oregon Business Review* in 1947, Pauline Maris observed that new planning statutes meant that counties would now be subject to development guidelines followed by cities. Developers had moved beyond corporate limits to avoid regulations, contributing to sprawling rural slums with evidence of open sewers along many county roads. The new laws, Maris believed, would provide "minimum land-use controls" in rural areas to conform with urban requirements. She added words of caution: *"Both acts are entirely permissive."* The language in the legislation included references to "shall" and "may provide for"; the ordinances applied only to future land uses and would have no regulatory effect on lands used for agriculture, horticulture, grazing, or growing timber. In addition, all county ordinances were subject to repeal through the initiative and refer-endum process.[7]

In its May 1948 "Progress Report," the Readjustment and Development Commission referred to sprawl on the urban fringe as a veritable "no-man's land where city ordinances do not apply and county courts have no juris-diction." The new real-estate subdivisions were made up of city dwellers who wanted to raise chickens, have more elbow room, and avoid the taxes required to support social services: "As a rule these fringe dwellers work in the city but decline to assume any of the municipalities' obligations." The report referred to one anonymous city where the fringe population equaled that within the incorporated limits. Beyond the city's boundaries, residents refused to build a sewer system despite the "deplorable sewage situation." Few counties had taken advantage of the legislature's grant of planning authority, according to the "Progress Report." Even without additional eyewitness reports, mere numbers suggest the magnitude of Oregon's unplanned growth during the postwar years. The state's population grew from just over 1 million people in 1940 to more than 1.5 million in 1950, an increase of nearly 40 percent and a rate surpassed only by California.[8]

Although the state's growth rate slowed to 16 percent in the 1950s and 18 percent in the 1960s, evidence of uncontrolled development was apparent everywhere, especially in the broad expanses of the Willamette Valley. While forward-thinking urban and county leaders worried about uncontrolled sprawl, a few state officials feared that the expanding suburbs would threaten Oregon's natural-resource industries. Oregon Department of Agriculture director E. L. Peterson warned in 1951 that population growth had the potential to threaten the state's material wealth "in its broad acres and green forests." Organized management of those resources "becomes imperative" in the wake of the swelling number of people moving to the state. Peterson noted that Oregon's future challenges raised an important question: "Can a people multiply and develop in an area rich in natural resources without themselves destroying the resources and natural beauty which is a part of it?" In keeping with his department's long-standing practice, Peterson advised "the use of demonstration and education" rather than legislative mandates to overcome such problems.[9]

But remonstrances against the conversion of agricultural land to residential, industrial, and other uses continued. In its annual report for 1956, the Department of Agriculture expressed fear that land lost to development would reduce agricultural production: "Oregon could well consider taking steps to minimize this tragic diversion." By the 1960s, it was also apparent that the 1947 legislation permitting counties to adopt zoning guidelines was proving an abysmal failure. McCall biographer Brent Walth argues that "Oregon was growing out of control by the mid-1960s," with overworked sewer systems, hit-and-miss zoning regulations, and horrendously convoluted traffic patterns. Opportunistic developers were ascendant everywhere. When the Lincoln County Chamber of Commerce declared Highway 101 along the northern coast the "twenty miracle miles," Republican governor Mark Hatfield referred to the chaotic string of beachfront towns that later became Lincoln City as the "twenty miserable miles." But Lincoln City was more than simple visual blight; county health officers reported numerous cases of untreated sewage pouring directly into estuaries and onto ocean beaches.[10]

State agricultural officials had good reason to worry over the loss of Willamette Valley farmland to industrial and residential development. Between the mid-1950s and 1965, subdivision and commercial expansion had reduced the valley's agricultural base by 500,000 acres, or roughly 20 percent, with most of the losses in Washington and Clackamas Counties. The speculative value of land for conversion to commercial purposes drove up real-estate prices and brought higher tax assessments, a "cycle of destruction," according to Walth, that fed upon itself. The legislature's attempt in 1961 to provide special tax status for farmland proved ineffective. As the future of Oregon's land-use practices would prove, Tom McCall's election to Oregon's governorship in 1966 was timely, linking place, policies, and a unique cast of political operatives at a unique moment in the state's history. While McCall was acting on several initiatives during his first months in office—the Willamette River cleanup, greenway initiatives, and a public-beaches bill—he made clear in his inaugural address that his administration's overarching theme would be Oregon's quality of life, the livability of his beloved state.[11]

Although Willamette Valley agricultural interests spearheaded support for Oregon's land-use system, concern about uncontrolled growth was widespread by the late 1960s. In a series of articles published in August 1968, *Oregonian* staff writer Stan Federman drew attention to the state's recreational land boom, especially events taking place in the seven counties along the Oregon coast. Retirement and second homes and recreational developments were the pivotal forces driving the state's hot real-estate market. While developers were enjoying a rosy "big land picture," Federman thought there was a need for stronger regulatory authority to control the runaway growth of subdivisions. Many state and local officials feared that "future Junkvilles may be the end result" of overheated land sales. The critical issues were real-estate sales without building restrictions, lacking water and sewer systems or roads, and with no provisions for social-support facilities such as schools and medical accommodations. Only Clackamas and

Lane Counties, Federman wrote, had enacted zoning and building controls. Sprawling subdivision sales were especially egregious in central Oregon's Lapine area; Deschutes County voters had turned down zoning restrictions in 1966, as its population grew by 32 percent in the decade.[12]

Unplanned and unrestricted development was already causing pollution problems in central Oregon, where state health officers found traces of domestic waste in residential wells. Federman reported that Deschutes County officials also were concerned that the practice of using dry wells to dispose of wastes had the potential to create a devastating groundwater problem. In the face of such looming troubles, however, the county was powerless to do anything because it lacked building and land-use enforcement authority. Although some counties were slowly beginning to adopt regulatory measures, Federman revealed that more promising land-use measures were being readied for the 1969 legislative session. "Unless counties do something about zoning," he predicted, "the state may do it for them." Salem State Senator Cornelius Bateman told Federman that unless Oregon took steps to control runaway land sales, "we shall run out of the physical capacity to live on the land."[13]

The frequent news-media references to Portland's booming metropolitan area reflect Washington County's 71 percent population increase during the 1960s and Clackamas County's nearly 47 percent growth during the same decade. From Portland to Eugene there was growing sentiment that the state legislature should take steps to prevent Oregon from becoming a replica of southern California. "A siege mentality," according to one writer, "the belief that Oregon's unprecedented growth . . . would transform it into another Los Angeles," drove the growing movement toward more stringent land-use controls. A major conference early in 1967, "The Willamette Valley: What is Our Future in Land Use?" led to the appointment of a legislative interim committee to study pressures on agricultural land. Representing rural constituents, the committee helped draft Senate Bill 10, Oregon's first move toward mandatory comprehensive land-use planning. The legislature passed Senate Bill 10 in 1969 and Governor McCall signed the measure into law. By all accounts SB 10 was a significant step toward requiring local and county governments to adopt planning and zoning ordinances. The statewide land-use-planning effort was further strengthened when Tom McCall was reelected in 1970 on a platform calling for a stronger version of SB 10. In the same election Oregon voters defeated an initiative measure to overturn the law.[14]

From the time he entered the governor's office until his death in 1983, McCall enjoyed great popularity with the press. Articulate, gifted at crafting the facile metaphor, and with an oversized ego, the governor was more aggressive than most of his contemporaries in speaking out on important issues of the day. Midway through his first term, McCall told a group of Los Angeles industrialists that Oregon had mostly avoided being "an overeager lapdog to the economic master." It had been "wary of smokestacks and suspicious of rattle and bang. . . . Oregon has wanted industry only when that industry was willing to want what Oregon is." McCall's most notorious remark, however, broadcast in a CBS television interview on January 12, 1971, would stay with him for the rest of his life: "Come visit us again and again. This [Oregon] IS a state of excitement. But for heaven's sake don't come here to live." An interviewer remarked years later that McCall possessed an agile mind and enjoyed testing his ideas "in the verbal marketplace of spontaneous dialogue." While his famous line inviting people to visit and then leave was tongue in cheek, McCall paid dearly for the comment, especially among the business community and in his own Republican Party. A short time later, the governor told a reporter that the reference was simply meant to draw attention to crowded and industrialized conditions elsewhere in the country, the "lack of breathing space and overtaxed recreational facilities [that] take the zest out of life."[15]

Despite the disparaging remarks of his more vocal critics, McCall never advocated a no-growth policy. Although his legislative programs moved the state in bold new directions in pollution control and land-use planning, the governor always articulated a common-sense approach to economic growth. After he left office, McCall proudly pointed to surveys indicating sharp increases in foreign trade, especially with the Far East; rising farm incomes; and a dramatic statewide expansion in new jobs. As he neared the end of his first term in office, Associated Oregon Industries named him Oregon's "Livability Governor," praising his "cooperation with business and industry, knowing that the health of one is the success of the other."

A handful of developers, including Omark Industries executive John Gray, provided important support for land-use planning. Addressing the National Association of Home Builders in early January 1973, Gray praised Oregon's pollution-control and environmental regulations and told the homebuilders that the central features of the state's new land-use bill being debated in the legislature would become law. The state's "penchant for purity," Gray observed, had promoted healthy growth, and Oregon's population was "growing at a rate faster than the national average." Home building permits were up and unemployment was below the national average. "If you attend your stewardship well," Gray concluded, "your care will be shown in the color of ink on your balance sheet, as well as in the color and shape of the land around us."[16]

As for the governor, Walth contends that McCall was more a conservationist than an environmentalist, preferring the policies that would protect Oregon's resources for future generations. Years later McCall reflected that he preferred a wise-growth rather than a no-growth policy, one that provided restraints on the market when it came to land. Despite the passage of SB 10, thousands of acres of Willamette Valley farmland continued to be lost to development, and few local governments were in compliance with the law. While city and county officials complained that the planning requirements were vague, land speculation continued apace and new housing and commercial development quickly overran local efforts to provide services. Those events further disturbed an already alarmed governor. As the tide began to run out on his second term, McCall took advantage of "Project Foresight," a two-year planning and design task force carried out under the auspices of the Willamette Valley Environmental Protection and Development Planning Council. Shortly after it was formed in 1970, the task force hired the San Francisco landscape architectural firm Lawrence Halprin and Associates to develop future planning scenarios for Oregon's big western valley. Released to the public in the fall of 1972, *The Willamette Valley: Choices for the Future* provided a powerful educational tool as Governor McCall and his strategists prepared for the 1973 legislative session.[17]

In his cover letter for the report, McCall asked important questions about the valley's future: "Will the valley fall prey to a now-familiar pattern of uncoordinated growth and urban sprawl? Or can its people, working in community, build a different future?" Using the Project Foresight report as an educational mechanism, the task force traveled across Oregon to meet with civic organizations in a variety of town-hall settings. Courting audiences with the motto: "We cannot change the past but we CAN INFLUENCE THE FUTURE," Project Foresight offered the public two future scenarios, one devoid of any planning mechanisms and a second scenario that contained urban growth, focused on filling in urban areas and revitalizing city cores, protected open space, developed integrated mass-transit systems, and curbed the growth of "bedroom communities." McCall added his voice to the task force, using his bully pulpit as a medium to spread the message about Oregon's land-use problems. In an interview with Walth, Hector Macpherson, author of the new land-use bill, praised McCall for bringing the public's attention to the issue: "Tom was a master with words. . . . He was not a nuts-and-bolts man. . . . But this is the kind of thing we needed. He prepared the public for the legislators back home and helped create that groundswell."[18]

Crisscrossing the state in the last months of 1972, McCall ended his land-use campaign with a late November conference in Portland in which he served as the principal speaker. He warned the assembled planners and local officials that roadblocks lay in the way of establishing comprehensive land-use planning in Oregon. Speculators and willful developers interested only in profits were the chief culprits. The state, he told the audience, "should not be a haven to the buffalo hunter mentality," to individuals so interested in money that "they rape the land and worship the dollar." But McCall did not stand alone in his tireless politicking for land-use legislation. The most important supporter of land-use regulations and the governor's chief legislative ally was Macpherson, a Republican who was an Albany dairy farmer and former Linn County commissioner.[19]

Vitally concerned since the early 1960s about protecting Oregon's agricultural land base, Macpherson worked with McCall's staff, taking Senate Bill 100 (the body of law that would govern the state's land-use practices) through several revisions in preparation for the state legislature's 1973 session. The *Oregonian* praised Macpherson's effort as "eminently worthwhile"—but warned that there would be many dissenters on the road ahead. The heart of Macpherson's legislation was a planning hierarchy involving local and state governments, with power distributed to each of fourteen regional districts. Meeting in Portland in November 1972, the League of Oregon Cities passed a resolution opposing the planning measure. Portland city commissioner (and future mayor) Francis Ivancie called the legislation "a trend toward colonial government" and referred to the district planning regions as "McCall's fourteen colonial units." Writing for the left-leaning *Oregon Times* magazine, Betty Merten defended the Macpherson bill because it would provide an excellent mechanism for statewide planning and protecting open spaces.[20]

With his lanky frame draped over the podium, Tom McCall addressed the opening session of the 1973 legislature, appealing for action to curb the "unfettered despoiling of the land," the most precious of Oregon's finite resources. In one of his most famous speeches, the governor called on lawmakers to approve the plan being forwarded to the legislature. Only effective land-use controls would bring an end to "sagebrush subdivisions, coastal condomania and the ravenous rampage of suburbia in the Willamette Valley." These "grasping wastrels of the land," he told the legislators, must be stopped in their relentless assault on Oregon's resource base and its open spaces. With Democrats in control of both the House and Senate, McCall relied on allies such as Portland's liberal Democrat Ted Hallock to push the land-use measure through the hearing and amendment process. The hearings themselves provided unusual testimony, with agricultural groups standing against land-use opponents such as forest-products giants Georgia-Pacific and Weyerhaeuser, Portland General Electric and Pacific Power and Light Company, the Oregon Home Builders Association, and Associated Oregon Industries. In the darkest moments of legislative wrangling, Hallock called upon L. B. Day, the irascible former head of the Department of Environmental Quality, to forge a panel of business groups to move the bill through the legislature.[21]

When the Senate approved a slightly modified version of the land-use bill by a vote of eighteen to ten, Hallock arranged with the House Democratic leadership to submit the legislation directly to the floor of the House—where it was approved without having to travel the gauntlet of conference-committee amendments. Even in victory, McCall worried that the land-use bill was "tremendously hedged," burdened with so many hidden caveats that it would prove difficult to implement. In the end, the new measure established the Land Conservation and Development Commission (LCDC) to oversee compliance with local and statewide land-use goals. More than two decades after the passage of Senate Bill 100, Brent Walth provided some perspective on the passage of the pioneering measure:

> No other act in McCall's administration left such a profound impact on Oregon, nor was any other as controversial. Land-use soon became a booming business for lawyers as local governments and private interests fought the new law in court.[22]

Among planners nationwide, the new land-use law further embellished Oregon's already progressive environmental reputation. In an October 1973 article, "The New American Land Rush," *Time* praised a trio of states—Vermont, Hawaii, and Oregon—for passing comprehensive laws to control development. The magazine's sixteen-page chronicle of haphazard suburban sprawl through the American countryside mentioned Oregon only one other time: Governor McCall's notorious invitation to visit the state but "don't come and live here." In a fit of local chauvinism, the *Oregonian* accused *Time* of snubbing Oregon and investing most of its attention in the East Coast and southwestern desert states. By lumping Oregon with Vermont and Hawaii, the magazine failed to acknowledge the truly innovative and comprehensive planning initiatives embodied in Senate Bill 100. "There is no other state in the nation, the *Oregonian* declared, that has so effectively established a legal remedy for fly-by-night developers."[23]

While the state's leading newspaper lamented *Time's* failure to acknowledge Oregon's achievements, officials at home were faced with the daunting task of implementing the complex and unique measure. Senate Bill 100 established a new state agency, the Department of Land Conservation and Development, which served as the administrative arm to implement planning strategies originating with LCDC. Much of the early planning work involved establishing urban-growth boundaries around the state's cities and towns to contain development. Beyond the urban centers, LCDC created exclusive agricultural and forest-use zones to protect land for natural-resource use. The objective of the rules was to restrict commercial and residential development beyond urban-growth boundaries. Although the governor's office directed counties and local governments to prepare comprehensive plans and zoning ordinances, lack of funding and staffing brought a flurry of requests for extensions and delayed compliance with the law. The unique feature of the new land-use legislation was its delegation of planning responsibility to the local level, where agencies were required to follow common statewide guidelines.[24]

The measure also engendered opposition, contributing to lengthy and conflicted debates over approaches to land-use planning. Preserving prime agricultural land became a point of controversy when citizens opposed a proposed thermal-power plant near Lebanon; ironically, Hector Macpherson favored the site because it would give a boost to Linn County's economy. One of the most vocal of the early planning opponents was former state representative (and Springfield realtor) B. J. Rogers. In a luncheon address to the Portland Board of Realtors, Rogers referred to Senate Bill 100 as the work of planners and environmentalists who were attempting to destroy the world of those who created "the wealth that sustains our society." "Poorly drafted" and "biased and vindictive in its intent," the legislation was unnecessary and should be repealed. "It is time for us to stop this NONSENSE," he thundered at the Realtors. Home builders, bankers, title people, and realtors, Rogers argued, were professionals capable of protecting the public from unwise legislation that would increase the cost of homes: "The truth of the matter is we are being planned to death."[25]

Since the inception of Oregon's planning system, William Moshofsky, then with the Georgia-Pacific Corporation, has been the most persistent critic of national and statewide planning. As company vice president for government affairs in the mid-1970s, he faced off with LCDC chairman L. B. Day in a series of exchanges over the adoption of land-use goals. Moshofsky expressed his early opposition to planning in a 1975 article in *Forest Industries*, the most widely read forest-products trade magazine. The Georgia-Pacific executive warned that land-use legislation granted huge discretionary authority to federal and state governments and that it was predicated on a no-growth philosophy. Planning legislation, Moshofsky argued, restricted the use of private property and shifted control to government agencies beyond the reach of most landowners. There was also a danger that such regulatory power would be used to " 'take' or impair private property rights without compensation." Because Oregon's program was still in its infancy, there was great apprehension over the legislation's vaguely drawn and conflicting goals, imprecise language, and endless red tape. As an antidote to federal and state regulation, Moshofsky recommended a limited governmental role, with most authority vested at the local level.[26] For more than a quarter of a century, Moshofsky would push a variety of initiatives promoting the view that environmental regulations were unconstitutional restrictions on private property and the free market.

Toward the close of Tom McCall's eight-year reign, Roger Williams wrote a laudatory appraisal of the governor's tenure for the *Saturday Review*. With his emphasis on livability and controlling growth, Williams declared, McCall might be "the nation's most progressive governor, . . . the heart and soul of his state's controlled-growth movement." Since the mid-1960s, Oregon had been a pacesetter in adopting environmental legislation, providing a good example of the nation's ability to regulate land use and manage its resources. There were many explanations for the state's environmental successes, according to Williams, but it was Oregon's unorthodox governor more than anyone else who deserved credit for the state's extraordinary round of environmental laws:

> McCall's real contribution to 'the livability thrust,' as one of his lieutenants calls it, has been leadership. Like him or not, almost everyone in Oregon would agree that McCall has articulated provocative public policy with a forcefulness and an honesty seldom seen at the higher levels of American politics. His speechmaking in behalf of controlled growth has ranged far afield: to Congress and to college and community groups.

In an article, published just as Oregonians were electing liberal Democrat Robert Straub to the governorship, Williams pointed to difficulties ahead— an understaffed and underfunded land-use agency and future legislative maneuvering to erode LCDC's regulatory authority.[27]

After he took office in early 1975, Robert Straub forcefully defended Senate Bill 100. In a tough-worded speech to Associated Oregon Industries (AOI), he warned that efforts to repeal land-use laws were "inconsistent with the goal of a livable Oregon, and a hostile act toward the people of the state and the future we must all share." Acknowledging that many in the audience supported Senate Bill 100, Straub told the AOI members that there were others who promoted "their own selfish interests" over that of livability. Although the governor mentioned no names, he was well aware that timber-industry and agricultural leaders, the Oregon State Grange, and Douglas County officials were pushing an initiative measure to repeal the state's land-use program. If LCDC were wisely administered, Straub believed, it would "insure the balanced growth and development of the state."[28]

There is little question that Oregon's land-use program exacerbated tensions between rural and urban parts of the state. Three initiative challenges to the law—1976, 1978, and 1983—showed strong support for land-use planning in the urban corridor from Portland south to Eugene. Predominantly ranching counties in southeastern Oregon and timber-dependent communities in Coos, Douglas, Curry, Josephine, and Jackson Counties opposed statewide planning. The first initiative to challenge comprehensive land-use planning was defeated 57 to 43 percent in 1976; the second test, in 1978, was turned back by 61 to 39 percent. A third measure to return all planning decisions to local authorities emerged in the depression-wracked early 1980s. Proponents of that initiative charged that Draconian land-use requirements were destroying local economies and slowing Oregon's recovery from the recession. Despite a badly slumping economy, voters again sustained the land-use planning system in 1982 by a large margin. Opponents of land-use planning began another initiative petition two years later but failed to collect enough signatures to place the measure on the ballot.[29]

As the Department of Land Conservation and Development (DLCD) worked with counties through the lengthy process of drafting plans that met statewide goals, the legislature periodically adjusted policies to make the system operate more efficiently. In addition to granting enforcement authority to LCDC in 1977, lawmakers took an important step in removing the logjam of cases before circuit courts when it created the Land Use Board of Appeals in 1979. In the next few months the board developed into a quasi-judicial appellate court with jurisdiction over land-use decisions. The legislature added another important modification in 1981 when it established formal guidelines for periodically reviewing DLCD plans and regulations, and in its 1983 session legislators directed LCDC to give more consideration to economic development. In 1985 lawmakers ordered LCDC to study ways to identify marginal farmlands, and two years later the legislature created the Ocean Resources Planning Program. At the outset of the 1990s, LCDC undertook an extensive review of the effectiveness of urban-growth boundaries in protecting farmland.[30] No state legislative session leaves Oregon's land-use planning system untouched.

Despite these adjustments, troubles persisted. What many outsiders considered a smoothly operating planning system actually involved a mix of legislative enactments, administrative decrees, and court decisions. Of all the planning issues that came before LCDC, rural land-use policy was the most controversial, the source of initiatives to overturn the system, and court challenges objecting to specific land-use decisions. Many of the disputes centered on state versus local control over decision making; the major contestants included landowners, developers, urban-based environmental groups, government planning professionals, and politicians. State and local interests debated policies involving agricultural and forest lands, sensitive habitats, and activities affecting underground and surface bodies of water. For the first ten years much of the opposition originated in eastern and southern Oregon counties. Marie Lester, a Union County commissioner, helped organize the Oregon Counties Land Use Coalition to provide a dissenting forum for those who opposed the new regulations. She conceded that the state needed some planning mechanism, but Lester strongly rejected the "one-size-fits-all" approach. "You can't plan our area, Eastern Oregon," she told a reporter, "the same way you plan Portland. Our plan isn't our plan—it was dictated by LCDC. It doesn't really fit Eastern Oregon."[31]

County officials also procrastinated in adopting zoning requirements to bring their jurisdictions in compliance with statewide goals. The most conspicuous delays occurred in agricultural and forest-dependent counties. When the legislature gave LCDC enforcement authority in 1977, however, county officials sped up the process of drafting planning measures.

Hood River and Deschutes Counties made headlines in 1977 and 1978 for refusing to comply with statewide planning objectives. In Hood River County, the LCDC exercised its newly delegated authority and ordered officials to cease the practice of subdividing orchard land for residential subdivisions. Although the county was two years in arrears in adopting guidelines to protect agricultural land, Hood River officials moved quickly to meet state-mandated requirements to ward off LCDC action. The commission imposed similar conditions on Deschutes County when it failed to implement policies to protect agricultural land.[32]

While rural counties were struggling with the compliance issue, Portland mayor Neil Goldschmidt sharply criticized proposed changes which would loosen the state's land-use policies in testimony before an October 1978 legislative interim committee hearing in Salem. He urged lawmakers to defer action until after the November election, in which voters faced the second initiative to repeal statewide land-use planning. Although voters turned back the effort, the criticisms and attacks continued. Speaking to the annual meeting of the League of Oregon Cities in 1979, Anthony Downs, an urban-affairs specialist with the Brookings Institution, warned that the state's land-use planning system had the potential to create a housing shortage and increase the price of homes. Although state land planning might be popular, Downs thought it would price low- and moderate-income families out of the market.[33] In the face of such criticism, legislators and the LCDC continued to allow flexibility in applying statewide goals at the county and local level.

Other visitors to Oregon left positive, even glowing images of Oregon's environmental accomplishments. Peter James, a British planner who traveled the state in 1978, compared Oregon to Ernest Callenbach's *Ecotopia*, a futuristic novel depicting the Pacific Northwest as an environmental paradise.[34] Writing for the *New Scientist*, James believed that Oregon could "lay some claim to being the most ecologically conscious area in the world." Its past provided evidence that citizens have been concerned with "quality, rather than quantity, of life," with the most recent confirmation the governorship of Tom McCall. The University of Nottingham professor listed the McCall administration's accomplishments: the cleanup of the Willamette River, preservation of coastal beaches, establishing estuarine

nature sanctuaries, protecting scenic areas, limiting air and water pollution, the "Bottle Bill," and the state's innovative land-use planning system.
The last measure, James observed, restrained the state's runaway growth
and had survived several initiative efforts and court challenges. James also
pointed to changes in citizen behavior—an emphasis on bicycling, volunteers cleaning up beaches, and well-earned pride in livable cities such as
Portland and Eugene.[35]

During its first decade of operation, the greatest challenge to Oregon's planning system occurred during the recession of the early 1980s when opponents charged that restrictive land-use policies were circumscribing the
state's recovery efforts. Conservative Republican Victor Atiyeh succeeded
Robert Straub as governor in 1979, and economic development became the
new administration's mantra—with a special emphasis on recruiting new
industries to the state. As a signator to Senate Bill 100 when he served in
the state legislature, the governor was no unabashed worshiper of unplanned
development; when the land-use law came up for reauthorization in 1981,
Atiyeh supported the measure. But in other respects, his administration
was equivocal, seeking legislative approval for funds to inventory lands for
development and pursuing strategies to loosen state regulations. To that
end, the governor criticized LCDC for establishing a regulatory environment that discouraged economic growth.[36]

In the midst of the recession, Governor Atiyeh appointed a prestigious
panel to inquire into the workings of Senate Bill 100. Headed by Board-
man farmer and former legislator Stafford Hansell, the task force traveled
the state taking testimony from more than 400 citizens. Some of the more
rancorous hearings took place in central and eastern Oregon in mid-July
1982. The LaGrande hearing provided overwhelming evidence of land-use
planning's negative economic effects. A large crowd attended the task force
hearing the following night in Bend, and the overwhelming sentiment was
to overhaul totally or abolish statewide planning. Deschutes County commissioner Abe Young called planning "an uncontrollable monster," because
supporters of the planning process such as 1000 Friends of Oregon caused
endless appeals and delays. Brooks Resources president William Smith, who

had directed the move to build Black Butte Ranch near Sisters, complained that LCDC's controls were preventing the kind of innovations that would help move the state out of recession. What developers faced, he told the panel, were lengthy and frustrating delays. Despite such testimony, the task-force report—delivered to Governor Atiyeh in mid-September—found no support for the belief that land-use planning was harming the state's over-all economic welfare.[37]

The report did not deter opponents of land-use planning, who were readying an initiative. Measure 6, which would appear on the November 6, 1982, ballot, asked a seemingly innocent question: "Shall the state's land use authority and goals be advisory only?" It was obvious to all that if the initiative passed, it would deprive LCDC of its regulatory authority. The financial contributors supporting the repeal of statewide planning included most of Oregon's leading corporate players: Associated Oregon Industries, Associated General Contractors, Boise Cascade, Georgia-Pacific, Weyer-haeuser, Louisiana-Pacific, and Publishers Paper. Planning supporters hired pollster Tim Hibbits, whose September opinion surveys showed solid backing for repealing Oregon's land-use laws. But the Hibbits poll also hinted that the public would invest great trust in former governor McCall's judgment.[38]

In failing health, suffering from an advanced stage of prostate cancer, McCall's regular KATU-TV evening commentaries limited his public role in opposing Measure 6. However, the still popular former governor took advantage of strategically timed speaking engagements (and his still notable rapport with the press) to speak out in favor of statewide planning. In a discreetly arranged media event on October 7 at Portland's University Club, where he was to introduce William Reilly of the Conservation Foundation, McCall delivered an impassioned speech calling upon Oregonians to defend the state's livability. In remarks that left few dry eyes in the audi-ence, he concluded:

> You all know I have terminal cancer—and I have a lot of it. But what you may not know is that stress induces its spread and induces its activity. Stress may even bring it on.

Yet stress is the fuel of the activist. This activist loves Oregon more than
he loves life. I know I can't have both very long. The trade-offs are all right
with me.

But if the legacy we helped give Oregon and which made it twinkle from
afar—if it goes, then I guess I wouldn't want to live in Oregon anyhow.[39]

McCall's address reversed the polling trends on Measure 6 and placed
the former governor in the national spotlight in the month before the election. The state's leading newspapers, the *Oregonian* and *Statesman-Journal,* hailed McCall's courage, and the Associated Press and United Press
International released stories on Oregon's upcoming election. McCall
appeared on the NBC evening news and the "Today" show, and CBS
reporter Terry Drinkwater delivered two commentaries on the former governor. In occasional appearances around the state, McCall continued to
hammer away at planning opponents, and he criticized state leaders for
apologizing for his years as governor: "To scapegoat the McCall administration which hasn't been in office for eight years," he said on one occasion, "is the refuge of weak, frightened men." Finally, in what turned into
one of the most expensive initiative campaigns to that date, voters defeated
the effort to repeal the state's planning system by a margin of 55 to 45 percent.[40] As with the other anti-planning measures, huge majorities in the
greater Portland area and in Marion and Lane Counties sustained the state's
land-use laws.

Through all the court challenges, initiatives, and legislative maneuvering,
Oregon's land-use experiment remained largely intact into the 1990s. From
the moment McCall signed the measure into law on May 29, 1973, the land-
use program had powerful supporters, none more influential than 1,000
Friends of Oregon. Anticipating that development interests would attack
the state's planning effort, McCall and other influential individuals, including Glenn Jackson, the powerful chairman of the Oregon Highway Commission, formed 1000 Friends in early 1975 to monitor the implementation
of Senate Bill 100. Eric Allen, editor of the *Medford Mail-Tribune;* J. W. For-
rester, Jr., publisher of the *Daily Astorian;* Thomas Vaughan, director of
the Oregon Historical Society; and Hector Macpherson served on the advi-

sory board. The purpose of the citizens' organization was to provide legal support for upholding the state's land-use laws. Financed through private gifts and donations, foundation grants, and membership dues, the organization lobbied the state legislature and initiated judicial review of LCDC planning directives. The organization's strength was its willingness to challenge the commission to follow its own guidelines, and, as the future would demonstrate, 1000 Friends was critical to the success of statewide land-use planning.[41]

On the twentieth anniversary of the land-use law, Portland State University's Carl Abbott and Deborah Howe judged the legislation a success because the system's core principles were still intact. Writing for the *Oregonian,* the two urban-studies scholars found "much to celebrate" about the nationally recognized program. They noted that the state's planning system had developed into "a sophisticated planning approach," reflecting changing legislative priorities and new challenges. Abbott and Howe also acknowledged that despite its obvious accomplishments, some legislators— many of them loosely identified with a new property-rights group, Oregonians in Action—wanted to dismantle the system and weaken its provisions. But the enemies of statewide land-use regulations faced a public that valued planning. Voters had already rejected several attempts to overturn the system, Abbott and Howe argued, and reputable observers were certain that the public remained solidly behind Oregon's planning system. The major problem for land-use supporters was a new generation of voters who were unaware of the struggles that took place in the 1970s.[42]

Despite threats that loomed in the future, Oregon's system remained one of a kind, especially in its comprehensive formula for statewide planning and its coordinated effort to involve state and local officials and citizens in the planning process. Hawaii had initiated comprehensive planning in the early 1960s and strengthened its policies in the 1970s, but subsequent legislative changes substantially weakened the law. Florida, Vermont, and a handful of other states followed suit in the early 1970s, but none of those efforts achieved the success of the Oregon system. As the state was celebrating two decades of planning, John DeGrove observed that only Oregon had implemented a truly comprehensive system for managing growth.

A student of statewide planning for more than two decades, DeGrove praised the scope and political support for the Oregon program in the face of several stiff challenges. Despite continued controversy and criticisms, DeGrove noted, "the system nonetheless has survived and in many ways thrived." Moreover, other states contending with uncontrolled growth were turning to Oregon for suggestions.[43]

Oregon's land-use planning system faced a problematic future during the 1990s, many of its problems related to the state's booming population growth. Renewed property-rights activism, especially legal challenges by Oregonians in Action, further complicated the state's ability to contain development. Oregon's population increased by more than 17 percent, to 3,421,399 people, between 1990 and 2000. That percentage increase, tenth highest in the nation for the decade, approached but did not surpass the state's peak decade of numerical growth during the 1970s and was less than that of Idaho and Washington.[44]

The Pacific Northwest's growth was especially apparent in the heavily populated urban counties surrounding the Seattle-Tacoma and Portland-Vancouver metropolitan areas. Portland's urban reach extended across the Columbia River to Clark County, east and south to Clackamas County, and west into Washington and Yamhill Counties. Developers, environmentalists, and planners struggled constantly with Portland's urban-growth boundary. City and county officials, state legislators, and the governor's office fought over ways to mitigate traffic congestion. Standard jokes poking fun at Seattle's notorious traffic tie-ups moved southward along I-5 to the Interstate Bridge spanning the Columbia River and to the Terwilliger curves in southern Portland, where frequent accidents brought traffic to a standstill. Complaints regularly surfaced about good farmland being lost to development, and critics once again evoked the memory of Tom McCall in their efforts to constrain runaway growth. By the mid 1990s, Oregon unemployment was at a twenty-year low and the state had one of the top rates of job growth in the nation.[45]

Oregon secretary of state Phil Keisling told a reporter in 1996 that the state's thriving economy proved Tom McCall's axiom that a clean environment and a healthy economy complimented each other. But the quality of life and livability "carrot" also created a Catch-22: overcrowding, pollution, and increasing traffic problems all placed strains on the attractions that drew industries and people to the state in the first place. In a special series entitled "Paradise Lost? Searching for Tom McCall's Oregon," published in the fall of 1996, the Salem *Statesman-Journal* reported that the Oregon that McCall had envisioned appeared to be eroding. The newspaper cited the proliferation of throwaway cans and bottles, increasing evidence of toxic substances in the Willamette River, and—most significantly—the new suburbs that continued to eat away at the valley's valuable farmland. Keizer resident Anna Lebold told a reporter that Oregon's land-use laws had worked well until the mid-1980s, when the state began to loosen its planning requirements: "The more people that come into Oregon, the more they relax the rules." Former governor Robert Straub thought there was still considerable residual support for maintaining the state's livability: "If you scratch a little under the surface of indifference among Oregonians, there lies tremendous support . . . for keeping the state a great place to live."[46]

Although Oregon's planning system had improved urban design and helped protect farm-and-forest land, *Statesman-Journal* writer John Henrikson reported that future growth in the Willamette Valley would tax anew the state's livability and "force some hard choices about how and where we live." Although the state's planning system had been in effect since 1973, Henrikson thought it would soon be getting its first real test. Planning supporters believed that the greatest threats were incremental, a slow erosion of the original principles, such as the appearance of "yuppie farms" on land zoned for exclusive farm or forest use. Speaking for the anti-planners, Bill Moshofsky of Oregonians in Action charged that the system was "very rigid, very inflexible, and very unreasonable." Still other groups thought that Oregonians in Action and the development community presented the greatest threat to the land-use system. Although Democratic governor John Kitzhaber continued to veto Republican-passed legislative measures that would weaken planning laws, Robert Liberty, executive director of 1000 Friends of Oregon, worried that special interests who profited from development would begin to prevail in Salem.[47]

One of the more obvious visual symbols of the inexorable loss of quality farmland in the Willamette Valley was the huge rectangular warehouses being constructed near Interstate-5 interchanges. Because they represented jobs and economic development, county planning commissions approved the warehouse plans, which ultimately passed muster with LCDC. The stretch of interstate highway between Stafford and Wilsonville had already been turned into a long line of commercial buildings by the early 1990s, and thereafter the big concrete boxes increasingly began to appear near freeway interchanges south of Wilsonville as well as near population centers in the Rogue Valley. Governor Kitzhaber cited the new factory-outlet stores and a huge warehouse at the Woodburn interchange as evidence that the state's planning system needed to be overhauled. Portland businessman John Russell, who regularly traveled the interstate, found the commercial development unsettling:

> I'm particularly troubled that scenic beauty just isn't on any governmental radar screen. There are no lobbyists for the aesthetics of Oregon's highways. Few besides Scenic America, which fights billboard clutter on our highways, are spending money to make life more pleasant for motorists. Governments are looking at efficiency and cost control and economic development—but these may not mean as much to future generations as what we are failing to do for the aesthetics of the Oregon experience.[48]

And the "boxes" along the state's most heavily traveled interstate reflected only part of the incremental sprawl that accompanied Oregon's booming economy in the 1990s.

On the twenty-fifth anniversary of Senate Bill 100, Oregon newspapers offered assessments of the state's land-use planning program. The Corvallis *Gazette-Times* wondered in May 1998 if citizens would continue to support the system in the face of the "powerful theme of private owners' rights to choose other uses for their land." Property-rights groups, it observed, had become better organized and would use growth pressures to promote their agenda and further jeopardize Oregon's land-planning mechanisms. But the *Gazette-Times*, echoing Abbott and Howe's concerns of five years earlier, found the greatest threat to Oregon's unique system to be a public unfamiliar with both the struggles to draft the comprehensive plans and the repeated efforts to undermine Senate Bill 100. Moreover, there was no longer anyone comparable to the charismatic Tom McCall to rally the public behind land planning.[49]

The most contentious land-use battle lines in the late 1990s centered on the issue of urban-growth boundaries. Although most of the state's municipalities experienced problems with growth, the political debates were more intense in the Portland metropolitan area, where fully developed subdivisions butted directly against open farmland and plowed fields. Development pressures in the lower Willamette Valley posed a severe test to the audacious planning vision of the McCall generation. Signs of change were everywhere in Multnomah, Clackamas, and Washington Counties as new commercial and housing developments continually pressed out into the countryside. Formally approved in 1979, Portland's urban-growth boundary embraced twenty-four municipalities in three counties. The only elected regional government in the United States, the Metropolitan Service District (Metro), is charged with managing the boundary, reviewing and coordinating comprehensive plans, and assuring that proposals are in compliance with the state's planning goals.[50]

It is widely recognized that the urban-growth boundary has benefitted the City of Portland. Because the boundary confines development to a specific area, the business community focused its attention on redevelopment, especially of properties along the riverfront and in old industrial areas such as northwest Portland's Pearl District. Clark County, Washington, with relatively lax land-use controls, has served as a safety valve of sorts, providing an avenue of opportunity for people seeking less expensive homes. Realtors and home-builder lobbyists have persistently attacked Metro for being too restrictive in making land available for commercial and residential development. Although there were complaints about crowding and density within the boundary, Portland proper had fewer people per square mile in 1998 than it did at the end of the Second World War. Election to the Metro board also became increasingly contentious, especially with the emerging political influence of groups such as Oregonians in Action.[51]

"Infilling" or concentrating development in the Portland metropolitan area has had a negative side: rapidly rising real-estate prices and the gentrification of older sections of the city have forced low-income families in the city's poorer districts to look elsewhere for reasonably priced dwellings. As Portland's flourishing economy roared through the 1990s, home values doubled and compact buildings filled the city's empty spaces. But the benefits of the boom were distributed unevenly; close-in neighborhoods became richer and whiter while poorer families were dispersed to areas with fewer services. The *Oregonian's* Brent Hunsberger observed that in their effort to persuade investors to develop old neighborhoods, city officials had neglected those who were displaced by the redevelopment of close-in districts. What took place during the 1990s was a striking shift of many African Americans to the city's outlying neighborhoods just as middle-class whites were revitalizing sections of northeast Portland. Such developments, according to Hunsberger, represented gentrification in its purest form: "Middle-class residents buy inexpensive homes in spotty neighborhoods and fix them up. Realtors, investors, and the media take note. Property values rise. Professional developers move in. The neighborhood takes on more of a middle-class to upper-class character."[52]

Despite critics who point to the hidden costs of gentrification, Portland has served as a laboratory for prudent urban growth. And planners point to the striking comparisons with other cities. Between 1990 and 1996 the Kansas City metropolitan area extended its spatial reach 70 percent while its population increased only 5 percent. In contrast, metropolitan Portland's built landscape expanded 13 percent, the same percentage as its population growth. Although the average building lot was reduced by half in twenty years (from 13,000 to 6,700 square feet), the cost of a medium-priced home more than doubled during the same period. Whereas the National Association of Home Builders once ranked Portland among the most affordable cities for housing, by the close of the twentieth century it was one of the most expensive in the nation.[53]

Despite rising housing costs, restricting urban sprawl in the Willamette and Rogue Valleys has had positive benefits—notably in protecting valuable farmland. Richard Brenner, director of the Department of Land Conservation and Development in the late 1990s, reported that Oregon had lost only 2 percent of its agricultural land between 1988 and 1992. In comparison, California lost 11 percent, Idaho 8 percent, and Washington 6 percent during the same period. "We do not have to choose between farmland and livable neighborhoods," Brenner contended, "we can have both if we use land efficiently and demand good design from our community leaders and community builders." Robert Liberty of 1000 Friends of Oregon compared the Willamette Valley with Utah's eighty-mile Wasatch Front, where ten-acre ranchettes dominate the landscape. In a talk to the City Club of Eugene in August 1998, Brenner cited sprawl as Oregon's greatest challenge: "If we can stop sprawl, Oregon may well become the most civilized place in the country." Extending urban services to sprawling developments is expensive and destructive to "sense of place." Sprawl, he told Eugene's civic leaders, was "ugly and lifeless" and offered little charm.[54]

Protecting Oregon farmland, however, was not always a selfless effort to preserve agrarian traditions. Certainly that was not the case of Holly-wood Video millionaire Mark Wattles and his infamous berry farm near Canby. But the Wattles incident was no singular, bizarre violation of Oregon's exclusive farm-use exemption. In a nearby case, Brian Ament, who operated a successful home-building business, constructed a less ambitious but still substantial 10,500-square-foot Victorian mansion on twenty acres zoned exclusively for farm use. To build the pink home (complete with turret and four-car garage), Ament submitted a farm plan stipulating that his agricultural endeavors would include raising sheep, growing Christmas trees, and cultivating shiitake mushrooms. When Clackamas County authorities inspected the "commercial" farm in the fall of 1998, they found empty greenhouses, weeds, sickly Christmas trees, and no sheep.[55]

Ament is one of thousands of wealthy "hobby" farmers who have signed agreements with counties to carry on farming activity, to build on small parcels of land, and to gain a break in taxes. Critics of Wattles, Ament, and other moneyed estate builders refer to such operations as "hobby farms" and even "phony farms," a disparaging reference to outright deception. Although the hobby-farm phenomenon brought public ridicule and scorn from critics, DLCD's Richard Brenner still believed that Oregon's accomplishments in preserving farmland were enviable when compared with other states. This was especially true after the 1993 legislature fashioned more stringent rules for building homes on agricultural land. The same legislature also prohibited the construction of golf courses, churches, and schools on prime farmland.[56]

Hobby farms were only one of the many challenges to Oregon's land-use planning system during the 1990s. Beginning with the 1995 legislature and continuing for the next several sessions, lawmakers were increasingly active in introducing legislation to overturn Oregon's land-use planning system. Of the many bills introduced in 1995, most failed in committee and Governor Kitzhaber vetoed the rest. The threat of the governor's veto power chastened the 1997 legislature into turning back measures that would weaken Senate Bill 100. Oregonians in Action backed a controversial bill that would map a new category for secondary lands, areas that were only marginally productive for agricultural and forestry purposes. When Kitzhaber threatened to veto the proposal, it died on the House floor. The secondary-lands issue surfaced again before the 1999 legislature when Newberg's senator Gary George pushed a property-rights measure that would allow building and development in marginal areas. While the lawmakers were considering a broad range of land-use bills emphasizing property rights, anti-tax activist Bill Sizemore was busy outside the legislature developing strategies to repeal the state's planning directives.[57] Although the 1999 session made no dramatic changes to Oregon's land-use planning system, Sizemore and his allies, Oregonians in Action, were orchestrating a broad-scale assault on the state's grand experiment.

Although Oregonians in Action has provided the most significant support for overturning Oregon's land-use laws, affiliated organizations, such as the Pacific Legal Foundation and its Northwest Center, have also actively challenged government regulations at all levels, especially those that infringe on private-property rights. Formed in 1992, the foundation's Northwest Center, which has challenged state and federal laws involved in what it calls "regulatory 'takings' of private property," has forced legal fights on behalf of property holders in the Columbia River Gorge National Scenic Area.[58] Passed in 1986, the Gorge legislation was an attempt to protect the beautiful river corridor from suburban sprawl, especially on the Washington side of the river where Clark County developers were pushing upriver and Skamania County was wide open to second-home development. The national scenic area act had its beginnings with the emergence of Portland-based Friends of the Columbia Gorge, which worked with the new bistate Columbia River Gorge Commission and the Forest Service to protect the scenic passageway. The commission and its supporters have fought property-rights groups at every step of the way to implement provisions protecting the beauty of the Gorge.[59]

But south of the Columbia River, Oregonians in Action has been the most active of the groups opposed to land-use restrictions. Reorganized in 1989 by its founder, Frank Nims (with the assistance of Bill Moshofsky), the association styles itself as a "non-profit organization devoted solely to fighting for property rights and against excessive land-use regulations." Its officers and members have been active on many fronts; Moshofsky failed in two efforts to win Oregon's First Congressional District seat. Nims and Moshofsky have viewed their organization as a counter to environmental groups such as 1000 Friends, which they believe had dominated the dialogue on land-use issues. Aside from sponsoring initiative measures to overturn the state's land-use laws, the group has also been heavily involved in state legislative races, supporting conservative Republican candidates who hold strong property rights beliefs. Larry George, the son of the Newberg senator and executive director of Oregonians In Action, represented the organization before the legislature and was active in pushing the property-rights agenda. Writing for the Corvallis-Albany *Mid-Valley Sunday,* columnist Russell Sadler observed that the new opponents of land-use planning were playing an old and familiar tune, offering alternatives that would limit mass transit, expand the highway system, extend urban-growth boundaries, and allow the development of secondary farmlands.[60]

Bill Moshofsky, who became OIA's vice president of governmental affairs, believed that LCDC's restrictive zoning mandates were stifling rural economies "and the rights of the land owners." The rigid "cookie cutter" urban-growth boundaries were exacerbating urban problems, "packing more people together, mandating smaller lots and apartments with little or no yard space, and increasing housing costs." Even worse, according to Moshofsky, were the "anti-car, anti-road policies" that were contributing to greater traffic congestion, especially in the lower Willamette Valley. His major grievance was the failure of governments to compensate owners when land-use restrictions prohibited individuals from developing their properties as they chose. The legislature and the governor should act, he argued, to "bring more realism and flexibility to zoning . . . and require just compensation for regulatory 'takings' of private land to provide public benefits."[61]

Oregonians in Action waged a successful campaign in 2000 to place an initiative on the November ballot that would amend the Oregon Constitution to require state and local governments to compensate landowners if land-use restrictions reduced the value of their property. In brief, the initiative would have made Oregon's land-use planning system inordinately expensive and impossible to administer. Measure 7, the smoking gun in the fight to kill Senate Bill 100, attracted opposition from the League of Women Voters, Governor John Kitzhaber, former governors Mark Hatfield and Neil Goldschmidt, the Oregon Business Association, several county commissioners and county sheriffs, the Association of Northwest Steelheaders, and traditional supporters of land-use planning such as the Nature Conservancy of Oregon, the Oregon Natural Resources Council, 1000 Friends of Oregon, and Friends of the Columbia Gorge. Proponents of Measure 7 included the Oregon Farm Bureau Federation, Associated Oregon Industries, Home Builders Association of Metropolitan Portland, Oregon Building Industry Association, Oregon Association of Realtors, and a Portland-based free-enterprise think tank, the Cascade Policy Institute.[62]

Because Bill Sizemore's Oregon Taxpayers United was pushing other initiative measures on the same ballot, OIA took over the campaign for Measure 7 and directed strategies to put before the public. In the months leading up to the November ballot, OIA's two political-action committees spent more than $880,000 promoting Measure 7. With ample support from other organizations and elected public officials, 1000 Friends charged that Measure 7 would destroy statewide land-use planning, open the gateway to unrestricted development, and vastly accelerate urban sprawl. According to Robert Liberty, OIA's effort was "based on the assumption that there's essentially no public good, no public interest. There's only a private financial interest." Russell Sadler contended that passing Measure 7 would breach urban-growth boundaries and spill small-parcel ranchettes across productive agricultural land. The initiative, he charged, was "not about 'just compensation.' It is about 'speculator compensation.'"[63]

When the votes were counted, the results confirmed pre-election polling surveys: Measure 7 was approved by nearly 53 percent of the voters. The "takings" proposal was defeated in only three counties, all in the Willamette Valley (Multnomah, Benton, and Lane). Southern and eastern Oregon counties approved the initiative by wide margins, many of them by more than 60 percent. After the election, Governor John Kitzhaber asked the state's attorney general for a legal opinion; the LCDC met in emergency session and announced that it would suspend its current work because of the uncertainties involved; and many local and county governments adopted ordinances to establish claims procedures under Measure 7. 1000 Friends of Oregon went into action, announcing that it would appeal any ordinances that waived land-use requirements to avoid litigation. Planners, lawyers, and city managers engaged in extensive e-mail discussion groups about the confusing, complex, and contradictory questions posed by the measure.[64]

Familiar persons from the past immediately filed suit to block the implementation of Measure 7. Audrey McCall, the widow of Tom McCall, and Hector Macpherson were joined by two sitting mayors in a complaint filed on November 24, 2000. The *McCall* plaintiffs argued that Measure 7 violated the "single question" requirement for constitutional amendments. The League of Oregon Cities also filed a complaint on December 5 raising essentially the same constitutional issues and asked the court for a preliminary injunction to prevent the measure from taking effect. The Marion County Circuit Court then consolidated the two cases and issued the preliminary injunction. The *McCall et al.* complainants succeeded in late February 2001 when Marion County Circuit Court judge Paul Lipscomb declared Measure 7 in violation of the Oregon constitution's separate-vote requirement. The state (the defendant in the case), and numerous interveners—including Oregonians in Action, the Pacific Legal Foundation, and Oregon Taxpayers United—appealed the case and the Oregon Court of Appeals sent the case to the state supreme court.[65]

The Oregon Supreme Court took thirteen months before rendering a unanimous finding on October 4, 2002, upholding the circuit court's determination that Measure 7 included more than one constitutional change. Property-rights groups reacted angrily; OIA legal counsel David Hunnicutt accused the justices of "spitting in the face of Oregon voters." Hunnicutt asserted that the issue would remain in the news "as long as we have a system in the state that forces the economic costs of regulations on a small group of landowners." An article on the organization's Web site expressed disappointment but not surprise, citing the decision as "a further blow to Oregon's initiative system and to voters' rights." Larry George called the Oregon court "a typical, liberal activist court" with no interest in caring "about the right of landowners to use their land."[66]

The Eugene *Register-Guard* expressed relief with the court decision, calling Measure 7 a cleverly disguised "stealth effort to undermine Oregon's landmark land use planning system." Voters, it contended, were "sucked in by the proposal's deceptive populist appeal" and distracted in the November election by more than two dozen ballot measures. The court ruling, according to the *Register-Guard*, prevented Oregonians in Action from "dismantling the state's land-use system." The *Oregonian* was also pleased to see Measure 7 "spiral into the dustbin set aside for grotesque attempts to mangle the Oregon Constitution." If the initiative had gone into effect, the newspaper claimed, it would have put "cities and counties in a crippling double-bind"—face bankruptcy from compensating property-rights claims or stop enforcing land-use regulations. In the coastal city of Newport, Lincoln County assistant counsel Rob Bovett thought the court's decision saved the county from countless cases of litigation. If the measure had survived, "it would have been a very dramatic thing . . . that would have led to years, maybe decades of litigation . . . to sort out all the questions."[67]

The Oregon Supreme Court's October decision did not end property-rights groups' attacks on the spirit of Senate Bill 100 and the state's multi-faceted land-use-planning system. In the November 2002 elections, David Hunnicutt failed in an aggressive campaign to win a seat on the Oregon Court of Appeals. The Oregon League of Conservation Voters, fearful that Hunnicutt would bring his anti-environmental agenda to bear on judicial decisions, spent $45,000 to expose his extreme ideological position on questions of property rights. OIA also sponsored Metro Service District ballot Measure 26–11, which would have ended the council's ability to manipulate population density within the urban growth boundary. With campaign support from 1000 Friends and the League of Conservation Voters, 26–11 was defeated. And OIA supported Kate Schiele, a property-rights advocate, as a candidate for Metro president. In a multi-candidate race, Schiele finished a distant second to David Bragdon.

Tom McCall envisioned Oregon as something more than an endless string of subdivisions, strip malls, and parking lots. If citizens had the collective will to act responsibly, he believed, the state could avoid a future with congested freeways, polluted airsheds, depressed urban centers, and ever diminishing open spaces. As Oregon moves into the twenty-first century, planning experts from across the country continue to praise the state for its accomplishments in protecting green spaces; keeping industrial, commercial, and residential development within urban-growth perimeters; and taking important strides to provide alternative modes of transportation. A Sierra Club report published in 1999 cited Oregon as a shining example for its efforts to protect farm- and forest land and for managing urban growth. The study praised the Metro council as an innovative model for coordinating land-use and transportation planning in the three-county Portland area. The report commended Metro, charged with these responsibilities since its inception in 1979, for its remarkable successes in the face of incredible growth pressures.[68] But the question remains whether Oregon voters will continue to have the political will to turn away those who want the exclusive right to manage and dispose of their property as they see fit.

EPILOGUE

THIS SPECIAL PLACE

I was shaped by the West and have lived most of a long life in it, and nothing would gratify me more than to see it, in all its subregions and subcultures, both prosperous and environmentally healthy, with a civilization to match its scenery.—WALLACE STEGNER

When a person holds property with the expectation of selling it for a profit, it is a speculative real estate investment. . . . Don't try to tell me it is a property right—FRANK DEMONTE[1]

Oregon writer Ken Kesey offered up a modern-day, real-life story in a brief personal memoir of his boyhood trek from eastern Colorado to Oregon during the war-hot summer of 1943. Kesey's father, who had enlisted in the Navy, was taking the family to Grandma and Grandpa's new farm in Coburg, Oregon, where they had recently moved "seeking richer dirt and brighter horizons." Crowded into the back seat with his brother, Kesey thought the drive stretched on interminably, the scenery just as "deserty" as the home the family had left behind. The westward-bound travelers passed "failed farms and ribby cattle" and side roads leading off to nowhere. On a sweltering August afternoon, the Keseys drove into Oregon across an arid and baked landscape. "Even the place names along our bleak route sounded bleak," Kesey wrote. "Burns. Hines. Wagontire. And the pass we were passing through? It was called Stinkingwater." The family crossed the Cascades after sundown and reached Coburg after midnight. After hellos and hugs, the Kesey boys were bedded down on the grandparents' porch. The writer concludes his personal Oregon Trail narrative with the following:

> At dawn I was awakened by the swelling of birdsong and bladder. In a grumpy blur I made my barefoot way down off the porch. I shuffled to a blue-black loom of bushes at the edge of a lawn. And, standing there, blinking and shivering and draining myself of warm sleep, I suddenly became aware of a wondrous thing. Those looming bushes were covered with berries, festoons of glistening, dew-beaded berries!—sweeter than soda pop and more numerous than the stars fading in the sky!
>
> As I stood filling my mouth with this wild bounty, I watched the light come up on the land that had produced it. Through rising mists I saw Grandma's vegetable garden, heavy with tomatoes and string beans . . . and Grandpa's sweet-corn patch with its long ears lifted, haphazard orchards drooping with plum and pear and fat Guernseys standing to their bags in clover, and further away, forest after emerald forest of timber, hemlocks and firs and pines, their points lifted like hope itself to the new day.
>
> "Maybe Grandma and them pioneers was right about this land after all," I found myself conceding through the blackberry juice. A concession I have never recanted, though nearly half a century has passed.[2]

Ken Kesey's story is a familiar one. With the exception of the Native people among us, Americans have lived at the edge of movement and change for more than three centuries. As the dominant culture pushed its way relentlessly from East to West, emigrant groups developed and elaborated a set of romantic myths that ended in a blissful land of milk and honey. At the end of the western trails were places where golden dreams came true, providential settings that glistened with hope for those seeking renewal, a wonderland filled with a raw and productive beauty that would defy the ages. Those great utopian myths have been part of the Oregon story for more than 150 years. Better than anyone else, the writer Wallace Stegner grasped the essence of that mythical world that pulled the nation westward, "where every day is payday, a Big Rock Candy Mountain where the handouts grow on bushes and the little streams of alcohol come trickling down."[3]

For most of its history, writes the *Oregonian's* Brian Meehan, Oregon represented an idea: "an Eden where people prospected not for gold but for a better life." Beginning with the Oregon Trail in the 1840s, the Oregon idea kept reinventing itself through successive generations. The Great Depression and New Deal programs attracted "busted farmers and laborers" to jobs on federal projects such as Bonneville Dam; the Second World War drew African Americans to Portland's shipyards; Oregon became a beacon during the 1960s and 1970s for young people seeking a bucolic existence living off the land; and finally during the 1990s cashed-out Californians headed north in search of a quieter, more bucolic life. Meehan contends that Oregon represented promise: "a place where the grass grew tall and green even in winter. Where steady rain nurtured crops and family ambitions." Even though "the dream didn't match reality," he contends, what really mattered was that people believed that Oregon offered something better.[4]

Wallace Stegner's lyrical voice and the *Oregonian's* celebration of twentieth-century Oregon obscure more than they reveal. To better understand the cluster of values, the symbols, and the cultural world surrounding those celebrated treks to the Pacific Northwest, it is important to recognize the hidden and often unacknowledged consequences of human activities over the last century and a half. I am reminded of the Interstate-5 signposts— "The Eden at the End of the Trail"—that celebrated the Oregon Trail sesquicentennial in 1993. In one respect the highway markers served as a stark reminder that one people's Eden can be another people's tragedy. Armed with ever more potent technologies, the newcomers imposed upon indigenous people and the regional landscape a markedly different cultural vision—one that led to the gradual, and in some cases spectacular, modification of ecosystems everywhere.

Today, native peoples and newcomers alike live amidst the awesome changes that have been wrought to Oregon's landscape since the forces of market capitalism first penetrated the region in the early nineteenth century. The technological reach of the industrial world that followed was revolutionary in the extreme and celebrated across the land as a sign of progress and forward advance. Because it was one of the last areas in the United States to undergo the process of resettlement, the region moved rapidly from subsistence to industrial modes of production. Through it all, the immigrants who flocked to the Northwest anticipated a place of boundless plenty: immeasurable stands of timber, rivers teeming with fish and harboring the potential for hydroelectric power, and land itself fertile, requiring only the application of human labor and technology to bring forth its bounty.

As the industrial revolution advanced from east to west across the United States, Americans developed fanciful and romantic ideas about natural and human-influenced landscapes. Nineteenth-century writers defined the great western outback as "natural," virgin, pristine, and unsullied by the human touch. Nature writer Michael Pollan refers to such thinking as an "absolutist ethic," a simplistic division of the world into the natural and cultural. Americans who write about nature, he contends, are not writing about humanly influenced landscapes. To see the world in simple dichotomies of nature versus civilization, Pollan points out, ignores the long-standing and vast reach of the human imprint on the world around us. More recently, global warming, holes in the ozone layer, and acid rain are powerful reminders that old metaphors about a "nature" that is separate from culture are no longer valid. If recent events indicate a shift in our collective thinking about culture and the natural world, we will acknowledge the wisdom of British writer Raymond Williams, who maintained that "the idea of nature contains, though often unnoticed, an extraordinary amount of human history." Historian William Cronon offers additional words of caution: "Nature is not nearly so natural as it seems. Instead, it is a profoundly human construction."[5]

The tendency to draw sharp distinctions between the cultural and natural worlds dates to the earliest European contacts with North America and has contributed to descriptions of the continent steeped in utopian imagery. Envisioned as a pristine wilderness unburdened by history, tradition, or culture, the North American landscape represented a blank tablet upon which Europeans and their descendants proceeded to build a civilization.[6] Such notions ignore demonstrated proof that cultural modifications to the landscape date back several millennia. I followed that line of reasoning in *Landscapes of Promise,* arguing that the Pacific Northwest provides some of the best documented evidence of human-induced environmental change anywhere in the world. In the words of one scientist, what the Northwest witnessed in the last 150 years was "an unprecedented acceleration" in the ever changing ecosystems of the region. We now know that those new and infinitely more intrusive cultural disturbances have been all-encompassing, bringing unprecedented changes to land, water, and even the atmosphere itself.[7]

The Oregon surroundings that we live with today—the deliberate introduction of a vast array of new species, the distant markets that shape and reshape our surroundings, and the hard work of simplifying local ecosystems into monocultural settings of wheat, ryegrass, and Douglas fir—are tremendously productive enterprises, turning out an enormous volume of material wealth. Those dramatically altered settings also possess a stark beauty of their own. While postwar writers have featured the attractiveness of the verdant green and highly productive Willamette Valley, observers in other parts of the state have contributed equally striking descriptions of the charm and the rich agricultural enterprises east of the Cascade Mountains. Shortly after the Japanese surrender, the Pendleton *East Oregonian* published a letter from an anonymous "visiting elderly lady" who provided a striking narrative representation of "second nature" after a drive through northeastern Oregon's wheat country:

> It was the beginning of wheat harvest, and the fields were yellow or red, according to the variety. Contrasted to this was the brown summer fallow, and above was the lovely blue sky flecked with fleecy clouds—and far in the distance the purple of the Blue Mountains. It thrilled me through and through as I viewed those vast fields going up hill and down dale over one hill after another for miles on end, from the Blue Mountains to the Columbia, and in my imagination, reaching to Walla Walla, Spokane and the Palouse country. I thought, "Surely here is one of the breadbaskets of the world."
>
> It was a scene impossible to picture but one to charm any resident of this part of Oregon or Washington, or any beauty lover.
>
> There is something so simple and yet so clean and wholesome in a handful of wheat. It has such wonderful potentialities for man and beast.[8]

Eastern Oregon's golden wheat fields, the blossoming fruit orchards in the Hood River and Rogue Valleys, and the lush greenery along the broad

Willamette represent what William Cronon calls "the artificial nature that people erect atop the first nature." While "natural" ecosystems represent localized floral and faunal adaptations, the ecosystems that have replaced them—second nature—manifest and reflect local aspirations and national and international market relations and prices. Although wheat fields possess an undeniable beauty, their real value rests in the utility of the harvested crop in the marketplace. As such, wheat fields and fruit orchards represent "structures of the *human* economy—second nature," in Cronon's words. It is fair to say that humans have been transcendent across most of Oregon's landscape: channelizing and damming streams, turning rich valley soils to commodity production, transforming public and private forests into commercial plantations, and creating an extensive built environment for residential and industrial purposes. Even our wilderness areas are profoundly cultural creations, reflecting human decisions at particular moments in time.[9]

Atlantic Monthly writer Robert Kaplan captured the attractiveness of the transcendent human imprint on Oregon's landscape when he visited the state in the mid-1990s. He portrayed the Willamette Valley as "arguably, the most richly endowed valley in the world for its size." Protected from ocean winds, blessed with a mild climate, and teeming with nuts and berries, the valley with its rich soil supported carpeted fields of grass and red clover. Marking a westerly loop to the coast and back, Kaplan praised the magnificent coastal forests and the well-spaced farms and suburbs between Eugene and Portland, home to two-thirds of Oregon's population. After three visits to Portland, Kaplan was impressed:

> With its neat trolley lanes, geometric parks, rustic flowerpots beside polymer and glass buildings, crowded sidewalk benches and cafes with modish awnings that hang from sandblasted stone and veiny marble fronts, Portland exudes a stagy perfection, as if it were not simply a city but a kind of open-air museum. "View corridors" regulated by municipal ordinances keep new construction from blocking downtown vistas of the Cascades and, in particular, Mt. Hood.[10]

Cleveland native Russell Sadler, who has made Oregon his home for nearly forty years, recalled his journey by rail to the state in 1965. Westbound to attend the University of Oregon in Eugene, Sadler remembers tender filet steaks and Idaho baked potatos served up in the dining car. When the train headed down the Washington side of the Columbia Gorge, he gazed out on "the diligently preserved Oregon side with its magnificent water-falls." As a freshman on the Eugene campus, Sadler reveled in living within an hour of ocean beaches and the Cascade Mountains and where "you didn't have to shovel snow all winter!" Although the rain during that first gloomy winter bothered him, he quickly realized "that Oregonians marched to a different drummer and were proud of it." He thought citizens were per-ceptive about the future because they had great passion for the outdoors: "camping, hiking, rafting, hunting, fishing, skiing, snowboarding, biking and mountain climbing." Those activities, he believed, created an appre-ciation for public space and that realization in turn made Oregonians "atten-tive public stewards."[11]

Despite a century or more of exuberant extractive activity—seining and scooping salmon from rivers, sluicing minerals from hillsides, and cutting some of the largest trees in North America—Oregon still possesses special attractions. Like Washington and California, Oregon offers dramatic geo-graphic diversity: several distinct climate zones, eleven mountain ranges, and spectacular differences in precipitation, with more than 100 inches falling in the Coast Range and true desert conditions on the lee side of Steens Mountain. While south coast residents sport palm trees on their front lawns, monthly summer frosts make it nearly impossible to grow tomatoes east of the Cascade Mountains. "What makes Oregon different," according to Portland State University's Carl Abbott, "is how Oregonians relate to the land, to the place where they live." He adds that what they treasure is not simply their own community, but the entire Pacific Northwest.[12] That affection for the region reminds me that my son and his family returned to the Northwest in the early 1990s after spending more than four years in Pennsylvania's beautiful Apple Valley, home to Penn State University. They never considered *not* returning to the Northwest as an option.

Troubling signs emerged in Oregon toward century's end, however, sug-gesting even greater divisions in a state that has never enjoyed a smoothly functioning polity. Issues of class and a growing rural-urban divide became more apparent, in part a reflection of the striking and historically significant changes that Oregon experienced during the 1990s. At the outset of the decade the spotted owl seemed to symbolize the state's stressed economy and an uncertain future that loomed on the horizon. Still the state's leading industry in 1990, Oregon's timber companies saw harvests plummet during the decade to a seventy-year low of 3.53 billion board feet in 1998. While the timber industry was finding its way toward a troubled future, however, the state's high-tech industry hit full stride, providing a powerful counterforce to the downward spiraling forest-products sector. Keith Thompson, who directed Oregon's Intel operations in the early 1990s, remembered that "everything we were doing was starting to explode." By mid-decade, high-tech manufacturing jobs surpassed Oregon's longtime staple, lumber and forest products employment, increasing a stunning 20 percent between 1990 and 1995. Even in the face of the declining timber economy, Oregon's unemployment dropped to a twenty-five-year low at mid-decade as the Willamette Valley's urban economies operated at full throttle.[13]

The Portland metropolitan area boomed, with most of the industrial expansion taking place in Clackamas and Washington Counties. Intel, LSI Logic, Fujitsu, Microelectronics, Wacker Siltronic, Mitsubishi Silicon America, and numerous smaller firms invested millions of dollars in the lower Willamette Valley. Other high-tech firms set up shop in Eugene, Corvallis, and Bend. While the Metropolitan Service District struggled with urban-growth issues, home building skyrocketed in greater Portland, along the Oregon coast, and in central Oregon's high-desert country. California refugees moved to the state at a record pace and, thanks to monied newcomers and increasing wealth in the high-tech sector, income-tax revenue soared. In fact, aided in part by voter approval in 1990 of a property-tax limitation measure—Ballot Measure 5—income-tax revenue surpassed property tax collections in 1994, increasing more than 70 percent during the decade. But Oregon's Silicon Forest industries are second-tier affairs and lack the truly creative undertakings that characterize California's Silicon Valley and the Puget Sound enterprises.[14]

Nineteen-ninety was a pivotal year for many Oregonians quite aside from the passage of Ballot Measure 5. Early in the new year the U.S. Fish and

Wildlife Service listed the spotted owl as threatened under the Endangered Species Act, a decision that set off a decade of rancor, litigation, and hard times for communities dependent on federal timber sales. Angry rural citizens descended on Portland's Courthouse Square to make public their frustration with the owl ruling. One protestor carried a sign that read: "Try paying your bills with an owl." When Eugene's 109-year-old Cone Lumber Company closed its hemlock sawmill in 1995, the firm's officials blamed cutbacks in federal logging sales for the decision to shut down its operations. Without timberland of its own, the company relied on purchasing western hemlock that grew on national forests above 3,500 feet. When the owl decision forced an end to such sales, R. B. Cone, the fourth generation of his family to operate the business, decided to close the mill. "You can just put the whole thing on environmentalists and the spotted owl," he told a Eugene *Register-Guard* reporter. Former lumberman L.L. "Stub" Stewart, who headed the now defunct Bohemia Lumber Company, criticized the environmental movement for going to such extremes to destroy Oregon's "fine old-time firms."[15]

Cone's decision to close the mill and Stewart's remark ignored the fact that the Cone Lumber Company operated with antiquated equipment and that Japanese firms were outbidding American buyers for hemlock logs. Although the long-term log supply was troublesome and little timber came from the national forests in 1995, Oregon continued to lead the nation in timber production. Restrictions on federal logging, however, were severe in some parts of the state. National forest and Bureau of Land Management sales that reached more than 4 billion board feet a year in the late 1980s plunged to near zero by 2003. Timber communities around the fringes of the Blue Mountains—Baker City, La Grande, Prineville, and Burns— were hurt the most, and statewide Oregon's lumber and wood products employment dropped from nearly 70,000 workers in 1988 to just over 50,000 in 1998. But the changes to Oregon's timber industry reflect region-wide shifts in the Northwest's natural-resource sectors. Since the 1970s, 500 forest-products mills have closed in Oregon, Washington, Alaska, and British Columbia. The fishing fleet in those same jurisdictions has declined by 5,000 boats, and more than 10,000 farms and ranches have gone out of business. During the same period, some of the strongest job growth has taken place in the recreation, travel, and tourism industry. Second-home purchases, huge trophy houses, gas-guzzling SUVs and pick-up trucks, and the increased paving of more of our common landscape are other symbols of the region's growing affluence.[16]

Regionwide, the economic boom of the 1990s disproportionately favored upper-middle and upper income groups, and people at the bottom of the income ladder actually saw a decline in their take-home pay. Seattle-based Northwest Environment Watch contended that widening income inequality contributes to social segregation, a weakening of shared values, increased economic anxiety, and frayed social cohesion. The organization's profile of the Northwest, *This Place on Earth 2002: Measuring What Matters,* aptly describes Oregon's divisive politics during the last decade.[17]

In the midst of Oregon's faltering resource economy, angry timber and agricultural interests attacked Oregon's first woman governor, Democrat Barbara Roberts (1991–1995), initiating three unsuccessful efforts to recall her from office. Since the election of Democrat governor Neil Goldschmidt in 1986, the Democratic Party has dominated statewide elective offices, once the near-exclusive province of moderate Republicans. But in the state legislature, Democrats have suffered a reversal of fortunes. Rural voters, who associate Democrats with environmentalism, have elected increasingly conservative, anti-tax Republicans who blame the Endangered Species Act and Oregon's land-use laws for the state's faltering rural economy. The collapse of the timber industry and the thriving high-tech economy sharpened already existing tensions between urban and rural Oregon during the 1990s. In addition to passing the tax-limitation measure, voters looked for scapegoats via the state's initiative system—launching what many have called Oregon's culture wars, threatening gay and lesbian rights and the legality of abortions, while granting terminally ill people the right to physician-assisted suicide. By 1995 voters had approved term limitations for state legislators, enacted tougher crime laws, passed another tax-limitation measure, and turned back a stringent riparian-protection initiative. Carrying on the state's tradition of reforming the electoral process, the Secretary of State's office has moved exclusively to vote-by-mail elections. Meanwhile, the region's sharply declining salmon runs and the federal government's decision to list several stocks as threatened or endangered have left rural residents with an additional sense that they are under siege to urban interests.[18]

When he left office in January 2003, Governor John Kitzhaber, who made frequent use of the veto to protect the state's land-use system and other environmental laws, was provoked to refer to the state as "ungovernable." The national recession that began in mid-2001 struck Oregon's high-tech sector especially hard, giving the state the nation's highest unemployment ranking for several months. The recession laid bare Oregon's class and rural-urban divisions and worsened its already divisive politics. Ashland's senator Lenn Hannon, a moderate Republican, feared that Oregon had lost its way: "We've lost our can-do attitude in Oregon. And we replaced it with a cannot do attitude."[19]

Shortly before the state's high-tech sector crashed, Chet Orloff, the executive director of the Oregon Historical Society, worried that Oregon had lost its sense of rootedness: "You don't have that sense of place, that groundedness and commitment." Writing for the *Oregonian* on the last day of 1999, Gail Kinsey Hill observed that the 1990s closed with a note of anxiety about what the state was becoming: "New people, new money, new technology, all rushing in without time for assimilation." Business mergers and corporate takeovers contributed further to citizens' worries: Enron purchased the venerable Portland General Electric Company in 1996; the Northwest's Thrifty Payless sold out to Pennsylvania-based Rite Aid; Cincinnati's Kroger Company bought Portland-based giant Fred Meyer; and the Weyerhaeuser Company acquired Willamette Industries, an Oregon-based Fortune 500 firm, including the company's vast 1.7 million acres of timberland in the United States.[20] It has not passed the notice of many Oregonians that both Enron and Rite-Aid wound up in bankruptcy court.

Amidst the unease shared by many Oregonians in the early twenty-first century are elements of citizen pride, especially in the state's rich heritage of federal lands: thirteen national forests embracing 15.6 million acres (equal in size to West Virginia); 2.1 million acres of federal wilderness areas; and 30 percent of Oregon's rivers carrying the federal designation "Wild and Scenic." The state has its own emerald-green gem, the 364,000-acre Tillamook State Forest, a freshly grown stand of timber "created" out of the devastating series of fires that ravaged the Tillamook country between 1933 and 1951. While the state Department of Forestry proudly boasts that the replanted Tillamook will be managed "to provide a wide range of social, economic and environmental benefits," the future promises tensions between local lumber mills and those who value the forest for its amenities. Oregonians are also justly proud of their "public" ocean beaches, preserved in a series of waysides and state parks extending from the Columbia River to the California border. Columnist Russell Sadler refers to this heritage as Oregon's public patrimony—"its public parks, its public forests, its rivers and beaches."[21]

Oregon continues to struggle with the residual effects of its long history of extractive activity and the unbridled development of its waterways. Logging practices in some areas continue to be problematic despite the state Department of Forestry's professions to the contrary. After several people died in landslide-related incidents following torrential rains in 1996, steep-slope logging became an instant political issue. Ignoring a host of scientific studies dating to the early 1970s that indicated a direct relationship among clear-cutting, logging roads, and landslides, Department of Forestry officials at first insisted that the agency had no authority to restrict timber harvests to protect life and property. Following a public hearing in March 1997, the Oregon Board of Forestry voted unanimously to ask loggers to voluntarily restrict steep-slope logging in slide-prone areas. But public pressure pushed the issue before the legislature later that year, and lawmakers passed an emergency provision to prohibit logging that poses a threat to public safety. The legislature attempted to add further restrictions in 1999, authorizing the Board of Forestry to draft permanent rules to reduce risks from landslides related to logging activities. Four years later, the Department of Forestry was still holding public hearings "to discuss draft forest practice rules on landslides."[22]

No symbol more powerfully represents the essence and meaning of the Northwest than its anadromous fishes, especially salmon. Salmon have had powerful cultural importance to Native American people for at least 10,000 years, providing spiritual, ceremonial, and cultural meaning and an essential protein for their diet. For more than fifty years people from all walks of life have commiserated about the declining number of salmon, engaging in noisy debates over the effects of dams, hatcheries, overgrazing, logging practices, urban pollution, and a host of other human-related environmental disturbances. As the region moves into the next millennium, fishery biologist Jim Lichatowich argues that the fate of the fish "has truly become everyone's problem." In many respects salmon may be our miner's canary, strongly suggesting that environmental problems are intimately interrelated. Logging-related landslides wreak havoc with mountain streams, agricultural operations deposit sediment in waterways, septic discharges and fertilizer runoff contribute non-point-source pollutants to rural and urban streams, and stripping streamsides of riparian vegetation raises water temperatures to levels that are harmful to anadromous fish.[23]

A casual reading of the daily newspapers or a glimpse at the evening news suggests the extent to which endangered salmon have become the focus of hot political debate. Science and politics are often at cross purposes, with some groups accusing public agencies of promoting "bad science." Policymakers, scientists, fishers, and average citizens have offered ingenuous evasions and pointed the finger of blame at singular causes for declining salmon runs. In the midst of the salmon crisis, conspiracy theories have abounded, especially when improved ocean conditions began to boost returns of hatchery fish beginning in 2000. Conservative political groups want to see hatchery fish counted with wild stocks to facilitate removing the runs from threatened and endangered listings. Conservative talk-show hosts have criticized federal and state fishery biologists for manufacturing the crisis to assure that salmon will continue to be listed. To do so, they argue, gives federal and state agencies leverage to increase environmental restrictions on private property.[24]

When scientists have weighed in on the debate, their ideas have often challenged some of the Northwest's most powerful economic interests. Proposals to remove the lower Snake River dams to restore the waterway's more natural flow ran counter to the powerful Columbia River barging companies, electric utilities, ports, agricultural groups (especially wheat farmers), and aluminum and other industries. When scientists proposed cutting back on the number of downstream-migrating smolt that were being transported around the dams, they argued that putting fish in trucks to cart them past the dams was based on economics, not biology. Jack Stanford, a University of Montana ecologist who served on a scientific panel in the late 1990s to study the issue of barging salmon past the dams, observed that the recovery of wild stocks could not be "accomplished by transportation, more and fancier hatcheries, better (turbine) screens, and continued mixed stock commercial fisheries." The answer to salmon recovery, he argued, depended on "improved habitat and river conditions." The Bonneville Power Administration (BPA) and the Northwest Power Planning Council favored transporting the salmon smolts because spilling the fish over the dams was expensive, wasting water that could otherwise be used to generate electricity. The Army Corps of Engineers and BPA contend that barging helps more fish survive the downsteam migration.[25]

For BPA and Columbia River managers, the real issues are hydropower costs. When the winter of 2000–2001 produced a deficient snowpack in the Northern Rockies, the ensuing power "crisis" revealed a federal agency willing to trade economics for fish. Because endangered salmon stocks required BPA to spill water over the dams to aid downstream-migrating fish, the agency lacked the generating capacity to satisfy all of its contracted obligations. To meet those commitments, BPA was forced to purchase electricity on the open market at much higher prices. As one critic pointed out, it was not that BPA was unable to purchase enough power to meet its contract, "it was that the agency couldn't find it at a price it wanted to pay." BPA's defenders charged that "flow augmentation" disrupted electrical-power generation at federal dams and raised the price of electricity to the public. Fisheries scientists pointed to the obvious: years of high runoff in the Columbia system produced much higher salmon returns. But when BPA tallied the "costs" to the public, it reported that ratepayers paid $1.5 billion for "salmon protection" for the fiscal year 2001–2002. According to that logic, Lewis and Clark law professor Michael Blumm observed, "if it weren't for salmon requirements, . . . the region's ratepayers would save lots of money." Blumm then asked: "Why is there no attribution of the annual 'costs' that BPA operations have inflicted on the salmon resource?"[26]

Returns of salmon species to the Northwest's coastal streams between 2000 and 2003 improved significantly and most fisheries biologists attributed the increases to a temporary cycle of abundant ocean foods rather than a major achievement in saving the fish. NOAA Fisheries, formerly the National Marine Fisheries Service, issued a preliminary report early in 2003 stating that none of the twenty-seven threatened and endangered anadromous fish populations should be removed from the endangered species list. NOAA Fisheries announced that it would issue a later ruling on whether both hatchery fish and wild salmon should be protected and that it would outline the role of hatcheries in the restoration process. The scientists contracted to NOAA believed that returns still were far below recovery targets and did not appear to be sustainable. Environmental groups criticized the preliminary report for continuing to rely on techno-fixes and not recommending tough decisions. The conservative Pacific Legal Foundation complained that NOAA had failed to decide if hatchery fish should be counted in determining endangered species listings.[27]

The survival of salmon is inextricably linked to larger issues involving quality of life in Oregon and the Pacific Northwest. The siren song of the cash register—pursuing unbridled market imperatives—continues to threaten Oregon's streams, forests, agricultural lands and the ability to provide stable community living. Because of the increasingly integrated nature of the national and global economy, the state has limited maneuverability in charting a course that will preserve its open spaces, strengthen its stewardship of valuable resources, and provide equal opportunity for all its citizens to share in its bounty. Twelve Catholic bishops representing 1.5 million parishioners in the binational Columbia River Basin issued a remarkable pastoral letter in early 2001 declaring that the region was threatened with environmental and social degradation. Defining the Columbia Basin as a "sacred landscape," the document called for the development of caring communities, because humans were "created as social beings who must exercise a certain responsibility toward our neighbors." The letter reminded parishioners that they held land in trust for present and future generations and that the idea of the common good meant that "community and individual *needs* take priority over private wants." And then, in a ringing declaration of the common good, the bishops insisted that "the right to own and use property is not . . . an absolute individual right" but a "right [that] must be exercised responsibly for the benefit of . . . the community as a whole." The bishops argued that a just society and a sustainable environment were not incompatible. A living wage, secure working conditions, good and affordable housing, health insurance, educational opportunities, and healthful surroundings pointed in the direction of a greater common good. Lives centered in privacy, charter schools, gated communities, and huge houses were not part of the bishops' statement.[28] Because I see an erosion of our attachment and affection for places everywhere, I believe the bishops' argument carries a powerful contemporary message for Oregonians. Like all Americans, we have become an increasingly rootless people, a collective behavior that has diminished the influence of place as a moral force in shaping our social, civic, and environmental values. Our overly acquisitive habits, our propensity to accumulate material things, and our fascination with the latest technology have blinded us to social and economic injustices and environmental damage in our communities. There is no better test of our collective will, I believe, than the stewardship we exercise toward each other and toward the world about us.

NOTES

PREFACE

1. Evan Eisenberg, *The Ecology of Eden* (New York: Alfred A. Knopf, 1998), 289.

PROLOGUE: A TIME TO REMEMBER

1. Henry R. Luce, *The American Century* (New York: Farrar and Rinehart, 1941), 27.

2. *New York Times*, August 15, 1945.

3. Eric F. Goldman, *The Crucial Decade—and After: America, 1945–1960*, enlarged edition (New York: Random House, 1960), 4.

4. *Coos Bay Times*, August 15, 1945.

5. *Astorian Evening Budget*, August 15 and 16, 1945.

6. Carlos A. Schwantes, *The Pacific Northwest: An Interpretive History*, revised and enlarged edition (Lincoln: University of Nebraska Press, 1996), 408–09; William G. Robbins, *Hard Times in Paradise: Coos Bay, Oregon, 1850–1986* (Seattle: University of Washington Press, 1988), 97; and Connie Hopkins Battaile, *The Oregon Book: Information A to Z* (Newport, Oregon: Saddle Mountain Press, 1998), 639.

7. James T. Patterson, *Grand Expectations: The United States, 1945–1974* (New York: Oxford University Press, 1996), 852; *Medford Mail-Tribune*, September 4, 1945; and *Pendleton East Oregonian*, August 30, 1945.

8. Portland *Oregonian*, September 2, 3, and 4, 1945.

9. *East Oregonian*, August 8, 13, and 14, 1945.

10. Howard Zinn, *A People's History of the United States: 1492–Present* (New York: Harper and Row, 1980), 398–434; and Studs Terkel, *"The Good War": An Oral History of World War Two* (New York: Pantheon Books, 1984). Also see Tom Brokaw, *The Greatest Generation* (New York: Random House, 1998).

11. Goldman, *The Crucial Decade*, 12.

12. Robbins, *Hard Times in Paradise*, 95–97.

13. David M. Kennedy, *Freedom From Fear: The American People in Depression and War, 1929–1945* (New York: Oxford University Press, 1999), 776–82; Richard White, *"It's Your Misfortune and None of My Own": A New History of the American West* (Norman: University of Oklahoma Press, 1991), 496–504; Schwantes, *The Pacific Northwest*, 415–16; and Robbins, *Hard Times in Paradise*, 102.

14. White, *"It's Your Misfortune and None of My Own"*, 496.

15. Vernon Jensen, *Lumber and Labor* (New York: Farrar and Rinehart, 1945), 276; Valerie (Wyatt) Taylor, interview with the author, March 27, 1984; and Schwantes, *The Pacific Northwest*, 413.

16. John M. Findlay, *Magic Lands: Western Cityscapes and American Culture After 1940* (Berkeley: University of California Press, 1992), 13–16; and Gerald D. Nash, *The American West Transformed: The Impact of the Second World War* (Bloomington: Indiana University Press, 1985), 203–05.

17. *Oregonian*, August 26, 1945.

18. Robbins, *Hard Times in Paradise*, 99–100; and Dow Beckham, interview with the author, March 28, 1984.

19. Beckham interview.

20. Schwantes, *The Pacific Northwest*, 328–29; and Roger Sale, *Seattle, Past to Present* (Seattle: University of Washington Press, 1976), 180–81.

21. Manly Maben, *Vanport* (Portland: Oregon Historical Society Press, 1987), 1–12; Carl Abbott, *Portland: Planning, Politics, and Growth in a Twentieth-Century City* (Lincoln: University of Nebraska Press, 1983), 126–27; and Schwantes, *The Pacific Northwest*, 331–32.

22. Maben, *Vanport*, 87–93; and Ellen A. Stroud, "A Slough of Troubles: An Environmental and Social History of the Columbia Slough" (M.A. thesis, University of Oregon, 1995), 1–3, 12–13.

23. Schwantes, *The Pacific Northwest*, 413; and Abbott, *Portland*, 126.

24. U.S. Bureau of the Census, *Sixteenth Census of the United States: 1940, Population*, Vol. 1: *Number of Inhabitants*, 886; and ibid., *Seventeenth Census of the United States: Census of Population, 1950*, Vol. 2: *Characteristics of the Population*, 37–120 to 37–129.

25. For accounts of the beginnings of the Hanford works, see Bruce Hevly and John M. Findlay, "The Atomic West: Region and Nation, 1942–1992," in *The Atomic West*, Hevly and Findlay, eds. (Seattle: University of Washington Press, 1998), 3–11; Ferenc M. Szasz, "Introduction," in *Working on the Bomb: An Oral History of WWII Hanford*, S. L. Sanger, ed. (Portland: Continuing Education Press, Portland State

University, 1995), 11–18; and Richard White, *The Organic Machine: The Remaking of the Columbia River* (New York: Hill and Wang, 1995), 81–88.

26. Jensen, *Lumber and Labor,* 276–78; *Coos Bay Times,* September 11, 19, 1942; and Robbins, *Hard Times in Paradise,* 95–106.

27. U.S. Forest Service, *Annual Report* (1940), 1–3; and ibid. (1942), 1–5.

28. Government Control of Local Forest Enterprise, and objections to Plan for a Forest Products Program of U.S. Forest Service, circa February 1943, box 27, National Forest Products Association Records (hereafter NFPA Records), Forest History Society, Durham, North Carolina.

29. Roosevelt to Harold Smith and Wayne Coy, May 19, 1942, in Edgar B. Nixon, ed., *Franklin D. Roosevelt and Conservation,* vol. 2 (Hyde Park, New York: Franklin D. Roosevelt Library, 1957), 550–54.

30. U.S. Forest Service, *Annual Report* (1943), 1–2; and ibid. (1944), 1–2.

31. Denis W. Brogan, *The American Character* (New York: Alfred A. Knopf, 1944), 25; and Works Projects Administration, Oregon, *Oregon: End of the Trail* (Portland: Binfords and Mort, 1940), v, vii, ix, 28.

32. *Sunset* 84 (January 1940), 6.

33. See Samuel P. Hays, *Beauty, Health, and Permanence: Environmental Politics in the United States, 1955–1985* (New York: Cambridge University Press, 1987), 2–3; and Brogan, *The American Character,* 25.

34. Rex Eastman, "Oregon Beckons New Industries," *The Spectator* 65 (January 1940), 10.

35. These ideas are borrowed from George McKinley and Doug Frank, "Stories on the Land: An Environmental History of the Upper Applegate and Illinois Valleys," a report prepared for the Bureau of Land Management, Medford District, Medford, Oregon, 1995, 165.

36. Eric Hobsbawm, *The Age of Extremes: A History of the World, 1914–1991* (New York: Pantheon Books, 1995), 200, 276.

37. Findlay, *Magic Lands,* 19–20.

38. Patterson, *Grand Expectations,* 61–70, 318.

1 / THE GREAT HOPE FOR THE NEW ORDER

1. E. R. Jackman, typescript of radio broadcast on KOAC, Corvallis, Oregon, November 15, 1945, in E. R. Jackman Papers, Oregon State University Archives, Corvallis.

2. The story and the quotation are in Daniel J. Kelves, *The Physicists: A History of a Scientific Community in America* (New York: Alfred A. Knopf, 1978), 333.

3. *Medford Mail-Tribune,* August 24, 1945.

4. Barry Commoner, *The Closing Circle: Nature, Man, and Technology* (New York: Alfred A. Knopf, 1971), 125–27; and Keith R. Benson, "The Maturation of Science in the Northwest: From Nature Studies to Big Science," in *The Great Northwest: The Search for Regional Identity,* William G. Robbins, ed. (Corvallis: Oregon State University Press, 2001), 144.

5. Vannevar Bush, *Science: The Endless Frontier* (Washington, D.C.: Government Printing Office, 1946), vii, 1–2, 5–6.

6. Portland *Oregonian,* September 7, 1945.

7. The General Motors advertisement is discussed in Truman E. Moore, *Nouveaumania: The American Passion for Novelty and How it Led Us Astray* (New York: Random House, 1975), 26–27, 35. Also see James T. Patterson, *Grand Expectations: The United States, 1945–1974* (New York: Oxford University Press, 1996), 70; and Eric Hobsbawm, *The Age of Extremes, 1914–1991* (New York: Pantheon Books, 1995), 265.

8. J. Ronald Oakley, *God's Country: America in the Fifties* (New York: Dembner Books, 1986), 9–10, 237.

9. Hobsbawm, *The Age of Extremes,* 270; and Patterson, *Grand Expectations,* 318–20.

10. Thomas R. Dunlap, *DDT: Scientists, Citizens, and Public Policy* (Princeton: Princeton University Press, 1981), 3–17, 37–39, 59–63; J. Brooks Flippen, "Pests, Pollution, and Politics: The Nixon Administration's Pesticide Policy," *Agricultural History* 71 (Fall 1997), 442; and Rachel Carson, *Silent Spring* (Boston: Houghton Mifflin Company), 1962.

11. Robert J. Samuelson, *The Good Life and Its Discontents: The American Dream in the Age of Entitlement, 1945–1995* (New York: Random House, 1995), xiii, xvi; and Moore, *Nouveaumania,* 7–9.

12. *Life* 28 (January 2, 1950), 28, 31, 85.

13. Henry R. Luce, *The American Century* (New York: Farrar and Rinehart, 1941), 27, 30; and Donald W. White, *The American Century: The Rise and Decline of the United States as a World Power* (New Haven: Yale University Press, 1996), 8–9, 11, 17.

14. Richard L. Neuberger, "The Land of New Horizons," *New York Times Magazine* (December 9, 1945), 18–23.

15. *Oregonian,* December 25, 1945; *Roseburg News-Review,* December 31, 1945; *Pendleton East Oregonian,* January 1 and 7, 1946; and *Mail-Tribune,* December 31, 1945.

16. *Oregonian,* August 26, 1945.

17. "Progress Report of Postwar Readjustment and Development Commission of the State of Oregon" (hereafter "Progress Report"), no. 14 (August 1944), 1–2. For a brief account of similar commissions in the other western states and the Earl Warren quote, see Gerald D. Nash, *The American West Transformed: The Impact of the Second World War* (Bloomington: Indiana University Press, 1985), 201–04.

18. "Progress Reports," no. 17 (November 1944), 1; and no. 22 (April 1945), 5.

19. For a sampling of the commission's ambitions, see "Progress Reports," no. 18 (December 1944), 2–4; no. 22 (April 1945), 13–14; no. 23 (May 1945), 2; no. 25 (July 1945), 9–11; and no. 32 (February 1946), 5–7.

20. Ibid., no. 15 (September 1944), n.p.

21. Ibid., no. 23 (May 1945), 13; and no. 32 (February 1946), 7–9.

22. Patterson, *Grand Expectations,* 71–73; and Oakley, *God's Country,* 10.

23. Bill McKenna, interview with the author, April 11, 1984.

24. Patterson, *Grand Expectations,* 141; *Oregonian,* September 7, 1945; and *Coos Bay Times,* February 13, 1943.

25. "Progress Reports," no. 15 (September 1944), n.p.; no. 22 (April 1945), 6–7; no. 23 (May 1945), 3.

26. *Bend Bulletin,* September 6, 17, and 20, 1945.

27. *Reedsport Port Umpqua Courier,* December 27, 1945; *Mail-Tribune,* September 5, 1945; and *East Oregonian,* August 31, September 1, 1945.

28. *News-Review,* January 2, 1946; and "Progress Report" (April 1947), 3.

29. Salem *Statesman,* August 26, 1945.

30. Nash, *The American West Transformed,* 213–16; Richard White, *"It's Your Misfortune and None of My Own": A New History of the American West* (Norman: University of Oklahoma Press, 1991), 514; and William G. Robbins, *Hard Times in Paradise: Coos Bay, Oregon, 1850–1986* (Seattle: University of Washington Press, 1988), 105.

31. R. W. Cowlin, "Some Economic Aspects of Oregon Lumber Manufacture," *Oregon Business Review* 4 (June 1945), 1, 9–11; and Roseburg *News-Review,* September 8, 1945.

32. Ibid. 9–11.

33. Ibid., 9–11; and Paul W. Hirt, *A Conspiracy of Optimism: Management of the National Forests since World War Two* (Lincoln: University of Nebraska Press, 1994), 44.

34. H. V. Simpson, "West Coast Lumber Enters a New Cycle," *Oregon Business Review* 4 (November 1945), 1, 3.

35. W. H. Horning, "Sustained-Yield Program of the O & C Lands," *Oregon Business Review* 5 (April 1946), 1–4.

36. *News-Review,* August 17 and September 3 and 27, 1945.

37. *Oregonian,* June 5, 1945; and "Progress Report," no. 25 (July 1945), 9–10.

38. *Pacific Northwest Development Association: By-laws and Articles of Incorporation* (Portland, 1945), n.p.; *Pacific Northwest Development Association Bulletin,* no. 9 (May 28, 1946), 2; and ibid., no. 12 (July 17, 1946), 2.

39. *Oregonian,* November 12, 1944; and *Bulletin,* September 20, 1945 and May 18, 1946.

40. *Bulletin,* May 18 and 20, 1946.

41. Ibid., May 20, 1946.

42. *Oregonian,* May 5, 1946.

43. Jarold Ramsey, "'New Era': Growing up East of the Cascades, 1937–1950," in *Regionalism and the Pacific Northwest,* William G. Robbins, Robert J. Frank, and Richard E. Ross, eds. (Corvallis: Oregon State University Press, 1983), 183, 200.

44. *Bulletin,* September 19, 1946. Data on the storage capacity of the reservoirs is in Works Progress Administration, *Present and Future Land Development in Oregon Through Flood Control, Drainage and Irrigation* (Salem: Oregon State Planning Board, 1938), 144.

45. The McCall quote is in Brent Walth, *Fire at Eden's Gate: Tom McCall and the Oregon Story* (Portland: Oregon Historical Society Press, 1994), 356.

46. "Progress Report," no. 32 (Februry 1946), 5-6.

47. *Mail-Tribune,* August 28, 1945; and *East Oregonian,* September 20, 1945.

48. J. R. Woodruff, "Portland Port Facilities," *Oregon Business Review* 4 (September 1945), 1, 4.

49. H. T. Shaver, "Upper Columbia River Traffic," ibid., 1, 4.

50. For a discussion of the earlier Columbia River development works, see William G. Robbins, *Landscapes of Promise: The Oregon Story, 1800–1940* (Seattle: University of Washington Press, 1997), 191–96, 200, and 242–44.

51. Ivan Bloch, "Pacific Northwest Horizons in the Postwar World," *Oregon Business Review* 5 (June 1946), 1–2.

52. Ibid., 2–4.

53. Ibid., 6; Sammons is quoted in Richard L. Neuberger, "The Cities of America: Portland, Oregon," *Saturday Evening Post,* March 1, 1947, p. 106.

54. "Progress Report" (January 1948), 2–5.

55. Ibid., 2–4; (August 1948), 10; and Charles McKinley, *Uncle Sam in the Pacific*

Northwest: Federal Management of Natural Resources in the Columbia River Valley (Berkeley: University of California Press, 1952), 7.

56. McKinley, *Uncle Sam in the Pacific Northwest,* 10; and "Progress Report," no. 45 (May 1947), 8.

57. "Progress Report," no. 45 (May 1947), 5–6.

58. Marc Reisner, *Cadillac Desert: The American West and Its Disappearing Water* (New York: Viking Penguin, 1986), 165–66, 172.

2 / INTO THE BRAVE NEW WORLD

1. Jim Lichatowich, *Salmon Without Rivers: A History of the Pacific Salmon Crisis* (Washington, D.C.: Island Press, 1999), 5–7.

2. Ibid.

3. Portland *Oregonian,* August 26, 1945.

4. John Gunther, *Inside U.S.A.,* 50th Anniversary Edition, Foreword by Arthur Schlesinger, Jr. (1946; New York: Book of the Month Club, 1997), 119–20.

5. The "rationalized" phrase is in Richard White, *The Organic Machine: The Remaking of the Columbia River* (New York: Hill and Wang, 1995), 77.

6. *Oregonian,* August 14, 1943. For a discussion of the alliance of groups and individuals who served as a catalyst for the Willamette Valley project, see William G. Robbins, "The Willamette Valley Project of Oregon: A Study in the Political Economy of Water Resource Development," *Pacific Historical Review* 47 (November 1978), 585–605.

7. *Oregonian,* May 5, 1945.

8. Roberta Ulrich, *Empty Nets: Indians, Dams, and the Columbia River* (Corvallis: Oregon State University Press, 1999), 3. For an excellent study of the Indians at Celilo Falls and the move to build the dam, see Katrine Barber, "After Celilo Falls: The Dalles Dam, Indian Fishing Rights, and Federal Energy Policy" (Ph.D. dissertation, Washington State University, 1999).

9. *Oregonian,* May 5, 1946.

10. Ibid., June 9, 1946.

11. Ibid., July 19, 1946.

12. Paul R. Needham, "Plans for Protection of Salmon Runs That Will Be blocked by Shasta Dam," box 39, E. E. Wilson Papers, Oregon State University Archives, Corvallis.

13. I offer a mild dissent here from Richard White who argues that fishery advocates "were not an imposing force." See White, *The Organic Machine,* 93. Also see

Anthony Netboy, *The Columbia River Salmon and Steelhead Trout* (Seattle: University of Washington Press, 1980), 78–84.

14. *Oregonian,* May 29, 1945.

15. Much of the information about the Columbia Basin Fisheries Development Association is in State of Oregon, *Report of the Interim Fisheries Committee to the Forty-fourth Legislative Assembly* (Salem, 1947), 1–2.

16. *Roseburg News-Review,* October 10, 1945, February 3, April 29, 1946.

17. Copy of a newspaper clipping, dated January 23, 1946, given to me by David Charlton.

18. *Report of the Interim Fisheries Committee,* 2.

19. House Subcommittee on Merchant Marine and Fisheries, *Hearings on Columbia River Fisheries,* 79th Cong., 2nd sess. August 14, 1946, 86–88.

20. Ibid., 88–89, 92–93.

21. Ibid., 97–98.

22. Mert Folts to Representative John A. Blatnik and members of the House Subcommittee on Appropriations, October 27, 1947. Copy in the author's possession. A copy of the letter is also in the David B. Charlton Papers, Oregon Historical Society, Portland, Oregon (hereafter Charlton Papers).

23. *Columbia River Fisheries,* 106–107.

24. Chessman's testimony is printed in *Report of the Interim Fisheries Committee,* 2–3.

25. "Columbia River Fisheries," 84.

26. Ibid., 82.

27. Ibid., 82, 85–86.

28. Ibid., 99–102.

29. Ibid., 102–103.

30. Ibid., 112–15.

31. *First Biennial Report, Willamette River Basin Commission* (Salem, 1946), 4–6.

32. Keith C. Petersen, *River of Life, Channel of Death: Fish and Dams on the Lower Snake* (Lewiston, Idaho: Confluence Press, 1995), 113.

33. David B. Charlton to Kenneth A. Reid, May 24, 1947, copy in the author's possession.

34. Netboy, *Columbia River Salmon,* 78–79; and *Oregonian,* June 25, 1947.

35. The quotation is in Petersen, *River of Life, Channel of Death,* 111–12.

36. Columbia Basin Inter-Agency Committee, *Minutes of the Tenth Meeting,* Walla Walla, Washington, June 25–26, 1947, n.p.

37. Ibid.; and *Oregonian,* June 27, 1947.

38. *Oregonian,* June 27, 1947.

39. Ibid.

40. Paul R. Needham, "Dams Threaten West Coast Fisheries Industry," *Oregon Business Review* 6 (June 1947), 1, 3–4.

41. Ibid., 4–6.

42. *Oregonian,* June 28 and July 24, 1949. The federal agency's decision is quoted in Netboy, *Columbia River Salmon,* 82.

43. The Fish and Wildlife Service is quoted in White, *The Organic Machine,* 96.

44. I am indebted to Portland State University historian William Lang for this information.

45. *New York Times,* May 31, 1948; and Manly Maben, *Vanport* (Portland: Oregon Historical Society Press, 1987), 104–105.

46. *Oregonian,* May 31, 1948.

47. "Progress Report of the [Oregon] Postwar Readjustment and Development Commission," no. 60, June 1948, 7; and White, *The Organic Machine,* 74–75.

48. William F. Willingham, *Army Engineers and the Development of Oregon: A History of the Portland District U.S. Army Corps of Engineers* (Washington, D.C.: Government Printing Office, 1983), 157–58.

49. *Oregonian,* June 2, 1948.

50. Ibid., June 12, 1948.

51. Ibid.

52. Richard L. Neuberger, "The Columbia," *Holiday* (June 1949), reprinted in Steve Neal, ed., *They Never Go Back To Pocatello: The Selected Essays of Richard Neuberger* (Portland: Oregon Historical Society Press, 1988), 30–45.

53. Ibid., 43–44.

54. For example, see Richard L. Neuberger, "The Great Salmon Mystery," *The Saturday Evening Post,* September 13, 1941, 20–21, 39, 41–42, 44; and Neuberger, "The Great Salmon Experiment," *Harper's* (February 1945), 229–36.

55. Willingham, *Army Engineers and the Development of Oregon,* 158–59; Gus Norwood, *Columbia River Power for the People: A History of the Bonneville Power Administration* (Portland: U.S. Department of Energy, Bonneville Power Administration, 1981), 160; and Columbia Basin Inter-Agency Committee, *The Columbia Basin Inter-Agency Committee and the Pacific Northwest* (Portland, 1949), 8.

56. *Your Columbia River: The Development of America's Greatest Power Stream* (Portland: Bonneville Power Administration, 1950), 5, 13, 20–21.

57. *Columbia River and Tributaries, Northwestern United States,* 81st Cong., 2nd sess., 1950, H. Doc. 531, vol. 1, 1–5, 12; and Willingham, *Army Engineers,* 158.

58. Norwood, *Columbia River Power for the People,* 160.

59. *Columbia River and Tributaries,* vol. 1, 74–75.

60. Ibid., 117–19.

61. Ibid., 7, 2863-64, 2916.

62. Ibid., 2917, 2919.

63. Ibid., 2919.

64. Petersen, *River of Life, Channel of Death,* 112–14; *Columbia Inter-Agency Committee and the Pacific Northwest,* 8.

65. David Charlton to C. Girard Davidson, February 6, 1950, box 18, Charlton Papers.

66. White, *The Organic Machine,* 90.

3 / BRINGING PERFECTION TO THE FIELDS

1. *Agriculture in Oregon* (Salem: State Department of Agriculture, 1965), 5.

2. *West Shore* 1, no. 4 (November 1885), 2; and Salem *Willamette Farmer,* June 20, 1874.

3. Oregon State Department of Agriculture, *Agriculture Bulletin,* no. 171 (September 1951), 2.

4. Portland Chamber of Commerce, *Farming in Oregon* (Portland: Agricultural Committee of the Portland Chamber of Commerce, 1945), 5, 64.

5. James E. Sherow, "Environmentalism and Agriculture in the American West," in *The Rural West Since World War II,* R. Douglas Hurt, ed. (Lawrence: University Press of Kansas, 1998), 58–61. Also see the editor's "Introduction," 2–4.

6. U.S. Department of Agriculture, Pacific Northwest Regional Committee on Post-War Programs, "Area Plans for Agriculture in the Post-War Period" (Portland, 1944), 12–14.

7. "Progress Report of the Post War Readjustment and Development Commission of the State of Oregon" (hereafter "Progress Report"), no. 66 (January 1949), 9–11; and Oregon State Department of Agriculture, *Agriculture in Oregon* (Salem: State Printing Department, 1946), 3–7.

8. Cooperative Extension Service, *Oregon's First Century of Farming: A Statistical Record of Agricultural Achievement and Adjustment* (Corvallis: Oregon State College, 1959), 2–10.

9. The quotations from the enabling legislation are in Judith D. Soule and Jon K. Piper, *Farming in Nature's Image: An Ecological Approach to Agriculture*

(Washington, D.C.: Island Press, 1992), 59–60. Also see Deborah Fitzgerald, "Beyond Tractors: The History of Technology in American Agriculture," *Technology and Culture* 32 (January 1991), 115–16; and Frieda Knobloch, *The Culture of Wilderness: Agriculture as Colonization in the American West* (Chapel Hill: University of North Carolina Press, 1996), 57.

10. Fitzgerald, "Beyond Tractors," 116; and Soule and Piper, *Farming in Nature's Image,* 60.

11. Oregon Agricultural Experiment Station, *Profile* (Corvallis, September 1976), 4–5, 13.

12. Oregon Department of Agriculture, *Agriculture in Oregon* (Salem, 1965), 8–9; *Biennial Report of the [Oregon]State Department of Agriculture,* (1966), 5, (1968), 9.

13. *Agriculture Bulletin,* no. 184 (December 1954), 8.

14. A brief history of the department can be found in *Biennial Report of the State Department of Agriculture* (1968), 4–6.

15. For the history of the grass-seed industry, see E. R. Jackman, "From a Shoestring to Ten Million," manuscript prepared for *Seed World,* October 17, 1941, copy in Manuscripts, 1940–1951, E. R. Jackman Papers, Oregon State University Archives, Corvallis (hereafter Jackman Papers); and Leonard R. Jernstedt, "Seed Crops: A Major Industry," *Agriculture Bulletin,* no. 171 (September 1951), 25–26.

16. "Progress Report" no. 47, May 1947 (Salem), 14; and Jernstedt, "Seed Crops," 26.

17. E. R. Jackman, "Oregon's Seed Industry," Statement to Columbia Basin Interagency [sic] Committee, February 8, 1950, copy in SGI, Manuscripts, 1940–1951, Jackman Papers; and Jernstedt, "Seed Crops," 26.

18. Jernstedt, "Seed Crops," 26; and E. R. Jackman to Marshall N. Dana, February 15, 1951, Jackman Papers.

19. Jackman, "Oregon's Seed Industry."

20. Frank S. Conklin, William C. Young III, and Harold W. Youngberg, *Burning Grass Seed Fields in the Willamette Valley: The Search for Solutions,* Extension Miscellaneous Publication 8397, February 1989, Oregon State University Extension Service, 1, 8; *Field Burning: Oregon State University Research,* Agricultural Experiment Station, Oregon State University, Corvallis, February 1973, 2; and Peter G. Boag, "The World Fire Created: Field Burning in the Willamette Valley," *Columbia* 5 (Summer 1991), 8.

21. Conklin, et al., *Burning Grass Seed Fields,* 1, 12; Boag, "The World Fire Created," 7; and *Oregon's Burning Issue,* prepared by Robert G. Davis for the Seed Growers of the Willamette Valley (Salem, 1980), 1–2.

22. Conklin, et al., *Burning Grass Seed,* 12; *Oregon's Burning Issue,* 1–2; and *Field Burning,* 1–2.

23. R. Wallace Rice to Gordon Walker, June 1, 1960, in Agricultural Experiment Station Records, Record Group 25 (hereafter AES Records), Oregon State University Archives, Corvallis, Oregon.

24. Oregon Seed Council, *Grass Seed: The Tiny Giant* (Salem: c. 1975), n.p.; Conklin, et al., *Burning Grass Seed Fields,* 25.

25. *Agricultural Field Burning in the Willamette Valley* (Corvallis: Air Resources Center, Oregon State University, January 1969), 8–10.

26. Boag, "The World Fire Created," 7–9; and Conklin, et al., *Burning Grass Seed Fields,* 12.

27. Conklin, et al., *Burning Grass Seed Fields,* 13; and Boag, "The World Fire Created," 10.

28. *Field Burning,* 1–2. The 1974 report is cited in *Oregon's Burning Issue,* 2–4.

29. Boag, "The World Fire Created," 11.

30. Conklin, et al., *Burning Grass Seed Fields,* 14–15.

31. Ibid., 40–41; and Ron Karten, "Growers Look for Green Markets," *Oregon Business* 7 (August 1984), 47–48.

32. Harold Youngberg, "Introduction," *Seed Production Research at Oregon State University* (1982), 1; and Ronald E. Welty, "Introduction," ibid. (1983), 2.

33. W. C. Young III, D. O. Chilcote, and H. W. Youngberg, "Post-Harvest Management Alternatives for Grass Seed Production," in ibid. (1983), 21–22; Youngberg, "Introduction," in ibid. (1985), 1; J. A. Kamm, "Reduction of Insecticide Activity by Carbon Residue Produced by Post-Harvest Burning of Grass Seed Fields," in ibid. (1988), 2; and W. C. Young III, T. B. Silberstein, and D. O. Chilcote, "Evaluation of Equipment Used by Willamette Valley Grass Seed Growers as a Substitute for Open-Field Burning," in ibid. (1992), 1.

34. *Oregon's Burning Issue,* 32–39; and Boag, "The World Fire Created," 11.

35. *Oregon's Burning Issue,* 64–65.

36. Boag, "The World Fire Created," 11.

37. Oregon Seed Council, "Oregon Grown Grass Seed"; and Oregon Department of Agriculture, "Willamette Valley Field Burning History."

38. Portland *Oregonian,* January 8, 1950.

39. "Oregon Starts Pilot Program on Starlings," *Agriculture Bulletin,* no. 204 (December 1959), 3–4.

40. "Operation Starling," ibid., no. 205 (March 1960), 5.

41. "Starling Studies Make Headway," ibid., no. 211 (September 1961), 5–6; and

"Starling Studies Make Headway But Now Go to Research Basis," ibid., no. 219 (September 1963), 5–6.

42. Larry Rymon, "A Critical Analysis of Wildlife Conservation in Oregon" (Ph.D. dissertation, Oregon State University, 1969), 269–71; and *Agriculture Bulletin,* no. 219 (September 1963), 7.

43. Rymon, "A Critical Analysis," 273–74; and Oregon State Department of Agriculture, *Biennial Report* (1958), 31.

44. *Agriculture Bulletin,* no. 219 (September 1963), 7.

45. Ibid; Donald Worster, *Nature's Economy: A History of Ecological Ideas,* second edition (New York: Cambridge University Press, 1985), 260; and Rymon, "A Critical Analysis," 277–79.

46. "Ragweed's On The Run," *Agriculture Bulletin,* no. 196 (December 1957), 8; and Oregon State Department of Agriculture, *Biennial Report* (1958), 15–16.

47. Oregon State Department of Agriculture, *Biennial Report* (1960), 20; ibid. (1966), 24; ibid. (1968), 22; and ibid. (1970), 20.

48. William B. Meyer, *Americans and Their Weather* (New York: Oxford University Press, 2000), 116, 158.

49. "Four Weather Modifiers Now Licensed Under New Oregon Act," *Agriculture Bulletin,* no. 180 (December 1953), 8.

50. Oregon Department of Agriculture, *Biennial Report* (1956), 61; and ibid. (1962), 51.

51. Isaiah Bowman, *Geography in Relation to the Social Sciences* (New York: Charles Scribner's Sons, 1934), 37.

52. Don McKinnis, "Irrigation and What's Ahead," *Oregon Agri-Record,* no. 237 (March 1968), 4–5.

53. See Donald J. Pisani, *To Reclaim a Divided West: Water, Law, and Public Policy, 1848–1902* (Albuquerque: University of New Mexico Press, 1992), xiv.

54. Oregon State Planning Board, *Present and Potential Land Development in Oregon Through Flood Control, Drainage and Irrigation* (Salem, July 1938), 156–57.

55. Eric A. Stene, *The Umatilla Project* (Bureau of Reclamation DataWeb, Bureau of Reclamation History Program, Denver, 1993), 2–7, and Christopher W. Shelley, "The Resurrection of a River: Re-watering the Umatilla Basin," (Center for Columbia River History 1999), 1–2.

56. "Water Conflict Resolved Cooperatively: The Umatilla Basin Project," (Confederated Tribes of the Umatilla Reservation, 1996), 1–2.

57. Shelley, "The Resurrection of a River," 5–6.

58. Ibid., 6–7.

59. Ibid., 7.

60. U.S. Bureau of Reclamation, *Klamath Project, Oregon and California,* 1; and Klamath River Basin Fisheries Task Force, *Upper Klamath River Basin Amendment to the Long Range Plan For the Klamath River Basin Conservation Area Fishery Restoration Program* (October 1992), 2–A-20.

61. For a more detailed account of the project, see William G. Robbins, *Landscapes of Promise: The Oregon Story, 1800–1940* (Seattle: University of Washington Press, 1997), 250–54.

62. Ibid., 252–53; and *Upper Klamath River Basin Amendment,* 2–A-20.

63. *Upper Klamath River Basin Amendment,* 2–A-4 to 2–A-5.

64. Ibid., 2–A-6 and 2–A-20 to 2–A-23.

65. *Oregonian,* June 10, 1998. Also see ibid., November 12, 2000.

66. Ibid., December 1, 1998.

67. Ibid., October 4, 1999. Also see William Kittredge, *Balancing Water: Restoring the Klamath Basin* (Berkeley: University of California Press, 2000), 137, 139.

68. *Oregonian,* September 21, 2000.

69. Klamath Falls *Klamath Herald and News,* January 17, 2001.

70. Ibid., December 20, 2000.

71. Ibid., April 9, 2001; Rebecca Clarren, "No Refuge in the Klamath Basin," *High Country News* 33 (August 13, 2001), 8; and *New York Times,* July 15, 2001.

72. Clarren, "No Refuge," 1.

73. *Oregonian,* February 13, 2002.

74. *Oregonian,* January 4, 2003; and Donald Worster, "A Dream of Water," *Montana The Magazine of Western History* 36 (Autumn 1986), 73.

4 / THE WONDER WORLD OF PESTICIDES

1. E. R. Jackman, "Weeds Hoist the White Flag," Manuscripts, 1940–1951, E. R. Jackman Papers, Oregon State University Archives, Corvallis.

2. Samuel P. Hays, *Beauty, Health, and Permanence: Environmental Politics in the United States, 1955–1985* (New York: Cambridge University Press, 1987), 3; and Linda Lear, *Rachel Carson: The Life of the Author of Silent Spring* (New York: Henry Holt, 1996), 118–19.

3. U.S. Department of Agriculture, Division of Plant Industry, *Annual Report* (1946), 39–42.

4. Ibid. (1946), 62; and ibid. (1968), 6.

5. John Wargo, *Our Children's Toxic Legacy: How Science and Law Fail to Protect Us from Pesticides* (New Haven: Yale University Press, 1996), ix; and Portland *Oregonian,* August 26, 1945.

6. *Oregonian,* ibid.; Edmund P. Russell III, "The Strange Career of DDT: Experts, Federal Capacity, and Environmentalism in World War II," *Technology and Culture* 40, no. 4 (1999), 793; and Thomas R. Dunlap, *DDT: Scientists, Citizens, and Public Policy* (Princeton: Princeton University Press, 1981), 59–63.

7. Robert Rieder to Certain County Agents, May 31, 1945, Extension Service Circular, Reel 12, Spray Warnings, Record Group 25, Agricultural Experiment Station Records (hereafter AES Records), Oregon State University Archives (hereafter OSU Archives), Corvallis.

8. Ernest C. Anderson, "Mosquito Control With DDT," July–September 10, 1945, Box 4, Record Group 158, College of Agricultural Sciences Dean's Office Records, (hereafter CASD Records), OSU Archives.

9. *Oregonian,* July 7, 1948.

10. See the following Extension Service memoranda: R. W. Lauderdale and C. R. Tanner, "Fly Control Problems on the Oregon State College Campus and Suggestions for the Future," December 31, 1952; B. G. Thompson, Filbert Spray Notice, August 15, 1947 and July 16, 1948; Robert Every, Hairy Vetch Weevil, June 9, 1948, and May 31, 1949; Don C. Mote and Robert W. "Every, Sheep Tick Control," circa 1947, and Cattle Lice Control, November 5, 1948, all in RG 25, all Reel 12, Spray Warnings, AES Records. Also see Wargo, *Our Children's Toxic Legacy,* 38.

11. "DDT Use Ends Banded Roller," undated newspaper clipping in Clippings File, AES Records.

12. "Livestock Spraying Very Effective," March 28, 1947, unidentified newspaper in ibid.; Mote and Every, "Sheep Tick Control," *Oregonian,* March 2, 1947; and Dunlap, *DDT,* 76.

13. The quotation is in Russell, "The Strange Career of DDT," 770–71.

14. Ibid., 773.

15. Mark Fiege, *Irrigated Eden: The Making of an Agricultural Landscape in the American West* (Seattle: University of Washington Press, 1999), 65.

16. F. C. Bishop, "The Tax We Pay to Insects," in *Science in Farming: Yearbook of Agriculture, 1943–1947* (Washington, D.C.: Government Printing Office, 1947), 613–15.

17. Mormon crickets are actually long-horned grasshoppers (*Anabrus simplex*), or simply big black grasshoppers. Communication from Glenn Fisher, Extension Entomologist, Oregon State University, October 30, 2000.

18. *Corvallis Gazette-Times,* May 23, 1947; and *Oregonian,* May 23, 1947.

19. *Oregonian,* May 24, 1947.

20. Ibid., May 27, 28, 29, 30, and June 1, 1947.

21. Ibid., May 31 and June 1, 1947.

22. Ibid., July 25, 1954.

23. *Biennial Report of the Department of Agriculture* (1948), 7, 11, 62, 66; and ibid. (1950), 8.

24. Ibid. (1952), 27–31.

25. Ibid. (1948), 11, 67; and (1952), 23.

26. Ibid. (1950), 6–7.

27. Wargo, *Our Children's Toxic Legacy,* ix.

28. Lear, *Rachel Carson,* 118, 305–06.

29. Leroy Childs, The Tampa, Florida, Meeting of the American Association of Economic Entomologists, December 1949, Spray Warnings, AES Records.

30. Ibid.

31. Ibid.

32. Lauderdale and Tanner, "Fly Control Problems."

33. On one occasion, the Oregon State College Agricultural Extension office circulated a 1947 progress report to county agents on experiments with DDT carried out by the Wisconsin Agricultural Experiment Station. The Wisconsin study involved livestock fed on DDT-treated silage and found no ill effects in subject cattle and sheep, although there were "moderate accumulations of DDT in body tissues and in milk." The levels of DDT in milk were so low "that they appear to be insignificant." Although the Wisconsin Experiment Station had no authority to establish DDT tolerance, its study concluded that the chemical could be used to control aphids on peas and borers in corn "without detrimental effects to livestock." See Robert Every to All County and Assistant County Agents, July 16, 1947, Spray Warnings, AES Records.

34. For a discussion of parallel developments at the national level, see Wargo, *Our Children's Toxic Legacy,* xi–xii.

35. Lauderdale and Tanner, "Fly Control Problems."

36. *Biennial Report of the Department of Agriculture* (1952), 23; ibid. (1954), 31; and J. Brooks Flippen, "Pests, Pollution, and Politics: The Nixon Administration's Pesticide Policy," *Agricultural History* 71 (Fall 1997), 445.

37. *Oregonian,* February 19, 1952.

38. Ibid., March 16, 17, 1954.

39. Ibid., May 14, 1954.

40. *Biennial Report of the Department of Agriculture* (1956), 31–33.

41. Ibid. (1958), 46–47.

42. Robert W. Every to Oregon County Extension Agents, October 3, 1962, CASD Records.

43. *Biennial Report of the Department of Agriculture* (1958), 16; and ibid. (1962), 13.

44. Ibid. (1962), 13–14; and ibid. (1964), 11.

45. H. H. Crowell, Onion Insect Pests Research at Oregon State College, March 16, 1961, CASD Records.

46. Paul O. Ritcher to Commissioner of Food and Drug Administration, January 15, 1960, ibid.

47. *Agriculture Bulletin* (March 1960), 12–13.

48. Ibid. (September 1960), 4; and ibid. (March 1961), 4. In a subsequent issue of the *Agriculture Bulletin* Patterson returned again to his charge that attaining zero tolerance in trace residues was "scientific fantasy," arguing that although the zero-tolerance requirement still remained, it now depended on the sensitivity of the method used. See ibid. (September 1961), 8.

49. Carson is quoted in Robert Gottlieb, *Forcing the Spring: The Transformation of the American Environmental Movement* (Washington, D.C.: Island Press, 1993), 81–82; John Opie, *Nature's Nation: An Environmental History of the United States* (New York: Harcourt Brace, 1998), 413; and Lear, *Rachel Carson*, 416–20.

50. Dunlap, *DDT, Scientists, Citizens, and Public Policy,* 97.

51. *Biennial Report of the Department of Agriculture, 1961–1962* (1962), 3, 13; ibid. (1964), 11; ibid. (1966), 25; Flippen, "Pests, Pollution, and Politics," 442; and "The Age of Effluence," *Time,* May 10, 1968, 52.

52. Orville Freeman, Secretary's Memorandum no. 1565, *U.S.D.A. Policy on Pesticides,* December 23, 1964, copy in Box 4/1/5/50, CASD Records.

53. *Oregonian,* June 6, 1966.

54. *Oregon Agri-Record,* no. 237 (March 1968), 13.

55. *Oregonian,* July 13, 1969.

56. *Oregon Agri-Record,* no. 243 (September 1969), 4.

57. *Oregonian,* July 13 and August 25, 1969, and January 20, 1970.

58. Ibid., August 24, 1969.

59. Ibid., August 25, 1969.

60. Ibid., October 19 and December 4, 17, 1969. At a Willamette Valley environmental conference held in Salem in November, Ian McHarg, a well-known landscape architect from the University of Pennsylvania, charged that DDT was "cutting into human beings' chances for survival." Calling for a ban on DDT, he

added that those who used the chemical should be required to show "burden of proof." See ibid., November 2, 1969.

61. Ibid., December 28, 1969.

62. Ibid., January 21, November 2, 1970.

63. U.S. Department of Agriculture, "Actions to Restrict DDT Uses," July 1970, copy in manuscript 2386, box 23, Oregon Environmental Council Records (hereafter OEC Records), Oregon Historical Society, Portland (hereafter OHS).

64. *Oregonian,* June 15, 1972; and Larry Williams to William Ruckleshaus, June 27, 1972, box 23, OEC Records.

65. Flippen, "Pests, Pollution, and Politics," 446–47; and *Oregonian,* June 15, 1972.

66. Flippen, "Pests, Pollution, and Politics," 445.

67. Ibid., 455–56.

68. Opie, *Nature's Nation,* 480.

69. Gottlieb, *Forcing the Spring,* 243–44; and Robert Gordon, "Poisons in the Fields: The United Farm Workers, Pesticides, and Environmental Politics," *Pacific Historical Review* 68 (February 1999), 51, 75–77.

70. *Oregonian,* December 5–7, 1999.

5 / PLANNING AND TECHNICAL EFFICIENCY IN THE FORESTS

1. David M. Kennedy, *Freedom From Fear: The American People in Depression and War, 1929–1945* (New York: Oxford University Press, 1999), 377–79.

2. Ibid., 783; and Charles McKinley, *Uncle Sam in the Pacific Northwest: Federal Management of Natural Resources in the Columbia River Valley* (Berkeley: University of California Press, 1952), 459–60.

3. Portland *Oregonian,* February 12, 1938. For a brief discussion of the National Resource Planning Board, see Kennedy, *Freedom From Fear,* 783–84.

4. *Coos Bay Times,* December 20, 1934, November 22, 1935, July 21, 1936; and E. I. Kotok, "Some Economic Problems in Pacific Coast Forestry," *Proceedings of the Seventeenth Annual Conference of the Pacific Coast Economic Association,* 17 (December 1938), 90.

5. Kennedy, *Freedom From Fear,* 785; and McKinley, *Uncle Sam in the Pacific Northwest,* 460.

6. McKinley, *Uncle Sam in the Pacific Northwest,* 38–44.

7. Portland *Oregonian,* April 17, 1945; and *Coos Bay Times,* April 14, 1945, April 5, 8, 1946, and April 3, 1947.

8. For a summary of Mason's ideas, see Elmo Richardson, *David T. Mason: Forestry Advocate* (Santa Cruz, California: Forest History Society, 1983). Also see William G. Robbins, *Lumberjacks and Legislators: Political Economy of the Lumber Industry, 1890–1941* (College Station: Texas A&M University Press, 1982), 139, 162.

9. Elmo Richardson, *BLM's Billion-Dollar Checkerboard: Managing the O & C Lands* (Santa Cruz, California: Forest History Society, 1980), 147; and Roy O. Hoover, "Public Law 273 Comes to Shelton: Implementing the Sustained-Yield Forestry Management Act of 1944," *Journal of Forest History* 22 (April 1978), 86–87, 101.

10. "Progress Report of the Post War Readjustment and Development Commission" (hereafter "Progress Report" (August 1947), 5.

11. Richardson, *BLM's Billion-Dollar Checkerboard,* 99–107; *Oregonian,* January 22, 1948; and Ellery Foster, "An Alternative Plan for Sustained Yield," a brief for presentation at U.S. Forest Service Sustained Yield Hearing, Quincy, California, March 8, 1948, Record Group 95, Records of the Forest Service (hereafter FS Records), Accession Number 10897 (hereafter AN), Federal Records Center, Seattle, Washington (hereafter FRC-Seattle).

12. *Oregonian,* June 25, 1948; and undated news clipping in FS Records, AN 10897.

13. *Crow's Pacific Coast Lumber Digest,* March 18, 1948, copy in FS Records, AN 10897.

14. *New York Times,* April 2, 1951.

15. Richardson, *BLM's Billion-Dollar Checkerboard,* 105–111.

16. *Oregonian,* January 22, 1956.

17. James E. Murray to Richard L. Neuberger, October 13, 1955 and attached document, "Report on the Smith River Salvage Program," in box 7, Richard L. Neuberger Papers, OHS (hereafter Neuberger Papers).

18. Richardson, *BLM's Billion-Dollar Checkerboard,* 144–47. The Eugene *Register-Guard* is cited in ibid., 147.

19. Paul W. Hirt, *A Conspiracy of Optimism: Management of the National Forests since World War Two* (Lincoln: University of Nebraska Press, 1994), 40–41.

20. Joseph Kinsey Howard, "Timber Troubles: How Can We Get Sustained Yield?" *The Pacific Spectator* (Spring 1950), typescript copy in FS Records, AN 19897.

21. Ibid.

22. Urling Coe, *Frontier Doctor: Observations on Central Oregon and the Changing West* (1940; Corvallis: Oregon State University Press, 1996), 237.

23. Nancy Langston, *Forest Dreams, Forest Nightmares: The Paradox of Old Growth in the Inland Northwest* (Seattle: University of Washington Press, 1995), 29.

24. Ferdinand A. Silcox to Regional Forester, September 15, 1934, box 4, Records of the Forest Service, Record Group 95, National Archives and Record Center, Washington, D.C. (hereafter FS Records-NA); *Bend Bulletin,* June 22, 1922; *Oregonian,* February 22, 1944.

25. U.S. Department of Agriculture, Forest Service, Pacific Northwest Forest Experiment Station, *Forest Statistics for Deschutes County, Oregon* (Portland, May 16, 1936), 3, 6.

26. Ibid., *Forest Statistics for Klamath County, Oregon,* November 1, 1934, 1, 3–6; and *Forest Statistics for Harney County,* May 7, 1936, 4–6.

27. *Forest Statistics for Harney County,* May 27 and June 3, 1914.

28. *Deschutes Pine Echoes,* August 1928, 5–6.

29. *Bulletin,* August 24, 1955.

30. Ibid., August 24, 1955, August 20, 1958, and June 9, 1960.

31. Barbara Arnold, "Central Oregon Fire Regimes," (Master's degree in Interdisciplinary Studies project paper, Oregon State University, 1999),18 (copy in the author's possession).

32. *Bulletin,* September 13, 1943, and January 1 and May 13, 1946.

33. Ibid., April 30, 1947.

34. Ibid., May 22, 1946, August 25, 1948; and Hans H. Plambeck, *The Population of Oregon, 1940–1970: Changes and Implications,* Prepared for the Rockefeller Foundation-sponsored project, "Man and His Activities as They Relate to Environmental Quality in Oregon," Oregon State University (May 1975), 35, 37.

35. *Bulletin,* November 19, 1945, and undated photocopy of an article that appeared in the *Bulletin* in the late 1940s, in the author's possession.

36. "Progress Report," no. 52 (December 1947); and *Bulletin,* August 25, 1948.

37. *Bulletin,* November 21, 1950.

38. Ibid., November 21, December 12, 1950, and August 22, 1951.

39. Ibid., November 22, 1950.

40. Ibid., June 29 and August 22, 1951.

41. Ibid., August 22, 1951.

42. Sheldon D. Erickson, *Occupance in the Upper Deschutes Basin, Oregon,* Research Paper no. 32, Department of Geography, University of Chicago, 1953, 99–102, 106–10.

43. *Bulletin,* August 22, 1951, September 10, 1952, and December 21, 1956.

44. Ibid., September 10, 1952; and Ronald L. Gregory, "Life in Railroad Logging

Camps of the Shevlin-Hixon Company, 1916–1950" (M.A. thesis, Oregon State University, 1997), 143.

45. *Bulletin,* October 26, 1955, June 4, 1957, and July 25, 1960.

46. *The Timberman* (February 1949), 56.

47. Denis C. LeMaster, *Mergers Among the Largest Forest Products Firms, 1950–1970,* Washington State University, College of Agriculture Research Center, Bulletin 854 (Pullman, 1977), 1.

48. *Oregonian,* August 29, 1958; *Coos Bay Times,* April 8, 1959; and Jerry Phillips, interview with the author, April 6, 1984.

49. *Oregonian,* February 29, May 27, June 7, July 11, 1956, March 25, 1984; *Coos Bay Times,* May 25, June 5, 7, 1956; and Harold Walton, interview with the author, April 4, 1984.

50. *Coos Bay Times,* May 1, 3, 1958; and April 8, 1959.

51. Ibid., June 8, 1950; April 4, 14, May 1, 1951; and January 21, 1952.

52. Stuart Moir, "Forests Seen as Major Factor in Northwest Economy," *Oregon Business Review* 10 (July 1951), 1, 4–5.

53. Ibid., 5–6.

54. Richard A. Rajala, *Clearcutting the Pacific Rain Forest: Production, Science, and Regulation* (Vancouver: University of British Columbia Press, 1998), 84–86.

55. Ibid., 167, 169–70.

56. *Oregonian,* October 19 and November 19, 1955.

57. Ibid., October 18, 1955.

58. "Progress Report," no. 40 (October 1946), 11; and ibid., no. 44 (February 1947), 13.

59. The committee's report is cited in the *Oregonian,* September 23, 1956.

60. Walter H. Lund, "Bright Future Forecast for Oregon Forest Industry," *Oregon Business Review,* 13 (October 1954), 4; and W. D. Hagenstein, "Looking Ahead in Douglas-Fir Forestry," ibid. 16 (July 1957), 1–3.

61. George H. Schroeder, "Rapid Strides Made in Oregon Forest Conservation," ibid., 13 (November 1954), 1–3.

62. *Oregonian,* September 30, 1956 and September 30, 1958.

63. U.S. Department of Agriculture, Forest Service, *Timber Resources for America's Future,* Forest Resource Report no. 14 (Washington, D.C.: Government Printing Office, 1958), ii–iii, 10.

64. Ibid., 108, 218–19.

65. Hirt, *A Conspiracy of Optimism,* 141.

66. Hill, "Georgia on My Mind," 18–20.

67. Ibid., 22–25.

68. LeMaster, *Mergers,* 1; and Tom Bates, "Clearcutting and Its Consequences: The Rape of the Siletz Gorge," *Oregon Times Magazine* 6 (June 1976), 14.

69. *Oregonian,* February 29, May 27, June 7, July 11, 1956; Wylie Smith interview with the author, April 16, 1984; Phillips interview; and Bates, "Clearcutting and Its Consequences," 14–15.

70. Louis Hamill, "Increasing the Allowable Cut From Federal Forests in Western Oregon," *Oregon Business Review* 19 (May 1960), 1, 4, 8.

71. Hirt, *A Conspiracy of Optimism,* 133–34.

72. Ibid., 150. The figures in table 5.1 are gleaned from the following sources: *Statistical Yearbook, 1952* (Portland: West Coast Lumbermen's Association, 1953), 7; *Statistical Yearbook, 1962* (Portland: West Coast Lumbermen's Association, 1964), 7; *Statistical Supplement to Facts, 1968* (Portland: Western Wood Products Association, 1969), 23; *Statistical Yearbook, 1972* (Portland: Western Wood Products Association, 1973), 25; and *Statistical Yearbook of the Western Lumber Industry* (Portland: Western Wood Products Association, 1981), 27.

73. James N. Tattersall, "The Oregon Economy Since World War II," *Oregon Business Review* 23 (July 1964), 3–5, 9.

74. Hirt, *A Conspiracy of Optimism,* 216.

6 / INTENSIVE FORESTRY AND CITIZEN ACTIVISM

1. "A Long Winter: Industry Experts Assess the Future of Oregon's Wood Products Industry," February 1982, 25, copy in box 4, Records of the Oregon Environmental Council (hereafter OEC Records), Oregon Historical Society, Portland (hereafter OHS).

2. Paul Hirt, *A Conspiracy of Optimism: Management of the National Forests since World War Two* (Lincoln: University of Nebraska Press, 1994) xlv–xlvii; William G. Robbins, *American Forestry: A History of National, State, and Private Cooperation* (Lincoln: University of Nebraska Press, 1985), 134; and Richard A. Rajala, *Clearcutting the Pacific Rain Forest: Production, Science, and Regulation* (Vancouver: University of British Columbia Press, 1998), 169–70.

3. J. E. Schroeder to Interested Persons and Organizations, "Forest Practice Rules—Review Draft," January 14, 1972, box 4, OEC Records.

4. Charles Coate to William P. Holtsclaw, February 9, 1972, ibid.

5. Anne W. Squier to J. E. Schroeder, February 24, and Schroeder to Squier, February 29, 1972, both ibid.

6. Meetings of the Oregon State Board of Forestry, Forest Practices Act Implementation Committee, January 5 and February 22, 1972, Oregon Department of Forestry, File D-205.02, Oregon State Archives, Salem, Oregon.

7. Schroeder to Eric W. Allen, November 6, and Allen to Schroeder, November 17, 1972, both box 4, OEC Records.

8. Schroeder To Whom It May Concern, March 7, 1974, ibid.

9. Charles A. Connaughton, "The Revolt Against Clearcutting," *Journal of Forestry* 68 (May 1970), 264–65.

10. "Clearcutting the Forest: One of Nature's Ways" (n.d., n.p.); and "Clearcutting: Vital Tool in Forest Management," internal information summary for Weyerhaeuser Company employees, both box 6, Industrial Forestry Association Records (hereafter IFA Records), manuscript 1906, OHS.

11. *Weyerhaeuser Management Viewpoint*, vol. 3, no. 4 (August 1971), 3–4, box 6, IFA Records. Unlike early logging companies that recklessly clear-cut large acreages, the Weyerhaeuser Company had always practiced conservative timber management, and it had initiated the tree farm movement in the 1940s.

12. W. D. Hagenstein, "Emotions Aside, Clearcutting is Silviculturally Sound Concept," *Forest Industries* 97 (December 1970), 26–28.

13. Senate Committee on Interior and Insular Affairs, "Clearcutting on Federal Timberlands," Report by the Subcommittee on Public Lands, 92nd Cong., 1st sess., March 1972, 2–9, copy in IFA Records.

14. W. D. Hagenstein, "Forest-based Industries Reaction to Report of Subcommittee on Public Lands of the Senate Interior Committee on Clearcutting Hearing," n.d., ibid.

15. Hagenstein, "Circular no. 556," to Subscribers, Industrial Forestry Association, June 19, 1972, and Statement of W. D. Hagenstein before Subcommittee on Forests, House Committee on Agriculture, June 20, 1972, ibid.

16. Jerry Franklin and Dean DeBell, "Effects of Various Harvesting Methods on Forest Regeneration," in *Even-age Management: A Symposium held August 1, 1972*, Richard K. Hermann and Denis Lavender, comps. and eds. (Corvallis: College of Forestry, Oregon State University, 1973).

17. *Wall Street Journal*, January 24, 1974.

18. Jerry Franklin, letter to *Wall Street Journal*, March 12, 1974, copy in box 6, IFA Records.

19. Steve Woodard to Governor Tom McCall, October 30, 1972, ibid.

20. Jason Boe to Steve Woodard, November 1, 1972, ibid.; and Roseburg *News-Review*, October 31, 1972.

21. Brent Walth, *Fire at Eden's Gate: Tom McCall and the Oregon Story* (Portland: Oregon Historical Society Press, 1994), 331–32.

22. Ibid., 159.

23. *Corvallis Gazette-Times,* October 30, 1975.

24. Ibid.; and Hirt, *A Conspiracy of Optimism,* 260–65.

25. Minutes of the Annual Meeting, Northwest Forest Pest Action Committee, November 1, 1954, Records of the Forest Service, Record Group 95, Access. no. 46460, Federal Records Center, Seattle, Washington (hereafter FS Record/AN).

26. W. S. Swingler, "Keeping Forest Insects in Their Place," reprinted from *American Forests* (February 1959), n.p., copy in box 9, IFA Records.

27. Cooperative Agreement between Oregon State Board of Forestry and the Forest Service for Control of Destructive Forest Insects and Diseases, January 2, 1961, copy in FS Records, AN48715; and Robbins, *American Forestry,* 217–20.

28. R. L. Furniss, "Protecting Forest Resources Against Insects," speech to the Portland Rotary Club, March 19, 1963, copy in box 9, IFA Records.

29. Royce Cornelius, "What Are the Responsibilities of Forest Pest Control?" reprinted from *Proceedings, Society of American Foresters,* Denver, Colorado, 1964, copy in box 8, ibid. Cornelius made a more extended argument about the differences between applying pesticides in agricultural and forest environments in "Where are We Going in Forest Pest Control?" *Journal of Forestry* (March 1966), reprint in box 9, ibid.

30. Minutes of the Meeting of the Executive Committee, Northwest Forest Pest Action Council, Seattle, Washington, October 19, 1965, copy in box 8, ibid.

31. Boyd E. Wickman, Richard R. Mason, and C. G. Thompson, *Major Outbreaks of the Douglas-fir Tussock Moth in Oregon and California,* U.S. Department of Agriculture, Forest Service General Technical Report, Pacific Northwest Forest and Range Experiment Station, PNW-5 (1973), 1–4, 13.

32. Ibid., 9–11; Burns Project, Douglas-fir Tussock Moth Control, Malheur and Ochoco National Forests, 1965, printed document in State and Private Forestry, Insect and Disease Control file, U.S. Forest Service, Region 6, Portland, Oregon; and *Biennial Report of the State Forester* (1966), 10.

33. Wickman, et al., *Major Outbreaks,* 13; and Narrative Report by Bob Bourhill, Northeast Oregon District, n.d., FS Records, AN148808.

34. Northwest Forest Pest Action Council, "Your Help is Needed to Protect Blue Mountain Forests from further Destruction by the Douglas-fir Tussock Moth," February 28, 1973; and William G. Hagenstein to W. H. Larson, February 24, 1973, both box 8, IFA Records.

35. T. A. Schlapfer to all offices and individuals who have received a copy of the Douglas-fir Tussock Moth Environmental Statement, February 26, 1973; Interagency Tussock Moth Steering Committee, "The Douglas-Fir Tussock Moth Problem in the Northwest," December 1973; and Northwest Forest Pest Action Council, Resolution no. 2, all ibid.

36. Department of Agriculture, Forest Service, USDA-USDI Environmental Statement, "Summary Sheet," n.d., ibid.; and U.S. Department of Agriculture, Forest Service—Pacific Northwest Region, *Environmental Monitoring Program: 1974 Cooperative Douglas Fir—Tussock Moth Control Project* (Portland: n.p., 1981), 5–6.

37. Minutes, Northwest Forest Pest Action Council Executive Committee Meeting, Portland, August 1, 1974, box 8, IFA Records.

38. These comments are reported in *The Wall Street Journal*, January 7, 1975.

39. Ibid.

40. William H. Larson to Theodore A. Schlapfer, January 14, 1975, box 9, IFA Records.

41. Nancy Langston, *Forest Dreams, Forest Nightmares: The Paradox of Old Growth in the Inland West* (Seattle: University of Washington Press, 1995), 3; and Paul F. Hessberg, Russel G. Mitchell, and Gregory M. Filip, *Historical and Current Roles of Insects and Pathogens in Eastern Oregon and Washington Forested Landscapes*, U.S. Department of Agriculture, Forest Service, Pacific Northwest Research Station, General Technical Report, PNW-GTR-327 (April 1994), 25.

42. A Report to the Regional Forester and the Forest Supervisors of the Blue Mountain Forests, *Restoring Ecosystems in the Blue Mountains*, U.S. Department of Agriculture, Forest Service, Pacific Northwest Region (July 1992), 1–8.

43. *Oregonian*, October 3, 1997.

44. Theo Colborn, Dianne Dumanoski, and John Peterson Myers, *Our Stolen Future: Are We Threatening our Fertility, Intelligence, and Survival? A Scientific Detective Story* (New York: Penguin Books, 1996), 113–14; Joel Primack and Frank von Hippel, *Advice and Dissent: Scientists in the Political Arena* (New York: Basic Books, 1974), 74; and Arthur W. Galston, "Herbicides: A Mixed Blessing," *Bioscience* (February 1979), 85. I am grateful to the Northwest Coalition for Alternatives to Pesticides (NCAP) in Eugene, Oregon for copies of these materials.

45. Primack and von Hippel, *Advice and Dissent*, 75; Thomas Whiteside, "The Pendulum and the Toxic Cloud," *The New Yorker* (July 1977), 30–32; and T. F. X. Collins and W. H. Williams, "Teratogenic Studies with 2,4,5–T and 2,4–D in the Hamster," copy in NCAP files.

46. Carol Van Strum, "Back to the Future: EPA Reinvents the Wheel on Repro-

ductive Effects of Dioxin," *Synthesis/Regeneration* 7–8 (Summer 1995), 2; and Jean Anderson to Judie Irons, November 20, 1972, box 23, OEC Records.

47. Larry Williams to Anderson, December 21, 1972; Anderson to Irons, January 2, 1973; and Anderson to Terry Lash, January 3, 1973, all box 12, OEC Records. Also see *Eugene Register-Guard,* April 23, 1973.

48. *Register-Guard,* April 23, 1973; Anderson to Williams, May 15, 1973; Granville Knight to Alice Pettus, April 23, 1973; and Anderson to Jerry Uhrhammer, May 15, 1973, all box 12, OEC Records.

49. MiMi Cutler for the Oregon Environmental Council, "Review of the Draft Environmental Impact Statement for the Proposed Siskiyou, Siuslaw, and Umpqua National Forests Herbicide Spraying Program for 1974–75," box 12 OEC Records.

50. "Statement of the Case," *Citizens Against Toxic Sprays, Inc.* v. *Bergland;* and Harold E. Hartman, "The Siuslaw Lawsuit," presented to the Pesticide Awareness Seminar, February 23, 1978, both box 10, IFA Records.

51. "Pesticides Awareness Seminar," in ibid.; and Donald R. Bliss to Dr. John Noell, June 21, 1977, copy in NCAP files.

52. Whiteside, "The Pendulum and the Toxic Cloud," 32–37.

53. Mike Sullivan, "History of the Herbicide Controversy, Oregon and Washington"; and "Forest Service News, Pacific Northwest Region," April 7, 1978, both box 6, IFA Records.

54. Sullivan, "History of the Herbicide Controversy"; Roy Murphy, "Alsea Woman Leads Fight Against Herbicides," *Oregonian,* December 2, 1979; and Wilbur P. McNulty to Federal Register Section, EPA, July 27, 1978, copies in NCAP files.

55. *Seattle Post-Intelligencer,* July 15, 1978; and A. Rupert Cutler to John R. McGuire, April 27, 1978, all box 10, IFA Records.

56. Interim Directive no. 1, Pesticide-Use Management, April 27, 1978; and Report of 2,4,5–T Task Group to the API/NFPA Forest Industry Chemicals Action Commitee, May 5, 1978, all ibid.

57. Jack Anderson, "Herbicide Causes Birth Defects," *Post-Intelligencer,* July 15, 1978, State of Oregon Department of Forestry News, May 30, 1978; and ibid.

58. Sullivan, "History of the Herbicide Controversy"; and Office of the Governor, News Release, September 6, 1978, both ibid.

59. Hill points out that her sample did not include the vast majority of miscarriages, because they usually occur during the first trimester of pregnancy and "most of these women never go to a doctor." Bonnie Hill, notes to the author, February 3, 2003. Also see "Direct Testimony of Bonnie Hill on the Possible Link Between

Spontaneous Abortions and Herbicide Applications in Alsea, OR," 2,4,5–T Cancellation Hearings (1980), 17–19, copy in NCAP files.

60. Hill, notes to the author. Also see *Oregonian,* July 18, 1978; and Murphy, "Alsea
Woman Leads Fight Against Herbicides."

61. "Direct Testimony of Bonnie Hill," 21–25; and *Oregonian,* July 18, 1978.

62. Murphy, "Alsea Woman Leads Fight Against Herbicides"; Hill, notes to the
author; "Direct Testimony of Bonnie Hill," 25–27; and *New York Times,* March 9,
1979.

63. Michael Sullivan to Roone Arledge, August 14, 1978; Charles Black to Elton
Rule, August 25, 1978; and Comments from CAST, August 25, 1978, all box 10, IFA
Records. In his "Notes from the Editor's Cuff," conservative media critic Reed Irvine
charged ABC with putting on a sloppy and superficial show. He added that 2,4,5–T
was made "the target of a major propaganda drive by communists, who attributed
to it all manner of problems." See *Accuracy in Media Report* 7 (October 1978), copy
in ibid.

64. Phil Keisling, "OSU's Herbicide Connection," Portland *Willamette Week* 6
(December 17, 1979), 1–6, copy in NCAP files.

65. W. D. Hagenstein to Members, Industrial Forestry Association, March 22,
1979; James Woods to William McCredie, March 26, 1979, all box 10, IFA Records;
and U.S. Environmental Protection Agency, Before the Administrator, In re: The
Dow Chemical, et al, FIFRA Docket no. 415 (1979), Direct Testimony of George
Streisinger; and FIFRA Docket no. 295, Prehearing Statement, Matthew Meselson,
both copies in NCAP files.

66. Ralph D. Hodges, Jr. to Don-Lee M. Davidson, Davidson Industries, December 11, 1979, box 10, IFA Records. Earlier in the year, Hodges urged Agriculture secretary Bob Berglund to conduct "an authoritative and dispassionate review of the
risks and benefits of forestry 2,4,5–T applications by bringing the Department's scientific resources to bear on the issue." See Hodges to Berglund, March 30, 1979,
ibid.

67. Hill, notes to the author; and Carol Van Strum, *A Bitter Fog: Herbicides and
Human Rights* (San Francisco: Sierra Club Books, 1983), 171–72. NCAP has two large
file boxes of the testimonies delivered at the cancellation hearings.

68. Van Strum, "Back to the Future," 3. Beginning in 1977 and continuing for
several years, the EPA was embroiled in a major fraud investigation of Industrial
Biotest Laboratory (IBT), a firm that had done thousands of tests for chemical manufacturers. EPA relied on IBT's pesticide safety tests before permitting the regis-

tration of chemicals for sale on the market. Ultimately, the Justice Department ordered the lab closed in 1978 when investigators discovered that thousands of scientific safety tests had been deliberately falsified. In October 1983, a federal jury convicted three former company officials of fabricating safety tests for two popular pesticides and two popular drugs. See United States Environmental Protection Agency, *Environmental News,* August 15, 1977; *The Sacramento Bee,* September 25, October 23, and December 4, 1983, all copies in NCAP files.

69. Brian Wall, *Log Production in Washington and Oregon: An Historical Perspective,* U.S. Forest Service, Resource Bulletin PNW-42, Pacific Northwest Forest and Range Experiment Station (Portland, 1972), 7–8, 29; John F. Beuter, K. Norman Johnson, and H. Lynn Scheurmann, *Timber For Oregon's Tomorrow,* Research Bulletin 19, Forest Research Laboratory, School of Forestry, Oregon State University (Corvallis, 1976), 1, 18, 43; and Russell Sadler, "John Beuter Reckons With Timber," *Willamette Week,* December 26, 1977.

70. Hirt, *Conspiracy of Optimism,* 272; and Catherine Caufield, "The Ancient Forest," *The New Yorker,* May 14, 1990, 82–83.

71. *Oregonian* (Special Issue: "Forests in Distress"), October 15, 1990.

72. Ibid.; and M. Lynn Corn, "Spotted Owls and Northwest Forests," *Congressional Research Service Issue Brief* (The Library of Congress), January 14, 1993, CRS-1.

73. Eric D. Forsman, "A Preliminary Investigation of the Spotted Owl" (M.S. thesis: Oregon State University, 1975).

74. Corn, "Spotted Owls and Northwest Forests," CRS-2; and *Oregonian,* October 15, 1990.

75. Caufield, "The Ancient Forest," 74–75.

76. Dan Goldy, "Stop the Federal Forest Follies!" *Oregon Business Review* 8 (May 1985), 91–93; and Goldy, "Federal Forest Follies, Part 2," ibid. 10 (January 1987), 63–64, 72.

77. Teresa Carp, "Another Blow for Timber?" ibid. 9 (November 1986), 16–17, 19; and Goldy, "Loggers Are An Endangered Species," ibid. 12 (March 1989), 43–46.

78. *Oregonian,* July 4, 1993.

79. Corn, "Spotted Owls and Northwest Forests," CRS-9; and *Oregonian,* July 4, 1993.

80. *Bend Bulletin,* August 11, 1995; and *Oregonian,* May 21, 1995.

81. *Seattle Post-Intelligencer,* February 20, 1994.

82. Kathie Durbin, "Logging and Landslides," *Oregon Quarterly* (Winter 1998), 20–21; and Corvallis *Mid-Valley Sunday,* February 14, 1999.

83. Durbin, "Logging and Landslides," 24–26.

7 / RICHARD NEUBERGER'S CONSERVATION POLITICS

1. Richard L. Neuberger, "Guarding Our Outdoor Heritage," *The Progressive* (January 1959), reprinted in *They Never Go Back to Pocatello: The Selected Essays of Richard Neuberger*, Steve Neal, ed. (Portland: Oregon Historical Society Press, 1988), 66–67.

2. Richard L. Neuberger, "The Columbia," *Holiday* (June 1949), reprinted in *They Never Go Back to Pocatello*, 31.

3. Richard Neuberger to The President, November 23, 1957, box 7, Richard L. Neuberger Papers, Oregon Historical Society, Portland (hereafter Neuberger Papers and OHS).

4. Senator Richard Neuberger, Press Release, October 17, 1957, ibid; and Bert E. Swanson and Deborah Rosenfield, "The Coon-Neuberger Debates of 1955," *Pacific Northwest Quarterly* 55 (April 1964), 56–57.

5. Swanson and Rosenfield, "The Coon-Neuberger Debates," 60–61.

6. Portland *Oregon Journal,* September 29 and 30, 1955.

7. Swanson and Rosenfield, "The Coon-Neuberger Debates," 66.

8. William F. Willingham, *Army Engineers and the Development of Oregon: A History of the Portland District U.S. Army Corps of Engineers* (Washington, D.C.: Government Printing Office, 1983), 161.

9. From the office of Senator Wayne Morse, March 15, 1955, copy in box 11, Neuberger Papers.

10. Richard L. Neuberger, "Hells Canyon, The Biggest Of All," *Harper's* (April 1939), reprinted in *They Never Go Back to Pocatello*, 14–15, 28–29.

11. Keith C. Petersen, *River of Life, Channel of Death: Fish and Dams on the Lower Snake* (Lewiston, Idaho: Confluence Press, 1995), 156–58; and Sara E. Dant Ewart, "Evolution of an Environmentalist: Senator Frank Church and the Hells Canyon Controversy," *Montana The Magazine of Western History* 51 (Spring 2001), 36–38.

12. Gus Norwood, *Columbia River Power For The People: A History of the Bonneville Power Administration* (Portland: U.S. Department of Energy, Bonneville Power Administration, 1981), 192–93; Petersen, *River of Life, Channel of Death,* 157; and Ewart, "Evolution of an Environmentalist," 38.

13. Petersen, *River of Life, Channel of Death,* 156–57.

14. Talk given by Richard Neuberger before the Kiwanis Club luncheon at Albany, Oregon, December 10, 1953, transcribed copy in box 1, Neuberger Papers.

15. Richard and Maurine Neuberger, "Washington Calling," October 4, 1955, copy in box 7, Neuberger Papers; and Richard L. Neuberger, "The Senator Behind the

Give-aways," *The Progressive* (February 1954), reprinted in *They Never Go Back to Pocatello*, 227.

16. Senator Neuberger's Remarks on the Occasion of Introduction of the Hells Canyon Bill, March 8, 1955, copy in box 11, Neuberger Papers.

17. Richard L. Neuberger, "What Hells Canyon Means to America's Future," press release, March 1955, copy in ibid.

18. Neuberger to Charles Sprague, April 14, 1955, box 1, Neuberger Papers.

19. Richard L. Neuberger, "Mistakes of a Freshman Senator," *American Magazine* (June 1956), reprinted in *They Never Go Back to Pocatello*, 242; and Ewart, "Evolution of an Environmentalist," 39.

20. Richard and Maurine Neuberger, "Washington Calling," June 1955, copy in box 11, Neuberger Papers; and Petersen, *River of Life, Channel of Death*, 118.

21. Carlos A. Schwantes, *The Pacific Northwest: An Interpretive History*, revised and enlarged edition (Lincoln: University of Nebraska Press, 1996), 494–95; *Columbia River Power for the People*, 196–97; Richard and Maurine Neuberger, "Washington Calling," August 15, 1955; Richard Neuberger to Henry Alderman, June 24, 1957; Neuberger to James Marr, July 15, 1957; Neuberger to Marr, November 27, 1957; and Statement of Richard Neuberger at Hearing on Preliminary Findings of Review of 308 Report on Columbia Basin, Portland, Oregon, November 15, 1957, all box 7, Neuberger Papers; and Ewart, "Evolution of an Environmentalist," 39.

22. Albert M. Day, "Report of the Fisheries Subcommittee, Columbia Basin Inter-Agency Committee," September 9, 1959, copy in box 3, Neuberger Papers; and Ewart, "Evolution of an Environmentalist," 41–42.

23. Neuberger to Lyle F. Watts, March 26, 1958, box 3, Neuberger Papers.

24. Anthony Netboy, *The Columbia River Salmon and Steelhead Trout: Their Fight for Survival* (Seattle: University of Washington Press, 1980), 94; and Tim Palmer, *The Snake River: Window to the West* (Washington, D.C.: Island Press, 1991), 189–90.

25. Schwantes, *The Pacific Northwest*, 495; and Neuberger to Editor of the *Oregon Journal*, November 30, 1957, box 1, Neuberger Papers.

26. Ewart, "Evolution of an Environmentalist," 42–43, 48. The Douglas opinion is quoted in Schwantes, *The Pacific Northwest*, 494.

27. Neuberger to Ed Chambers, September 12, 1956, box 6, Neuberger Papers.

28. Neuberger to Lt. Gen. Samuel D. Sturgis, Jr., October 25, 1956; Neuberger to Major General Emerson C. Itschner, December 7, 1956; and Itschner to Neuberger, January 17, 1957, all ibid.

29. Statement of Richard Neuberger at Hearing of Corps of Engineers on Channel Modification for Columbia River, Astoria to Portland-Vancouver, June 5, 1958, ibid.

30. Neuberger to the President, November 23, 1957, box 7, Neuberger Papers.

31. Ibid.

32. A brief summary of Klamath termination is in the *Oregonian,* December 14, 1999. Also see Donald L. Fixico, *The Invasion of Indian Country in the Twentieth Century: American Capitalism and Tribal Natural Resources* (Niwot: University Press of Colorado, 1998), 79–102.

33. Neuberger to Malcolm Bauer, August 31, 1957, box 1, Neuberger Papers.

34. Fixico, *The Invasion of Indian Country,* 92–94.

35. Robert W. Sawyer, "Klamath Timber Should Be in National Forests," *American Forests* (March 1958), 25–26, 47. Neuberger had both essays printed in the *Congressional Record,* 85th Cong., 2d sess., 1958, Vol. 104, 5537–39, copies in box 1, Neuberger Papers.

36. Bill Jenkins, "Klamath Water Big Cog in Oregon's Prospects," in *American Forests* (March 1958), 24, 46–47, copy in ibid.

37. Portland *Oregonian,* April 6, 1958.

38. Neuberger to Malcolm Bauer, June 2 and June 30, 1958, box 1, Neuberger Papers.

39. Fixico, *The Invasion of Indian Country,* 95; and "Passage of Klamath Reservation Bill Hailed as Conservation Milestone by Oregon Senator, News Release," October 8, 1958, copy in box 3, Neuberger Papers.

40. Fixico, *The Invasion of Indian Country,* 96–97.

41. Neuberger to J. Herbert Stone, November 3; Neuberger to James F. Doyle, November 3; Neuberger to Walter H. Lund, November 13, 1958; and Neuberger to Lawrence E. Slater, June 29, 1959, all box 7, Neuberger Papers.

42. Richard L. Neuberger, "How Oregon Rescued a Forest," *Harper's Magazine* (April 1959), 48–52, reprinted in *They Never Go Back to Pocatello.*

43. William Kittredge, *Balancing Water: Restoring the Klamath Basin* (Berkeley: University of California Press, 2000), 100–101.

44. *Oregonian,* June 28, 1955.

45. Ibid., September 9, 11, 1955; and Neuberger to James Kezer, September 2, 1955, box 7, Neuberger Papers.

46. *Oregonian,* November 11, 23, and 26, 1955. Neuberger is quoted in the issue for November 26.

47. *Oregonian,* November 8, 1955.

48. Ibid., December 22, 1955; and Neuberger to Luvilia M. Richards, December 30, 1955, box 7, Neuberger Papers.

49. Neuberger's open contempt for McKay is revealed in a letter to Karl Onthank, president of the Federation of Western Outdoor Clubs: "Although it does not seem possible, McKay constantly gets worse and worse!" See Neuberger to Onthank, October 10, 1955, box 3, Neuberger Papers.

50. *Oregonian,* March 22, 1956. Senator Wayne Morse also testified at the sub-committee hearing, speaking to the unique character of the upper McKenzie— its excellent fishing, the white-water rapids, and the "crystal clarity" of Clear Lake (see ibid). For a critique of McKay, see Philip Shabecoff, *A Fierce Green Fire: The American Environmental Movement* (New York: Hill and Wang, 1993); and Elmo Richardson, *Dams, Parks, and Politics: Resource Development and Preservation in the Truman-Eisenhower Era* (Lexington: University Press of Kentucky, 1973).

51. *Oregonian,* March 27 and 29, 1956.

52. *Eugene Register-Guard,* March 27, 1956.

53. Neuberger to Jerome K. Kuykendall, November 28, and Neuberger to Robert Frazer, November 30, 1956, both box 7, Neuberger Papers.

54. *Oregonian,* January 31, February 9, and December 3, 1957.

55. Ibid., July 22, 1958. Even without the big Beaver Marsh project, the city of Eugene enjoyed inexpensive electrical power for several decades. The Federal Power Commission reported in 1967 that Eugene residents used more electricity per home during the previous year than any other U.S. city—and Eugene also enjoyed the lowest electrical rates in the nation. See ibid., April 6, 1967.

56. *Oregonian,* April 17, 1949.

57. Ibid.

58. Thomas M. Robins to Colonel Theron D. Weaver, May 29, 1951, box 6, Thomas M. Robins Papers, University of Oregon Library, Eugene; and *Oregonian,* June 13, 1951.

59. *Oregonian,* August 11, 1951; January 13, 14, and 16, 1952.

60. Ibid., July 16, 1952, and November 9 and 11, 1954.

61. Ibid., June 7 and 8, 1955.

62. Thomas R. Cox to Neuberger, June 9; Robert Y. Thornton to Neuberger, June 13; Neuberger to Cox, June 15; and Rollin E. Bowles to Neuberger, June 17, all 1955, box 7, Neuberger Papers.

63. Jerome S. Bischoff to Neuberger and Neuberger to Bischoff, both March 5; and J. L. Waud to Neuberger, February 22, 1956, all ibid.

64. William L. Stollmach to Neuberger, March 16; Neuberger to Stollmach, April 16; and Neuberger to Stollmach, May 5, 1956, all ibid.

65. Neuberger to Rollin E. Bowles, March 9, 1956; and Erskine B. Wood to Neuberger, March 30, 1956, both ibid.

66. *Oregonian,* March 29, June 2, 1956; Neuberger to Russell M. Dickson, May 10; and Neuberger to Herbert Lundy, June 23, 1956, both box 7, Neuberger Papers.

67. *Oregonian,* June 26, 1956.

68. Neuberger to Lundy, June 23; Mrs. George Scharnal to Neuberger, June 28; Howard W. Turner to Neuberger, July 17, 1956, both box 7, Neuberger Papers; and *Oregonian,* June 14, 1956.

69. Richard Neuberger, news release, September 18, 1956, box 7, Neuberger Papers. copy in box 7, Neuberger Papers.

70. *Oregonian,* December 9, 1956.

71. Robert H. Short to Neuberger, April 17, 1958 in box 7, Neuberger Papers. Neuberger is quoted in the *Oregonian,* November 28, 1958.

72. *Oregonian,* November 27, 1958; and Neuberger to Thomas W. Delzell, November 29, 1958, box 7, Neuberger Papers.

73. Albert Day, "Report of the Fisheries Subcommittee, Columbia Basin Inter-Agency Committee," September 9, 1959, copy in box 3, ibid; and *Oregonian,* September 24, October 15, 17, 1959.

74. Neuberger to Bill Aldridge, December 7, box 7; Neuberger to Cordelia Murphy, December 3, 1959, box 3, both in Neuberger Papers; and *Oregonian,* June 3, 1960.

75. *Oregonian,* July 19, 1963, and March 11, 2001.

76. Ibid., March 15, 2000.

8 / TOM McCALL AND THE STRUGGLE FOR THE WILLAMETTE

1. R. E. Dimmick and Fred Merryfield, *The Fishes of the Willamette River System in Relation to Pollution,* Bulletin Series no. 20, Engineering Experiment Station Oregon State College (Corvallis: June 1945), 53.

2. Lewis Mumford, *The City in History: Its Origins, Its Transformations, and Its Prospects* (New York: Harcourt Brace, 1961), 459–60.

3. Portland *Oregonian,* September 12, 1926; and Public Health Section of the City Club of Portland, "Stream Pollution In Oregon," *The Pacific Engineer* 5 (May 1927), n.p.

4. George W. Gleeson, *The Return of A River: The Willamette River, Oregon,* Oregon Water Resources Research Institute (WRRI-13), Oregon State University, June 1972, 13, 16, 22.

5. *Oregonian,* October 20, 1935.

6. *Oregonian,* December 1, 1936 and March 2, 1937; and Gleeson, *Return of a River,* 21–22.

7. *Oregonian,* January 21, 1945.

8. Fred Merryfield and W. G. Wilmot, *1945 Progress Report on Pollution of Oregon Streams,* Bulletin Series no. 19, Engineering Experiment Station (Corvallis: Oregon State College, June 1945), 7, 11–15.

9. Merryfield and Wilmot, *1945 Progress Report,* 47–48.

10. Dimmick and Merryfield, *Fishes of the Willamette River,* 53–54.

11. Undated *Oregonian* clipping in box 18, David B. Charlton Papers, Oregon Historical Society, Portland (hereafter Charlton Papers).

12. Ibid.

13. "Cleaning up the Willamette," unidentified mimeographed document in the author's possession, 51–53.

14. Undated *Oregonian* clipping in box 18, Charlton Papers; and *Oregonian,* January 2, 1948.

15. David B. Charlton to Harold Wendell, December 6, 1947, box 18, Charlton Papers; and *Oregonian,* December 21, 1948.

16. *Oregonian,* December 21, 1948.

17. Charlton to Oregon State Game Commission, January 11, 1949, box 18, Charlton Papers.

18. J. W. Smith to William A. Bowes, February 7, 1949; and Harold B. Say to Charlton, October 29, 1948, both ibid.

19. Walter J. Murphy to Charlton, January 19, 1949, ibid.

20. Charles Sprague to Charlton, November 11, 1949, ibid.

21. *Oregonian,* December 12, 1949.

22. Ibid., February 18, 1950.

23. The University of Michigan study is quoted in Gleeson, *Return of a River,* 24 and 26.

24. *Oregonian,* May 10, 1950.

25. Ibid., March 23 and 28, 1952.

26. Ibid., July 17, 1952.

27. Ibid., July 18, 1952.

28. Ibid., July 19, 1952.

29. Ibid., September 10, 11, and 12, 1958.

30. Ibid., September 12, 1958.

31. Ibid., September 15, 30, and October 18, 1958.

32. Ibid., October 18, 1958.

33. Ibid., February 20, 1959.

34. Ibid., August 6, 1959.

35. Portland *Oregon Journal,* September 1, 1965 and January 11, 1972.

36. George Abed, *Water Pollution Control Policy in the Willamette Basin,* vol. 5, *Willamette Basin Land Use Study* (Salem: Oregon Department of Commerce, n.d.), 1–4.

37. Gleeson, *Return of a River,* 30–33; *Oregon Journal,* September 1, 1965; and *Oregonian,* August 16, 1965.

38. Gleeson, *Return of a River,* 59.

39. "Cleaning up the Willamette," 54.

40. Gleeson, *Return of a River,* 73.

41. Brent Walth, *Fire at Eden's Gate: Tom McCall and the Oregon Story* (Portland: Oregon Historical Society Press, 1994), 77–112 and 149–59. I am indebted to Brent Walth for his fine assessment of *Pollution in Paradise.*

42. Ibid., 141–46.

43. Ibid., 147–48.

44. Ibid., 172–73.

45. Ibid., 181–82; and "Cleaning up the Willamette," 57–58.

46. Gleeson, *Return of a River,* 49–57.

47. Walth, *Fire at Eden's Gate,* 325–31.

48. *Oregonian,* August 6, 1970.

49. "Cleaning up the Willamette," 58–59.

50. *Oregonian,* December 19, 2000.

51. Ibid., October 10, 1971.

52. Ethel Starbird, "River Restored: Oregon's Willamette," *National Geographic* 141 (June 1972), 817–19.

53. David B. Charlton to Frank P. Terraglio, May 9, 1975 (copy in the author's possession); and Charlton to Loren Thompson, June 28, 1974, box 18, Charlton Papers.

54. *Oregonian,* March 15, 1973.

55. Oregon Department of Environmental Quality, *Willamette River Basin Task Force: Recommendations to Governor John Kitzhaber* (December 1997), 4–5.

56. Ibid., 6, 18.

57. *Oregonian,* December 28, 1997.

58. *Willamette River Basin Task Force,* 23–36.

59. *Oregonian,* November 8, 1998.

60. Ibid., March 14, 1999, and December 17, 2000.

61. Ibid., March 1, 1999.

62. Ibid., March 10, 1999.

63. Ibid., May 5, 2001.

64. Ibid., April 18, 1999.

65. Ibid., October 20, 2000.

66. Ibid., December 17, 2000.

67. Ibid., December 18, 2000.

68. Ibid., December 19, 2000.

69. Ibid., and *Corvallis Gazette-Times,* April 23, 2001.

9 / ECOLOGIES OF SPRAWL: THE LAND-USE NEXUS

1. Quoted in Gerrit Knaap, "Land Use Politics in Oregon," in *Planning the Oregon Way: A Twenty-Year Evaluation,* Carl Abbot, Deborah Howe, and Sy Adler, eds. (Corvallis: Oregon State University Press, 1994), 5.

2. Patricia Nelson Limerick, *The Legacy of Conquest: The Unbroken Past of the American West* (New York: W. W. Norton, 1987), 55; and Knaap, "Land Use Politics," 5.

3. I made this argument in "Introduction: In Search of Western Lands," in *Land in the American West: Private Claims and the Common Good,* William G. Robbins and James C. Foster, eds. (Seattle: University of Washington Press, 2000), 3–4.

4. Portland *Oregonian,* January 23, 1997.

5. Ibid., December 15, 1998.

6. These figures are listed in the on-line version of the *Oregon Blue Book.*

7. Carl Abbott and Deborah Howe, "The Politics of Land-Use Law in Oregon: Senate Bill 100, Twenty Years After," *Oregon Historical Quarterly* 94 (Spring 1993), 5; "Introduction," in Abbott, et al., eds., *Planning the Oregon Way,* xi; and Pauline Maris, "Rural Planning and Zoning Laws in Oregon," *Oregon Business Review* 8 (September 1947), 1–2.

8. "Progress Report of Postwar Readjustment and Development Commission of the State of Oregon" (hereafter "Progress Report"), no. 59 (May 1948), 8; and Hans H. Plambeck, *The Population of Oregon, 1940–1970: Changes and Implications* (Corvallis: Oregon State University, 1975), 32.

9. Oregon Department of Agriculture, *Agriculture Bulletin* (September 1951), 1.

10. *Biennial Report of the Department of Agriculture* (1956), 14; and Brent Walth, *Fire At Eden's Gate: Tom McCall and the Oregon Story* (Portland: Oregon Historical Society Press, 1994), 243–44.

11. Walth, *Fire At Eden's Gate,* 242, 244–45.

12. *Oregonian,* August 25–26, 1968; and Plambeck, *Population of Oregon,* 34.

13. *Oregonian,* August 27, 1968.

14. Plambeck, *Population of Oregon,* 34; Mark Hoy, "The Great Land Experiment," *Northwest Magazine, Oregonian,* June 14, 1987, 10; and Abbott and Howe, "Politics," 5–6. To punish the governor for signing SB 10, conservatives launched a short-lived "Recall McCall" effort. See Walth, *Fire At Eden's Gate,* 249.

15. *Oregonian,* February 11, 1971; and Charles Deemer, "A Conversation with Tom McCall," *Oregon Business* (September 1981), 16.

16. Deemer, "A Conversation with Tom McCall," 16–18; and John Gray, "Land Use: Profits *AND* Environmental Integrity," Address to the National Association of Home Builders, January 8, 1973, box 21, Records of the Oregon Environmental Council (hereafter OEC Records), Oregon Historical Society, Portland.

17. Walth, *Fire At Eden's Gate,* 354–55; Abbott and Howe, "Politics," 6; and Lawrence Halprin and Associates, *The Willamette Valley: Choices for the Future* (Salem: Willamette Valley Environmental Protection and Development Planning Council, 1972).

18. Halprin and Associates, *Choices for the Future,* 1, 57; and Walth, *Fire At Eden's Gate,* 355.

19. Walth, *Fire At Eden's Gate,* 353–54; and Abbott and Howe, "Politics," 6–7.

20. *Oregonian,* November 24, 1972; and Betty Merten, "Beyond the Valley of the Malls," *Oregon Times* 2 (December 1972), 14.

21. Walth, *Fire At Eden's Gate,* 356–57, 360–61; and Abbott and Howe, "Politics,"7.

22. Walth, *Fire At Eden's Gate,* 361.

23. *Time,* October 1, 1973; and *Oregonian,* October 1, 1973.

24. Abbott and Howe, "Politics," 8; and *Oregonian,* November 4, 1973.

25. *Oregonian,* May 15, 1974; and B. J. Rogers to the Portland Board of Realtors, February 6, 1974, box 21, OEC Records. The final version of Senate Bill 100 dropped language that specified certain areas off limits to development. Private-power interests dropped their opposition to the bill when L. B. Day successfully amended the legislation to exempt the siting of nuclear power facilities from land-use restrictions. See Walth, *Fire At Eden's Gate,* 360.

26. William J. Moshofsky, "Warning! Land Use Planning Ahead," *Forest Industries* 102 (July 1975), 26–27.

27. Roger M. Williams, "Oregon: The Fight for 'Survival,'" *Saturday Review* (November 16, 1974), 10–12, 14–15.

28. *Oregonian,* September 24, 1975.

29. Abbott and Howe, "Politics," 8; and Hoy, "The Great Land Experiment, 10.

30. Edward J. Sullivan, "The Legal Evolution of the Oregon Planning System," in Abbott, et al., eds., *Planning the Oregon Way,* 54, 56, and 60–61; and Knapp, "Land Use Politics in Oregon," 10.

31. James R. Pease, "Oregon Rural Land Use: Policies and Practices," in Abbott, et al., eds., *Planning the Oregon Way,* 164-65; and Hoy, "The Great Land Experiment,"14.

32. Sullivan, "Legal Evolution," 55; and *Oregonian,* July 8, 1978.

33. *Oregonian,* October 27, 1978, and November 13, 1979.

34. Ernest Callenbach, *Ecotopia: The Notebooks and Reports of William Weston* (Berkeley: Banyan Tree Books, 1975).

35. Peter James, "Ecotopia in Oregon?" *New Scientist* 81 (January 4, 1979), 28–30.

36. Matthew L. Slavin, "Irreconcilable Differences: Economic Development and Land Use Planning in Oregon," in Abbott, et al., eds., *Planning the Oregon Way,* 83–84; and *Oregonian,* May 20, 1982.

37. Abbott and Howe, "Politics," 8; and *Oregonian,* July 21 and September 16, 1982.

38. Walth, *Fire At Eden's Gate,* 456–57. As this note indicates, I am deeply indebted to Brent Walth for his summary of the events of September and October 1982.

39. Ibid., 458–60.

40. Ibid., 460–63.

41. Hoy, "The Great Land Experiment," 14; *Oregonian,* January 9, 1975; and Gerrit Knaap, "Land Use Politics in Oregon," 10.

42. Carl Abbott and Deborah Howe, "A Landmark Decision," *Oregonian,* June 13, 1993.

43. John M. DeGrove, "Following in Oregon's Footsteps: The Impact of Oregon's Growth Management Strategy on Other States," in Abbott, et al., *Planning the Oregon Way,* 227, 243.

44. Larry Swisher, "Northwest Faces Issues of Growth," *Corvallis Gazette-Times,* January 7, 2000.

45. "Agriculture," *Oregon Business* (January 1995), 43; and *Statesman-Journal,* September 8, 1996.

46. *Statesman-Journal,* September 8, 1996.

47. Ibid., consolidated special issue, "Paradise Lost? Searching for Tom McCall's Oregon," originally published in four weekly installments, September and October, 1996.

48. *Oregonian,* May 1 and 31, 1998.

49. *Gazette-Times,* May 31, 1998.

50. *Oregonian,* May 31, 1998; and Sullivan, "Legal Evolution," 59–60.

51. *Oregonian,* January 8 and May 31, 1998.

52. Ibid., December 16, 1998.

53. Ibid., March 21, 1999.

54. Ibid., January 8, 1998; *Gazette-Times,* May 31, 1998; and *Eugene Register-Guard,* September 6, 1998.

55. *Oregonian,* December 14, 1998.

56. Ibid., December 14 and 17, 1998.

57. Ibid., December 17, 1998 and February 28, 1999.

58. <*http://www.pacificlegal.org/NorthwestCenter.asp*>

59. Carl Abbott, Sy Adler, and Margery Post Abbott, *Planning a New West: The Columbia River National Scenic Area* (Corvallis: Oregon State University Press, 1997), 14, 82–83, 118–23.

60. <http://oia/oia2.html>; and Corvallis-Albany *Mid-Valley Sunday,* March 13, 1999.

61. *Oregonian,* January 19, 1999.

62. *Oregonian,* December 25, 2000.

63. *Oregonian,* December 12, 2000; and Russell Sadler, "Measure Threatens Oregon Farms," *Mid-Valley Sunday,* August 21, 2000.

64. This information is gleaned from the following Web sites: <*http://www.lcd.state.or.us/perspectives/measure7.html*>; and <*http://www.orcities.org/members/m7/m7page.html*>

65. <*http://www.lcd.state.or.us/perspectives/measure7.html*>; <*http://www.orcities.org/members/m7/m7page.html*>; and <*http://www.publications.ojd.state.or.us/S48450.html*>

66. *Register-Guard,* October 5, 2002.

67. *Register-Guard,* October 5, 2002; *Oregonian,* October 5, 2002; and Newport *News-Times,* October 5, 2002.

68. *Solving Sprawl: The Sierra Club Rates the States* (San Francisco: The Sierra Club, 1999), 1–3, 13.

EPILOGUE: THIS SPECIAL PLACE

1. Wallace Stegner, *Where the Bluebird Sings to the Lemonade Springs: Living and Writing in the West* (New York: Random House, 1992), xv. Delmonte quote: Letter to the editor; Corvallis *Gazette-Times,* January 27, 1997.

2. Ken Kesey, "My Oregon Trail," *Vis a Vis* (United Airlines In-Flight Magazine, March 1989), 66–70.

3. Stegner, *Where the Bluebird Sings to the Lemonade Springs,* xix.

4. Portland *Oregonian,* December 19, 1999.

5. Michael Pollan, *Second Nature: A Gardener's Education* (New York: Dell, 1991), 5, 220, 222, 228; Raymond Williams, *Problems in Materialism and Culture* (London: Verso, 1980), 67; and William Cronon, "Introduction: In Search of Nature," in *Uncommon Ground: Toward Reinventing Nature,* William Cronon, ed. (New York: W. W. Norton, 1995), 25.

6. For a discussion of these questions, see Gerhard Strohmeier, "Wild West Imagery: Landscape Perception in Nineteenth-Century America," in *Nature and Society in Historical Context,* Mikulas Teich, Roy Porter, and Bo Gustafsson, eds. (Cambridge: Cambridge University Press, 1997), 257–73.

7. Helmut K. Buechner, "Some Biotic Changes in the State of Washington, Particularly during the Century 1853–1953," *Research Studies of the State College of Washington* 21 (1953), 154.

8. *Pendleton East Oregonian,* September 1, 1945.

9. William Cronon, *Nature's Metropolis: Chicago and the Great West* (New York: W. W. Norton, 1991), xvii, 56–57, 62, 266–67.

10. Robert D. Kaplan, *An Empire Wilderness: Travels into America's Future* (New York: Vintage Books, 1998), 330–31.

11. Corvallis-Albany *Mid-Valley Sunday,* May 27, 2001.

12. *Oregonian,* December 19, 1999.

13. Ibid., December 31, 1999.

14. Ibid.

15. M. Lynne Corn, "Spotted Owls and Northwest Forests," Congressional Research Service, *CRS Brief* (January 14, 1993), 1–4; *Oregonian,* January 12, 2003; and *Eugene Register-Guard,* April 27, 1995.

16. *Register-Guard,* April 27, 1995; *Bend Bulletin,* August 11, 1995; and *Oregonian,* July 25, 1999.

17. *This Place on Earth 2002* (Seattle: Northwest Environment Watch, 2002), 13.

18. *Oregonian,* December 31, 1999.

19. Ibid., January 12, 2003.

20. Ibid., December 31, 1999.

21. *Oregon Blue Book, 1999–2000* (Salem: Office of the Secretary of State, 2000), 189; Gail Wells, *The Tillamook: A Created Forest Comes of Age* (Corvallis: Oregon State University Press, 1999); and *Mid-Valley Sunday,* May 27, 2001.

22. Kathie Durbin, "Slide at Stump Acres," *Oregon Quarterly* (Winter 1998), 20–26.

23. Jim Lichatowich, *Salmon Without Rivers: A History of the Pacific Salmon Crisis* (Washington, D.C.: Island Press, 1999), 7.

24. *Oregonian,* March 19, 2000.

25. Ibid., March 5, 1998; and *Corvallis Gazette-Times,* December 8, 1999.

26. *Oregonian,* March 2, 13, 2001, and September 24, 2002.

27. *Oregonian,* February 26, 2003.

28. *Christianity Today,* March 16, 2001; and An International Pastoral Letter by the Catholic Bishops of the Region, "The Columbia River Watershed: Caring for Creation and the Common Good" (draft copy in the author's possession).

BIBLIOGRAPHY

GOVERNMENT DOCUMENTS, REPORTS, PUBLICATIONS

Abed, George. *Water Pollution Control Policy in the Willamette Basin.* Vol. 5, *Willamette Basin Land Use Study.* Salem: Department of Commerce, n.d.

Beuter, John F., K. Norman Johnson, and H. Lynn Scheurmann. *Timber For Oregon's Tomorrow.* Research Bulletin 19. Forest Research Laboratory, Oregon State University. Corvallis, 1976.

U.S. Congress. Senate. *Clearcutting on Federal Timberlands.* Report by the Subcommittee on Public Lands to the Committee on Interior and Insular Affairs. 92nd Cong., 1st sess., 1972.

Columbia Basin Inter-Agency Committee. *The Columbia Basin Inter-Agency Committee and the Pacific Northwest.* Portland: n.p., 1949.

————. *Minutes of the Tenth Meeting.* Walla Walla, Washington, June 25–26, 1947.

Corn, M. Lynn. "Spotted Owls and Northwest Forests." *Congressional Research Service Issue Brief.* CRS-1, January 14, 1993.

Gleeson, George. *The Return of a River: The Willamette River, Oregon.* Oregon Water Resources Research Institute, WRRI-13. Corvallis: Oregon State University, 1972.

Halprin, Lawrence, and Associates. *The Willamette Valley: Choices for the Future.* Salem: Willamette Valley Environmental Protection and Development Planning Council, 1972.

McKinley, George, and Doug Frank. *Stories on the Land: An Environmental History of the Upper Applegate and Illinois Valleys.* Report prepared for the Bureau of Land Management, Medford District, Oregon, 1955.

Merryfield, Fred, and W. G. Wilmot. *1945 Progress Report on Pollution of Oregon Streams.* Bulletin Series No. 19. Engineering Experiment Station, Oregon State College. Corvallis, 1945.

Oregon State Department of Agriculture. *Biennial Reports,* 1946–1966.

————. *Agriculture Bulletin.* 1951–1963

Oregon Department of Agriculture. *Agriculture in Oregon.* Salem, 1946.

———. *Agriculture in Oregon.* Salem, 1965.

———. *Oregon Agri-Record.* 1968–1969.

———. *First Biennial Report, Willamette River Basin Commission.* Salem, 1946.

———. Progress Reports of the Postwar Readjustment and Development Commission of the State of Oregon. 1943–1949.

———. *Report of the Interim Fisheries Committee to the Forty-Fourth Legislative Assembly.* Salem, 1947.

———. State Planning Board. *Present and Potential Land Development in Oregon Through Flood Control, Drainage and Irrigation.* Salem, 1938.

———. *Willamette River Basin Task Force: Recommendations to Governor Kitzhaber.* December 1997.

———. Works Projects Administration. *Oregon: End of the Trail.* Portland: Binfords and Mort, 1940.

U.S. Congress. House. *Columbia River Fisheries.* Hearings Before the Subcommittee on Merchant Marine and Fisheries. 79th Cong., 2d sess., August 14, 1946.

———. *Columbia River and Tributaries, Northwestern United States.* Vol. 1. 81st Cong., 2d sess., 1950. H. Doc. 531.

U.S. Department of Agriculture. Agricultural Research Service. *Field Burning.* February 1973.

———. Forest Service. *Annual Reports, 1940–1943.*

———. *Forest Statistics for Deschutes County, Oregon.* Portland, 1936.

———. *Forest Statistics for Harney County, Oregon.* Portland, 1936.

———. *Forest Statistics for Klamath County, Oregon.* Portland, 1934.

———. Hessberg, Paul F., Russel G. Mitchell, and Gregory M. Filip. *Historical and Current Roles of Insects and Pathogens in Eastern Oregon and Washington Forested Landscapes.* General Technical Report, Pacific Northwest Research Station. PNW-GTR-327. 1994.

———. Pacific Northwest Regional Committee on Post-War Programs. "Area Plans for Agriculture in the Post-War Period." Portland, 1944.

———. Report to the Regional Forester and Forest Supervisors of the Blue Mountain Forests. *Restoring Ecosystems in the Blue Mountains.* Pacific Northwest Region. 1992.

———. Wall, Brian. *Log Production in Washington and Oregon: An Historical Perspective.* Pacific Northwest Forest and Range Experiment Station. Resource Bulletin PNW-42. 1972.

———. Wickman, Boyd, Richard Mason, and C. G. Thompson. *Major Outbreaks of the Douglas-fir Tussock Moth in Oregon and California.* General Technical Report, Pacific Northwest Forest and Range Experiment Station. PNW-5 (1973).

————. *Timber Resources for America's Future.* Forest Resource Report No. 14. Washington, D.C.: Government Printing Office, 1958.

U.S. Department of Commerce. Bureau of the Census. *Sixteenth Census of the United States: 1940, Population.* Vol. 1. *Number of Inhabitants.* Washington, D.C.: Government Printing Office, 1941.

————. *Seventeenth Census of the United States: 1950, Population.* Vol. 2. *Characteristics of the Population.* Washington, D.C.: Government Printing Office, 1951.

U.S. Department of Interior. *Your Columbia River: The Development of America's Greatest Power Stream.* Portland: Bonneville Power Administration, 1950.

West Coast Lumbermen's Association. *Statistical Yearbook, 1952.* Portland, 1953.

————. *Statistical Yearbook, 1962.* Portland, 1964.

Western Wood Products Association. *Statistical Supplement to Facts, 1968.* Portland, 1969.

————. *Statistical Yearbook, 1972.* Portland, 1973.

————. *Statistical Yearbook of the Western Lumber Industry.* Portland, 1981.

INTERVIEWS AND PRIVATE COMMUNICATIONS

Beckham, Dow. March 28, 1984.

Hill, Bonnie. Notes to the author. February 3, 2003.

McKenna, Bill. April 11, 1984.

Phillips, Jerry. April 6, 1984.

Smith, Wylie. April 16, 1984.

Taylor, Valerie (Wyatt). March 27, 1984.

Walton, Harold. April 14, 1984.

BOOKS

Abbott, Carl, Deborah Howe, and Sy Adler, eds. *Planning the Oregon Way: A Twenty-Year Evaluation.* Corvallis: Oregon State University Press, 1994.

Abbott, Carl. *Portland: Planning, Politics, and Growth in a Twentieth-Century City.* Lincoln: University of Nebreska Press, 1983.

Battaile, Connie Hopkins. *The Oregon Book: Information A to Z.* Newport, Oregon: Saddle Mountain Press, 1998.

Bowman, Isaiah. *Geography in Relation to the Social Sciences.* New York: Charles Scribner's Sons, 1934.

Brogan, Denis W. *The American Character.* New York: Alfred A. Knopf, 1944.

Brokaw, Tom. *The Greatest Generation.* New York: Random House, 1998.

Bush, Vannevar. *Science: The Endless Frontier.* Washington, D.C.: Government Printing Office, 1946.

Callenbach, Ernest. *Ecotopia: The Notebooks and Reports of William Weston.* Berkeley: Banyan Tree Books, 1975.

Colburn, Theo, Dianne Dumanoski, and John Peterson Myers. *Our Stolen Future: Are We Threatening our Fertility, Intelligence, and Survival? A Scientific Detective Story.* New York: Penguin Books, 1996.

Commoner, Barry. *The Closing Circle: Nature, Man, and Technology.* New York: Alfred A. Knopf, 1971.

Cronon, William. *Nature's Metropolis: Chicago and the Great West.* New York: W. W. Norton, 1991.

Dunlap, Thomas R. *DDT: Scientists, Citizens, and Public Policy.* Princeton, N.J.: Princeton University Press, 1981.

Erickson, Sheldon. *Occupance in the Upper Deschutes Basin, Oregon.* Research Paper No. 32. Department of Geography. Chicago: University of Chicago, 1953.

Fiege, Mark. *Irrigated Eden: The Making of an Agricultural Landscape in the American West.* Seattle: University of Washington Press, 1999.

Findlay, John M. *Magic Lands: Western Cityscapes and American Culture After 1940.* Berkeley: University of California Press, 1992.

Fixico, Donald L. *The Invasion of Indian Country in the Twentieth Century: American Capitalism and Tribal Natural Resources.* Niwot: University Press of Colorado, 1998.

Goldman, Eric. *The Crucial Decade—and After: America, 1945–1960.* Enlarged edition. New York: Random House, 1960.

Gottlieb, Robert. *Forcing the Spring: The Transformation of the American Environmental Movement.* Washington, D.C.: Island Press, 1993.

Gunther, John. *Inside U.S.A.* 1946. 50th Anniversary Edition. Foreword by Arthur Schlesinger, Jr. New York: Book of the Month Club, 1997.

Hays, Samuel P. *Beauty, Health, and Permanence: Environmental Politics in the United States, 1955–1985.* New York: Cambridge University Press, 1987.

Hevly, Bruce, and John M. Findlay, eds. *The Atomic West.* Seattle: University of Washington Press, 1998.

Hirt, Paul W. *A Conspiracy of Optimism: Management of the National Forests Since World War Two.* Lincoln: University of Nebraska Press, 1994.

Hobsbawm, Eric. *The Age of Extremes: A History of the World, 1941–1991.* New York: Pantheon Books, 1995.

Jensen, Vernon. *Lumber and Labor.* New York: Farrar and Rinehart, 1945.

Kaplan, Robert. *An Empire Wilderness: Travels into America's Future.* New York: Vintage Books, 1998.

Kennedy, David M. *Freedom From Fear: The American People in Depression and War, 1929–1945.* New York: Oxford University Press, 1999.

Kevles, Daniel J. *The Physicists: A History of a Scientific Community in America.* New York: Alfred A. Knopf, 1978.

Kittredge, William. *Balancing Water: Restoring the Klamath Basin.* Berkeley: University of California Press, 2000.

Knobloch, Frieda. *The Culture of Wilderness: Agriculture as Colonization in the American West.* Chapel Hill: University of North Carolina Press, 1996.

Langston, Nancy. *Forest Dreams, Forest Nightmares: The Paradox of Old Growth in the Inland Northwest.* Seattle: University of Washington Press, 1995.

LeMaster, Dennis C. *Mergers Among the Largest Forest Products Firms, 1950–1970.* Pullman: Washington State University, 1977.

Lichatowich, Jim. *Salmon Without Rivers: A History of the Pacific Salmon Crisis.* Washington, D.C.: Island Press, 1999.

Limerick, Patricia Nelson. *The Legacy of Conquest: The Unbroken Past of the American West.* New York: W.W. Norton, 1987.

Luce, Henry. *The American Century.* New York: Farrar and Rinehart, 1941.

Maben, Manly. *Vanport.* Portland: Oregon Historical Society Press, 1987.

McKinley, Charles. *Uncle Sam in the Pacific Northwest: Federal Management of Natural Resources in the Columbia River Valley.* Berkeley: University of California Press, 1952.

Meyer, William B. *Americans and Their Weather.* New York: Oxford University Press, 2000.

Moore, Truman E. *Nouveaumania: The American Passion for Novelty and How it Led Us Astray.* New York: Random House, 1975.

Mumford, Lewis. *The City in History: Its Origins, Its Transformations, and Its Prospects.* New York: Harcourt Brace, 1961.

Nash, Gerald D. *The American West Transformed: The Impact of the Second World War.* Bloomington: Indiana University Press, 1985.

Neal, Steve, ed. *They Never Go Back to Pocatello: The Selected Essays of Richard Neuberger.* Portland: Oregon Historical Society Press, 1988.

Netboy, Anthony. *Columbia River Salmon and Steelhead Trout.* Seattle: University of Washington Press, 1980.

Nixon, Edgar B., ed. *Franklin D. Roosevelt and Conservation.* Vol. 2. Hyde Park, New York: Franklin D. Roosevelt Library, 1957.

Norwood, Gus. *Columbia River Power for the People: A History of the Bonneville Power Administration.* Portland: U.S. Department of Energy, Bonneville Power Administration, 1981.

Oakley, J. Ronald. *God's Country: America in the Fifties.* New York: Dembner Books, 1986.

Opie, John. *Nature's Nation: An Environmental History of the United States.* New York: Harcourt Brace, 1998.

Palmer, Tim. *The Snake River: Window to the West.* Washington, D.C.: Island Press, 1991.

Patterson, James T. *Grand Expectations: The United States, 1945–1974.* New York: Oxford University Press, 1996.

Petersen, Keith C. *River of Life, Channel of Death: Fish and Dams on the Lower Snake.* Lewiston, Idaho: Confluence Press, 1995.

Pisani, Donald J. *To Reclaim a Divided West: Water, Law, and Public Policy, 1848–1902.* Albuquerque: University of New Mexico Press, 1992.

Primack, Joel, and Frank von Hippel. *Advice and Dissent: Scientists in the Political Arena.* New York: Basic Books, 1974.

Pyne, Stephen J. *Fire in America: A Cultural History of Wildland and Rural Fire.* Princeton, N.J.: Princeton University Press, 1982.

Rajala, Richard A. *Clearcutting the Pacific Rain Forest: Production, Science, and Regulation.* Vancouver: University of British Columbia Press, 1998.

Reisner, Marc. *Cadillac Desert: The American West and Its Disappearing Water.* New York: Viking Penguin, 1986.

Richardson, Elmo. *BLM's Billion-Dollar Checkerboard: Managing the O & C Lands.* Santa Cruz, California: Forest History Society, 1980.

_____. *Dams, Parks, and Politics: Resource Development and Preservation in the Truman-Eisenhower Era.* Lexington: University Press of Kentucky, 1973.

_____. *David T. Mason: Forestry Advocate.* Santa Cruz, California: Forest History Society, 1983.

Robbins, William G. *American Forestry: A History of National, State, and Private Cooperation.* Lincoln: University of Nebraska Press, 1983.

_____. *Hard Times in Paradise: Coos Bay, Oregon, 1850–1986.* Seattle: University of Washington Press, 1988.

_____. *Landscapes of Promise: The Oregon Story, 1800–1940.* Seattle: University of Washington Press, 1997.

_____. *Lumberjacks and Legislators: Political Economy of the Lumber Industry, 1890–1941.* College Station: Texas A&M University Press, 1982.

Robbins, William G., and James C. Foster, eds. *Land in the American West: Private Claims and the Common Good.* Seattle: University of Washington Press, 2000.

Sale, Roger. *Seattle, Past and Present.* Seattle: University of Washington Press, 1976.

Samuelson, Robert J. *The Good Life and its Discontents: The American Dream in the Age of Entitlement, 1945–1995.* New York: Random House, 1995.

Schwantes, Carlos A. *The Pacific Northwest: An Interpretive History.* Revised and enlarged edition. Lincoln: University of Nebraska Press, 1996.

Shabecoff, Philip. *A Fierce Green Fire: The American Environmental Movement.* New York: Hill and Wang, 1993.

Soule, Judith D., and Jon K. Piper. *Farming in Nature's Image: An Ecological Approach to Agriculture.* Washington, D.C.: Island Press, 1992.

Stegner, Wallace. *Where the Bluebird Sings to the Lemonade Springs.* New York: Random House, 1992.

Terkel, Studs. *"The Good War": An Oral History of World War Two.* New York: Pantheon Books, 1984.

Van Strum, Carol. *Bitter Fog: Herbicides and Human Rights.* San Francisco: Sierra Club Books, 1983.

Walth, Brent. *Fire At Eden's Gate: Tom McCall and the Oregon Story.* Portland: Oregon Historical Society Press, 1994.

Wargo, John. *Our Children's Toxic Legacy: How Science and Law Fail to Protect Us from Pesticides.* New Haven: Yale University Press, 1996.

Wells, Gail. *The Tillamook: A Created Forest Comes of Age.* Corvallis: Oregon State University Press, 1999.

White, Donald W. *The American Century: The Rise and Decline of the United States as a World Power.* New Haven: Yale University Press, 1996.

White, Richard. *"It's Your Misfortune and None of My Own": A New History of the American West.* Norman, University of Oklahoma Press, 1991.

_____. *The Organic Machine: The Remaking of the Columbia River.* New York: Hill and Wang, 1995.

Willingham, William F. *Army Engineers and the Development of Oregon: A History of the Portland District U.S. Army Corps of Engineers.* Washington, D.C.: Government Printing Office, 1983.

Worster, Donald. *Nature's Economy: A History of Ecological Ideas.* 1977. New York: Cambridge University Press, 1985.

Zinn, Howard. *A Peoples' History of the United States.* New York: Harper and Row, 1980.

ARTICLES AND CHAPTERS

Abbott, Carl, and Deborah Howe. "The Politics of Land-Use Law in Oregon: Senate Bill 100, Twenty Years After." *Oregon Historical Quarterly* 94 (Spring 1993), 4–35.

Bates, Tom. "Clearcutting and Its Consequences: The Rape of the Siletz Gorge." *Oregon Times Magazine* 6 (June 1976), 13–18.

Benson, Keith R. "The Maturation of Science in the Northwest: From Nature Studies to Big Science." In *The Great Northwest: The Search for Regional Identity,* edited by William G. Robbins. Corvallis: Oregon State University Press, 2001.

Bishop, F.C. "The Tax We Pay to Insects." In *Science in Farming: Yearbook of Agriculture, 1943–1947.* Washington, D.C.: Government Printing Office, 1947.

Bloch, Ivan. "Pacific Northwest Horizons in the Postwar World." *Oregon Business Review* 5 (June 1946), 1–6.

Boag, Peter G. "The World Fire Created: Field Burning in the Willamette Valley." *Columbia* 5 (Summer 1991), 3–11.

Carp, Teresa. "Another Blow for Timber?" *Oregon Business Review* 9 (November 1986), 16–17, 19.

Caufield, Catherine. "The Ancient Forest." *The New Yorker,* May 14, 1990. 46–84.

Clarren, Rebecca. "No Refuge in the Klamath Basin." *High Country News* 33 (August 31, 2001), 1, 8–11.

Connaughton, Charles A. "The Revolt Against Clearcutting." *Journal of Forestry* 68 (May 1970), 264–65.

Cowlin, R.W. "Some Economic Aspects of Oregon Lumber Manufacture." *Oregon Business Review* 4 (June 1945), 1, 9–11.

Durbin, Kathie. "Slide at Stump Acres." *Oregon Quarterly* (Winter 1998), 20–26.

Ewart, Sara E. Dant. "Evolution of an Environmentalist: Senator Frank Church and the Hells Canyon Controversy." *Montana The Magazine of Western History* 51 (Spring 2001), 36–52.

Fitzgerald, Deborah. "Beyond Tractors: The History of Technology in American Agriculture." *Technology and Culture* 32 (January 1991), 114–126.

Flippen, J. Brooks. "Pests, Pollution, and Politics: The Nixon Administration's Pesticide Policy." *Agricultural History* 72 (Fall 1997), 442–56.

Franklin, Jerry, and Dean DeBell. "Effects of Various Harvesting Methods on Forest Regeneration." In *Even-Age Management: A Symposium,* edited by Richard K. Hermann and Denis Lavender. Corvallis: College of Forestry, Oregon State University, 1973.

Goldy, Dan. "Federal Forest Follies, Part 2." *Oregon Business Review* 10 (January 1987), 63–64, 72.

_____. "Loggers Are An Endangered Species." *Oregon Business Review* 12 (March 1989), 43–46.

_____. "Stop the Federal Forest Follies." *Oregon Business Review* 8 (May 1985), 91–93.

Hagenstein, W. D. "Emotions Aside, Clearcutting Is Silviculturally Sound Concept." *Forest Industries* 97 (December 1970), 26–28.

_____. "Looking Ahead in Douglas-Fir Forestry." *Oregon Business Review* 16 (July 1957), 1–3.

Hamill, Louis. "Increasing the Allowable Cut From Federal Forests in Western Oregon." *Oregon Business Review* 19 (May 1960), 1, 4–8.

Hill, Robert. "Georgia on My Mind." *Oregon Business Review* 31 (June 1982), 18–25, 72.

Hoover, Roy O. "Public Law 273 Comes to Shelton: Implementing the Sustained-Yield Forestry Management Act of 1944." *Journal of Forest History* 22 (April 1978), 86–101.

Horning, W. H. "Sustained-Yield Program of the O & C Lands." *Oregon Business Review* 5 (April 1946), 1–4.

Hoy, Mark. "The Great Land Experiment." *Northwest Magazine Section.* Portland *Oregonian,* June 14, 1987, 9–16.

Karton, Ron. "Growers Look for Green Markets." *Oregon Business Review* 7 (August 1984), 47–48.

Kesey, Ken. "My Oregon Trail." *Vis a Vis* (United Airlines In-flight Magazine), (March 1989), 66–70.

Kotok, E.I. "Some Economic Problems in Pacific Coast Forestry." *Proceedings of the Seventeenth Annual Conference of the Pacific Coast Economic Association* 17 (December 1938), 88–92.

Lund, Walter. "Bright Future Forecast for Oregon Forest Industry." *Oregon Business Review* 13 (October 1954), 1, 3–4.

Maris, Pauline. "Rural Planning and Zoning Laws in Oregon." *Oregon Business Review.* 6 (September 1947), 1–2.

Merten, Betty. "Beyond the Valley of the Malls." *Oregon Times Magazine* 2 (December 1972), 13–14.

Moir, Stuart. "Forests Seen as Major Factor in Northwest Economy." *Oregon Business Review* 10 (July 1951), 1, 4–5.

Moshofsky, William J. "Warning! Land Use Planning Ahead." *Forest Industries* 102 (July 1975), 26–27.

Needham, Paul R. "Dams Threaten West Coast Fisheries Industry." *Oregon Business Review* 6 (June 1947), 1, 3–4.

Neuberger, Richard L. "The Land of New Horizons." *New York Times Magazine* (December 1945), 18–23.

_____. "The Cities of America: Portland, Oregon." *Saturday Evening Post* (March 1, 1947), 22–24.

_____. "The Great Salmon Experiment." *Harper's* (February 1945), 229–36.

_____. "The Great Salmon Mystery." *The Saturday Evening Post* (September 13, 1941), 20–21, 39–44.

Ramsey, Jarold. "New Era: Growing up East of the Cascades." In *Regionalism and the Pacific Northwest,* edited by William G. Robbins. Corvallis: Oregon State University Press, 1983.

Robbins, William G. "The Willamette Valley Project: A Study in the Political Economy of Water Resource Development." *Pacific Historical Review* 47 (November 1978), 585–605.

Russell, Edmund P. III. "The Strange Career of DDT: Experts, Federal Capacity, and Environmentalism in World War II." *Technology and Culture* 40, no. 4 (1999), 770–96.

Schroeder, George H. "Rapid Strides Made in Oregon Forest Conservation." *Oregon Business Review* 13 (November 1954), 1–3.

Shaver, H. T. "Upper Columbia River Traffic." *Oregon Business Review* 4 (September 1945), 1,4.

Sherow, James E. "Environmentalism and Agriculture in the American West." In *The Rural West Since World War II,* edited by R. Douglas Hirt. Lawrence: University Press of Kansas, 1998.

Simpson, H. V. "West Coast Lumber Enters a New Cycle." *Oregon Business Review* 4 (November 1945), 1–3.

Starbird, Ethel A. "River Restored: Oregon's Willamette." *National Geographic* 141 (June 1972), 816–35.

Swanson, Bert E., and Deborah Rosenfield. "The Coon-Neuberger Debates of 1955." *Pacific Northwest Quarterly* 55 (April 1964), 55–66.

Szasz, Ferenc M. "Introduction." In *Working on the Bomb: An Oral History of WWII*

Hanford, edited by S. L. Sanger. Portland: Continuing Education Press, Portland State University, 1995.

Tattersall, James N. "The Oregon Economy Since World War II." *Oregon Business Review* 23 (July 1964), 1–11.

Whiteside, Thomas. "A Reporter at Large: The Pendulum and the Toxic Cloud." *The New Yorker* (July 25, 1977), 30–46.

Williams, Roger M. "Oregon: The Fight for 'Survival.'" *Saturday Review* (November 16, 1974), 10–15.

Woodruff, J. R. "Portland Port Facilities." *Oregon Business Review* 4 (September 1945), 1, 4.

Worster, Donald. "A Dream of Water." *Montana The Magazine of Western History* 36 (Autumn 1986), 72–74.

PAMPHLETS

Air Resources Center. Oregon State University. *Agricultural Field Burning in the Willamette Valley.* Corvallis, 1969.

Davis, Robert G. *Oregon's Burning Issue.* Salem, 1980.

Conklin, Frank S., William C. Young III, and Harold W. Youngberg. *Burning Grass Seed Fields in the Willamette Valley: The Search for Solutions.* Extension Miscellaneous Publication 8397. Oregon State University Extension Service. Corvallis, 1989.

Dimmick, R.E., and Fred Merryfield. *The Fishes of the Willamette River System in Relation to Pollution.* Bulletin Series No. 20, Engineering Experiment Station, Oregon State College. Corvallis, 1945.

Oregon Agricultural Experiment Station. *Field Burning: Oregon State University Research.* Corvallis, 1973.

_____. *Profile.* Corvallis, 1976.

Oregon Seed Council. *Grass Seed: The Tiny Giant.* Salem, 1975.

Oregon's First Century of Farming: A Statistical Record of Agricultural Achievement and Adjustment. Federal Cooperative Extension Service. Corvallis, n.d.

Pacific Northwest Development Association: By-laws and Articles of Incorporation. Portland, 1945.

Pacific Northwest Development Association Bulletin. Portland, 1946.

Plambeck, Hans H. *The Population of Oregon, 1940–1970: Changes and Implications.* Prepared for the Rockefeller Foundation Project, "Man and His Activities as Related to Environmental Quality," Oregon State University. Corvallis, 1975.

Portland Chamber of Commerce. *Farming in Oregon.* Portland, 1945.

Solving Sprawl: The Sierra Club Rates the States. San Francisco: The Sierra Club, 1999.

This Place on Earth. Seattle: Northwest Environment Watch, 2002.

PERIODICALS AND NEWSPAPERS

Astorian Evening Budget. 1945.

Bend Bulletin. 1922–2002.

Coos Bay Times. 1934–1959.

Corvallis Gazette-Times. 1947–2003.

Christian Science Monitor. 1969.

Christianity Today. 2001.

Crow's Pacific Coast Lumber Digest. 1948.

Deschutes Pine Echoes. 1928.

Eugene Register-Guard. 1956–2003.

Klamath Herald and News. Klamath Falls. 2000–2003.

Life. 1945–1960.

Medford Mail-Tribune. 1945.

Mid-Valley Sunday. Corvallis/Albany. 2000–2003.

Newport News-Times. 2002.

New York Times. 1945–2003.

Oregon Business Review. 1945–1990.

Oregon Journal. Portland. 1965, 1972.

Oregonian. Portland. 1939–2003.

Oregon Statesman. Salem. 1945–1946.

Pendleton East Oregonian. 1945–1946.

Reedsport Port Umpqua Courier. 1945.

Roseburg News-Review. 1945–1946, 1972.

Seattle Post-Intelligencer. 1994.

Seed Production Research at Oregon State University. 1982–1992.

Statesman-Journal. Salem. 1996–2003.

Sunset. 1940.

The Spectator. 1940.

The Timberman. 1949.

Wall Street Journal. 1974–1975.

West Shore. 1885.

Willamette Farmer. Salem. 1874.

THESES, DISSERTATIONS, MANUSCRIPTS

Agricultural Experiment Station Records. Record Group 25. Oregon State University Archives. Corvallis.

Barber, Katrine. "After Celilo Falls: The Dalles Dam, Indian Fishing Rights, and Federal Energy Policy." Ph.D. dissertation, Washington State University, 1999.

Charlton, David B. Papers. Oregon Historical Society. Portland.

College of Agricultural Sciences Dean's Office Records. Record Group 158. Oregon State University Archives. Corvallis.

Forsman, Eric D. "A Preliminary Investigation of the Spotted Owl." M.S. thesis, Oregon State University, 1975.

Gregory, Ronald L. "Life in Railroad Logging Camps of the Shevlin-Hixon Company, 1916–1950." M.A. thesis, Oregon State University, 1997.

Industrial Forestry Association. Records. Oregon Historical Society. Portland.

Jackman, E. R. Papers. Oregon State University Archives. Corvallis.

National Forest Products Association. Records. Forest History Society. Durham, North Carolina.

Neuberger, Richard. Papers. Oregon Historical Society. Portland.

Northwest Coalition for Alternatives to Pesticides. Records. Eugene, Oregon.

Oregon Environmental Council. Records. Oregon Historical Society. Portland.

Robins, Thomas M. Papers. University of Oregon Library. Eugene.

Rymon, Larry. "A Critical Analysis of Wildlife Conservation in Oregon." Ph.D. dissertation, Oregon State University, 1969.

Stroud, Ellen A. "A Slough of Troubles: An Environmental and Social History of the Columbia Slough." M.A. thesis, University of Oregon, 1995.

USDA. Forest Service. Miscellaneous Files. Region 6. Pacific Northwest Region. Portland, Oregon.

U.S. Forest Service Records. Record Group 95. Federal Records Center, Seattle, Washington, and National Archives and Records Administration, Washington, D.C.

Wilson, E. E. Papers. Oregon State University Archives. Corvallis.

INDEX

Power Commission, 223; oppor-
tunism of, 224; supports The Dalles
as ocean port, 225; on hydropower
development, 226; federal purchase
of Klamath timber, 228, 230; and
Klamath Tribe, 230, 231; and Beaver
Marsh Project, 232–236; and Pelton
project, 240–43, 244; and Round
Butte project, 245, 246; death, 246
Newberg, 253
Newberg Pool, 274, 277
New Brunswick, 135
New Deal, 23, 86, 147, 218, 316
New Scientist, 297
"New South," 174
Newsweek, 147
Newton, Michael, 203
New York City, 4, 94
New Yorker, 198, 205
New York Times, 3, 66, 110, 152
New York Times Magazine, 26
Niemela, Henry, 55
Nims, Frank, 309
Nixon, Richard, 140, 141
NOAA Fisheries, 328
non-point source pollution, 274
North Bend, 167–68
Northern California Lumbermen's
Association, 151
northern spotted owl. *See* spotted owl
Northern Spotted Owl v. Hodel (1987), 207
North Unit irrigation project, 38–39,
40, 159
Northwest Environmental Advocates,
275
Northwest Environment Watch, 323
Northwest Forest Pest Action Council,
188, 190, 191, 192

Northwest Forest Plan, 212
Northwest Forest Research Council, 210
Northwest Power Planning Council, 327
Northwest Power Supply Company, 237
Norton, Gale, 111
"no-till" agriculture, 125

Oakes, Ivan, 60
Ocean Resources Planning Program,
296
O'Connell, Sean, 92
old-growth forests, 13, 34, 169, 170, 175;
spotted owl dependence on, 206,
207
1000 Friends of Oregon, 298, 300–301,
303, 312
Ontario, 29
Onthank, Karl, 236
Ordnance, 6
Oregon: mythic origins of, xvii; mythical promise of, xviii, 316; as component of global superpower, xix;
promise in postwar era, 8, 26; as
land of abundance, 15; and markets
for natural resources, 17; as home-
building market, 33; as leading
lumber-producing state, 34, 149,
210, 322; forest bounty of, 37; rural
electricity in, 45–46, 81; agricultural
abundance of, 79; rural population
decline, 81; seed-grass industry in,
85–94; limits use of DDT, 140; timber dependency of, 160; lumber
industry's political power in, 169;
lumber production of, 176; forest
practices of, 178; challenges FPC
authority, 239; anti-pollution fight
gains national attention, 270–71;